In Situ Archaeological Conservation

IN SITU ARCHAEOLOGICAL CONSERVATION

Proceedings of meetings

April 6–13, 1986

Mexico

Miguel Angel Corzo, *Conference Coordinator*
The Getty Conservation Institute

Henry W.M. Hodges, *Senior Editor*

INSTITUTO NACIONAL DE ANTROPOLOGIA E HISTORIA DE MEXICO

THE GETTY CONSERVATION INSTITUTE

Conference Coordinator Miguel Angel Corzo, The Getty Conservation Institute

Senior Editor Henry W.M. Hodges

Publication Credits

Editing and coordination: Irina Averkieff, The Getty Conservation Institute
Design: Zempel Group, Inc., Los Angeles, California
Typography: American Typesetting, Inc., Reseda, California
Color separations: Heinz Weber Inc., Los Angeles, California
Printing: Southern California Graphics, Los Angeles, California
Photography: Guillermo Aldana pp. xi, xiv–xxiv, 70–71, 150–151, 153, 187, 203; Luciano Cedillo Alvarez pp. 91, 93, 95; John M. Coles pp. 34, 35, 38; Agustín Espinosa cover, pp. xiv, xv; Maria Luisa Franco pp. 171–174; Malika Hachid pp. x, xi; Lars-Uno Johansson p. 135; J.P. Mallory pp. 22–23, 39–41, 44, 49–51; Eduardo Matos Moctezuma pp. 159, 160, 162, 164, 165; P.J.R. Modderman pp. 42, 43; Sergio Arturo Montero pp. 102–105; Pacatnamu Project pp. 72, 74–77; James A. Spriggs pp. 106–110, 112; Richard Todd pp. xii, xiii, 74, 75; Peter Van Geersdaele pp. 117, 118.

The photograph on the cover shows a portion of a wall painting in the Temple of the Paintings, Bonampak, Chiapas, Mexico.

The small symbol that appears on the cover and elsewhere in this book is a plumed serpent design from a cylindrical clay stamp found in Mexico City and now in the National Museum of Mexico.

Library of Congress Catalog Card Number 86-083099
International Standard Book Number 0-941103-03-X

Printed in the United States of America

Instituto Nacional de Antropología e Historia de Mexico

The National Institute of History and Anthropology of Mexico (INAH) is an entity of the Mexican federal government and a part of the Ministry of Education. INAH is responsible for the dissemination of information on and protection of the country's pre-Hispanic and colonial cultural heritage. INAH oversees numerous and diverse sites, monuments, national and regional museums, and cultural objects. A school of anthropology, a school of conservation, and an active publications program fall under its auspices. INAH actively organizes conferences, meetings, exhibits, and other cultural events both in Mexico and abroad that are related to its mandate.

The Getty Conservation Institute

The Getty Conservation Institute (GCI), an operating program of the J. Paul Getty Trust, was created in 1982 to enhance the quality of conservation practice in the world today. Based on the belief that the best approach to conservation is interdisciplinary, the Institute brings together the knowledge of conservators, scientists, and art historians. Through a combination of in-house activities and collaborative ventures with other organizations, the Institute plays a catalytic role that contributes substantially to the conservation of our cultural heritage. The Institute aims to further scientific research, to increase conservation training opportunities, and to strengthen communication among specialists.

Contents

PROLOGUE

A large part of the world's cultural heritage has been revealed through archaeological excavation. And every day, in every country, more artifacts and architectural remains from past civilizations are being unearthed. Archaeology may be seen as a destructive activity. To gain evidence from the past, the strata of accumulated human occupation must be removed. As a result, unearthed archaeological remains are exposed to new, more aggressive environments.

During the last hundred years, archaeological field techniques have been refined, but the methods used to preserve excavated materials have not kept pace with these developments. In situ archaeological conservation is still in its formative stages. This is why the Instituto Nacional de Antropología e Historia (INAH) and the Getty Conservation Institute (GCI) organized an international conference in Mexico City, in April 1986, focusing on the problems of preserving sites and artifacts left in situ.

These issues are central to the activities of both institutions. INAH is responsible not only for the exploration of Mexico's archaeological sites, but also for the supervision, conservation, and restoration of its archaeological, historical, and artistic monuments, and associated objects. The GCI, an operating program of the J. Paul Getty Trust, serves the international conservation community by undertaking activities that enhance conservation practice and that address the urgent needs of the profession.

By publishing these papers presented at the conference, INAH and the GCI hope to provide a survey of the current state of knowledge of in situ archaeological conservation.

Dr. Enrique Florescano
Director General
Instituto Nacional de Antropología e Historia, Mexico

Luis Monreal
Director
The Getty Conservation Institute, Los Angeles

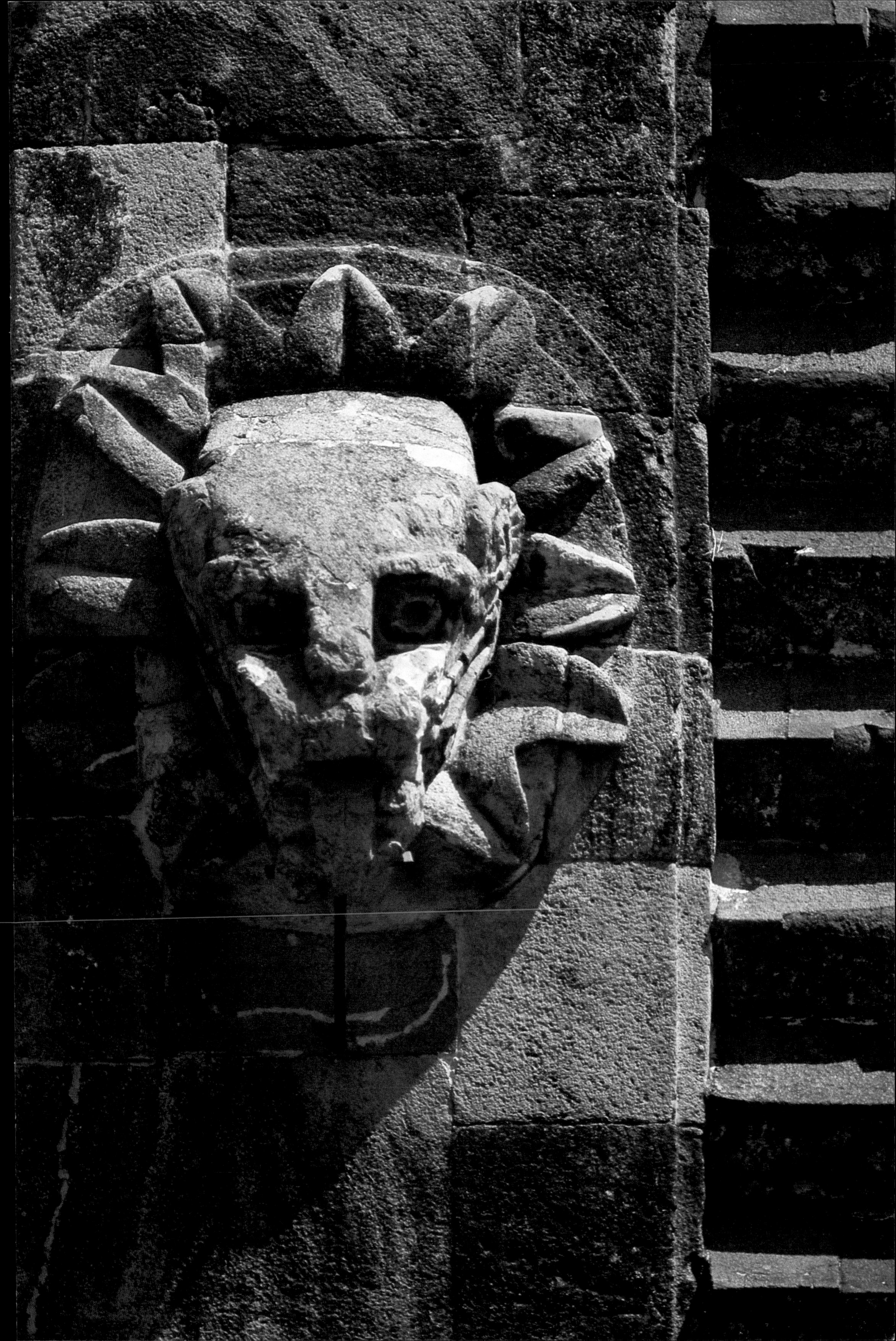

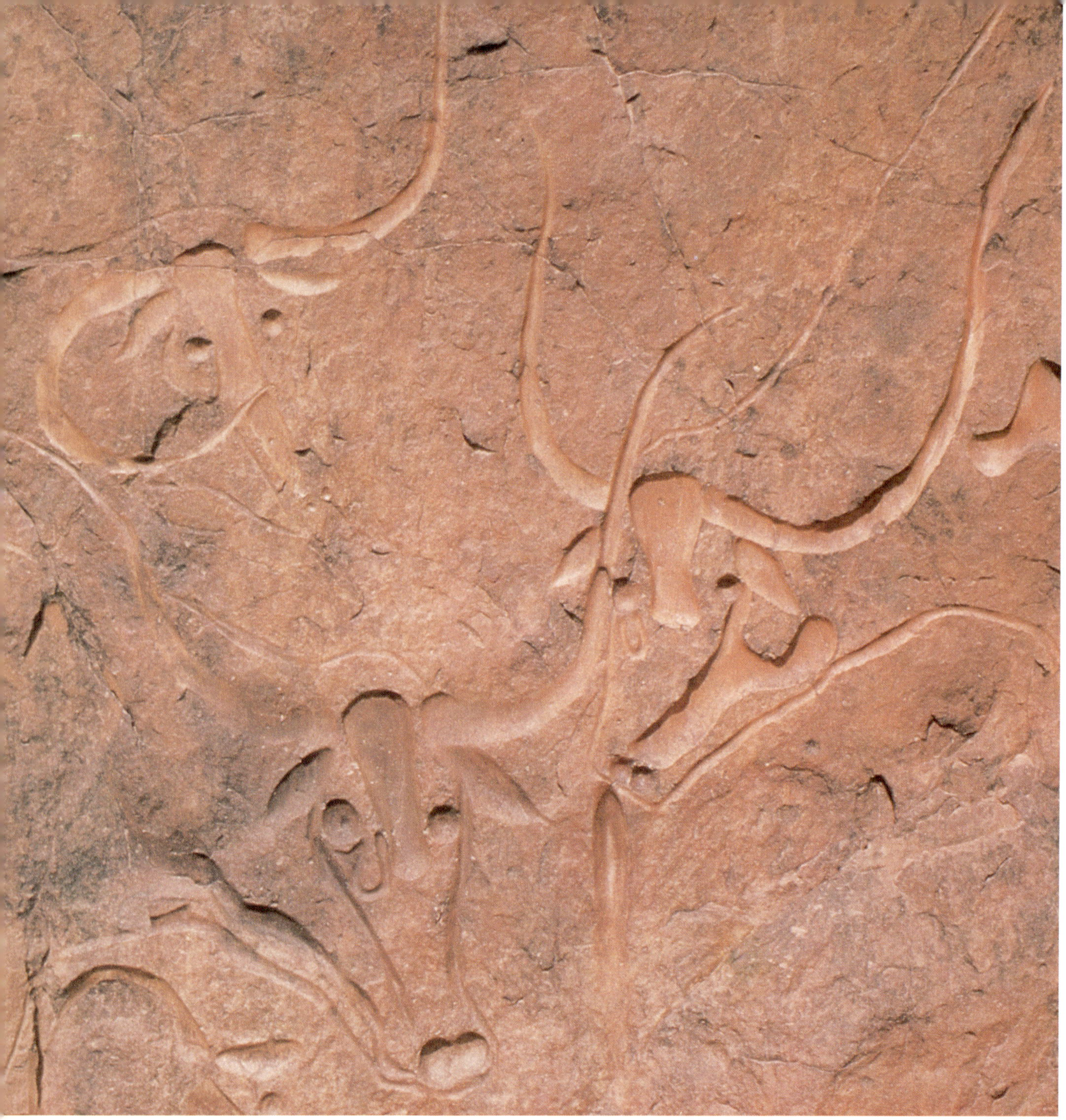

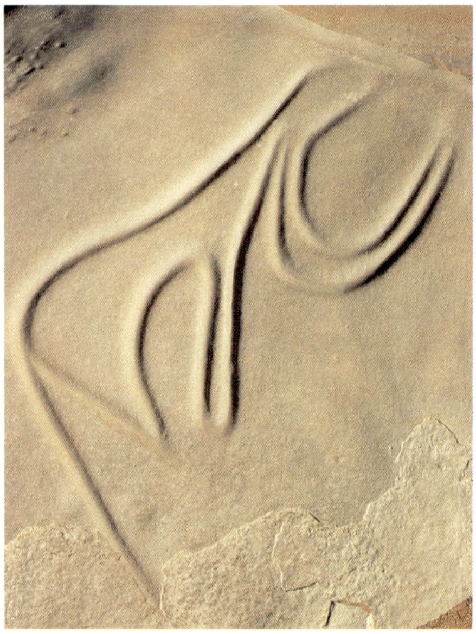

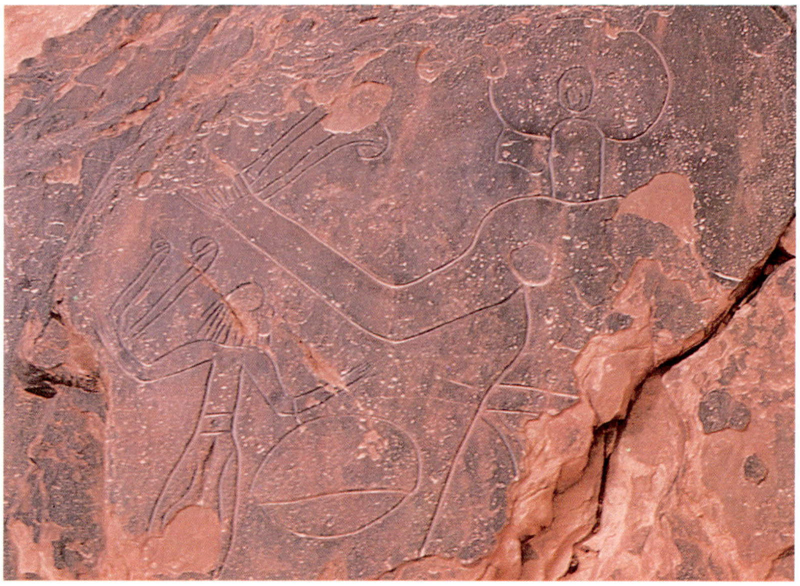

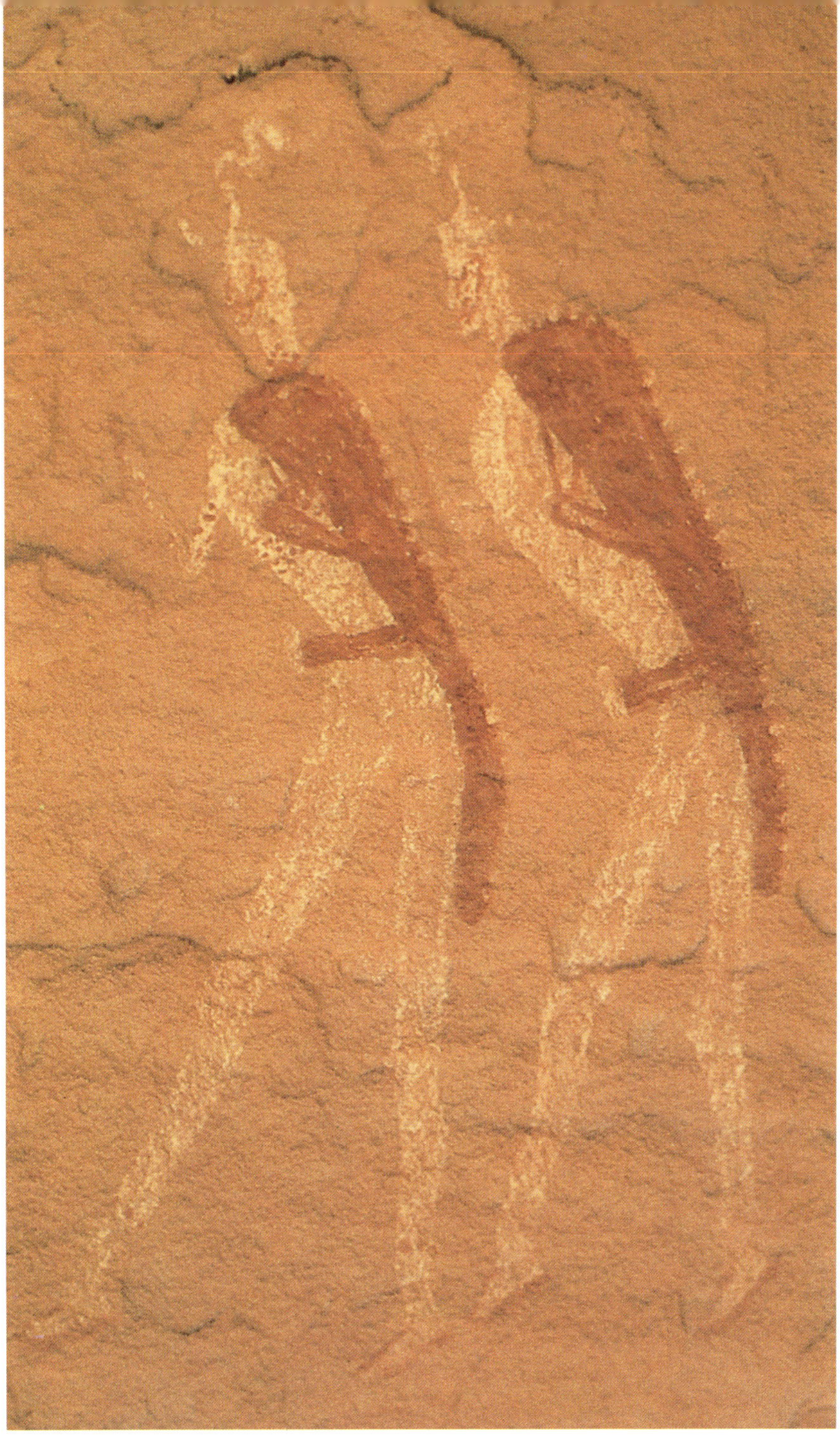

Plate 2 (opposite, above). Low-relief rock art. Tassili n'Ajjer, Algeria.

Plate 3 (opposite, below left). Rock art. Saharan Atlas, Algeria.

Plate 4 (opposite, below right). Rock art. Saharan Atlas, Algeria.

Plate 5 (left). Rock painting. Tassili n'Ajjer, Algeria.

Plate 6 (below). Landscape. Tassili n'Ajjer, Algeria.

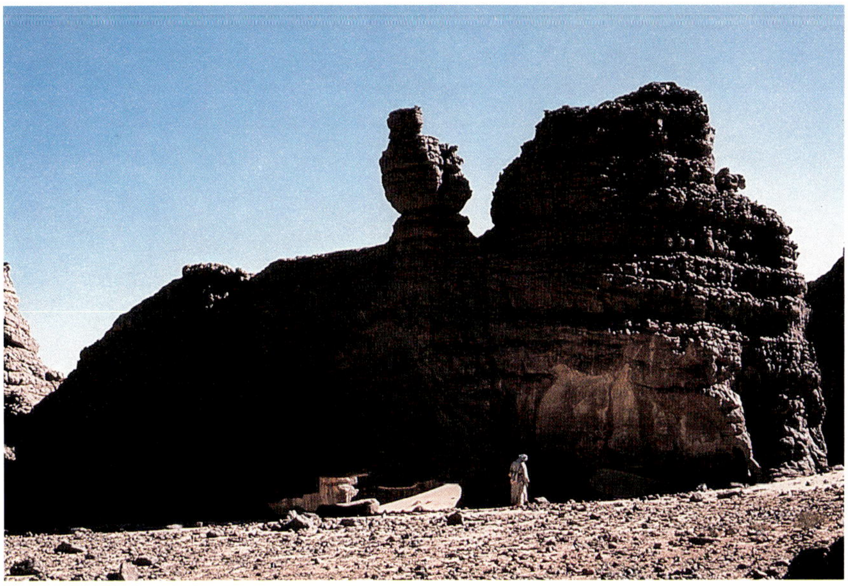

Plate 7 (above). Miniature crown with textile covering (left) and reed armature (right). Pacatnamu, Peru.

Plate 8 (center). Miniature textiles. Pacatnamu, Peru.

Plate 9 (below). Miniature loincloths. Pacatnamu, Peru.

Plate 10 (opposite). Elaborate textile from the Major Quadrangle. Pacatnamu, Peru.

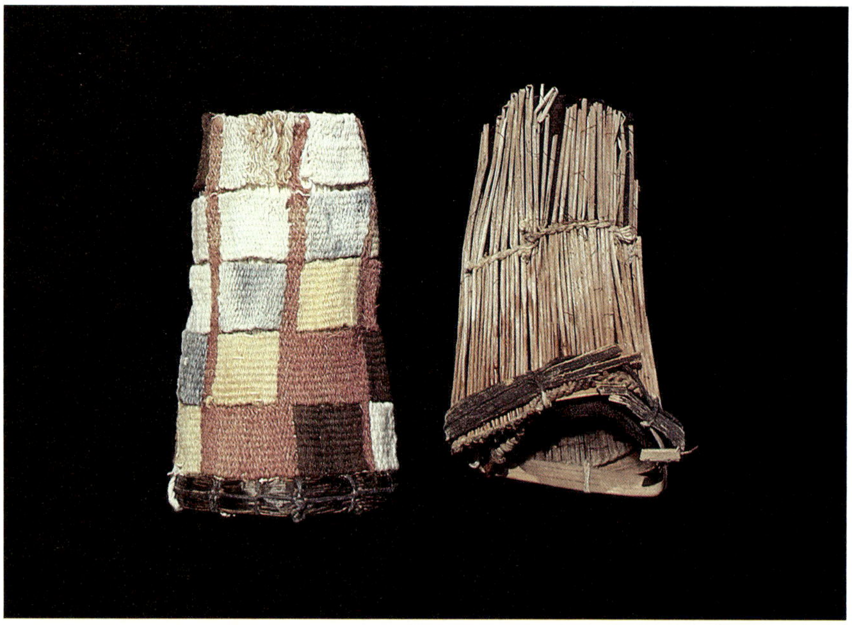

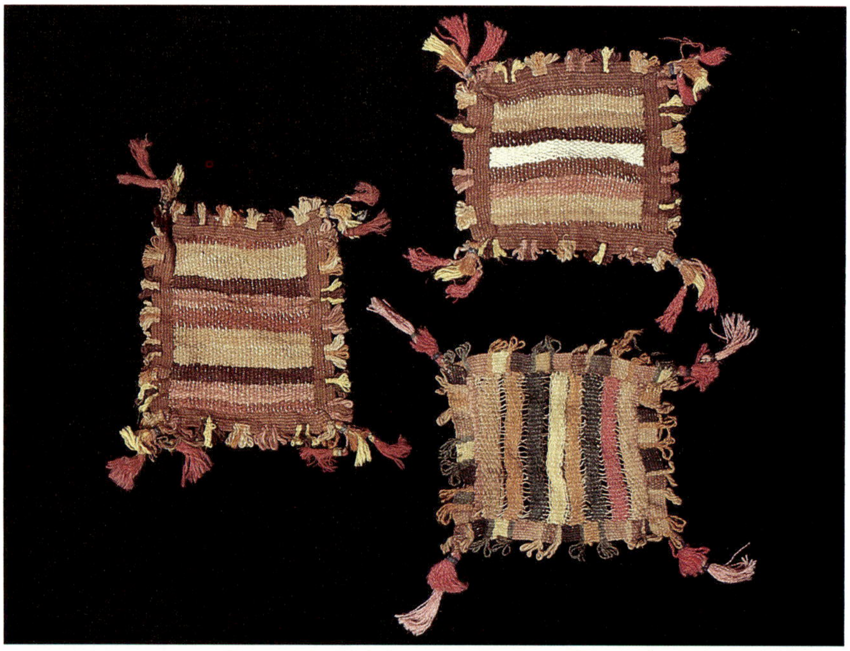

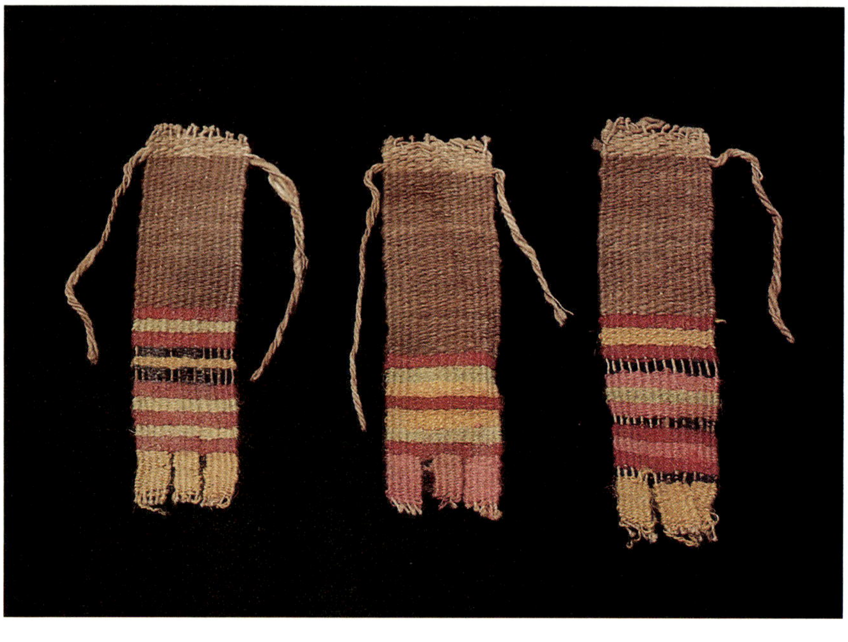

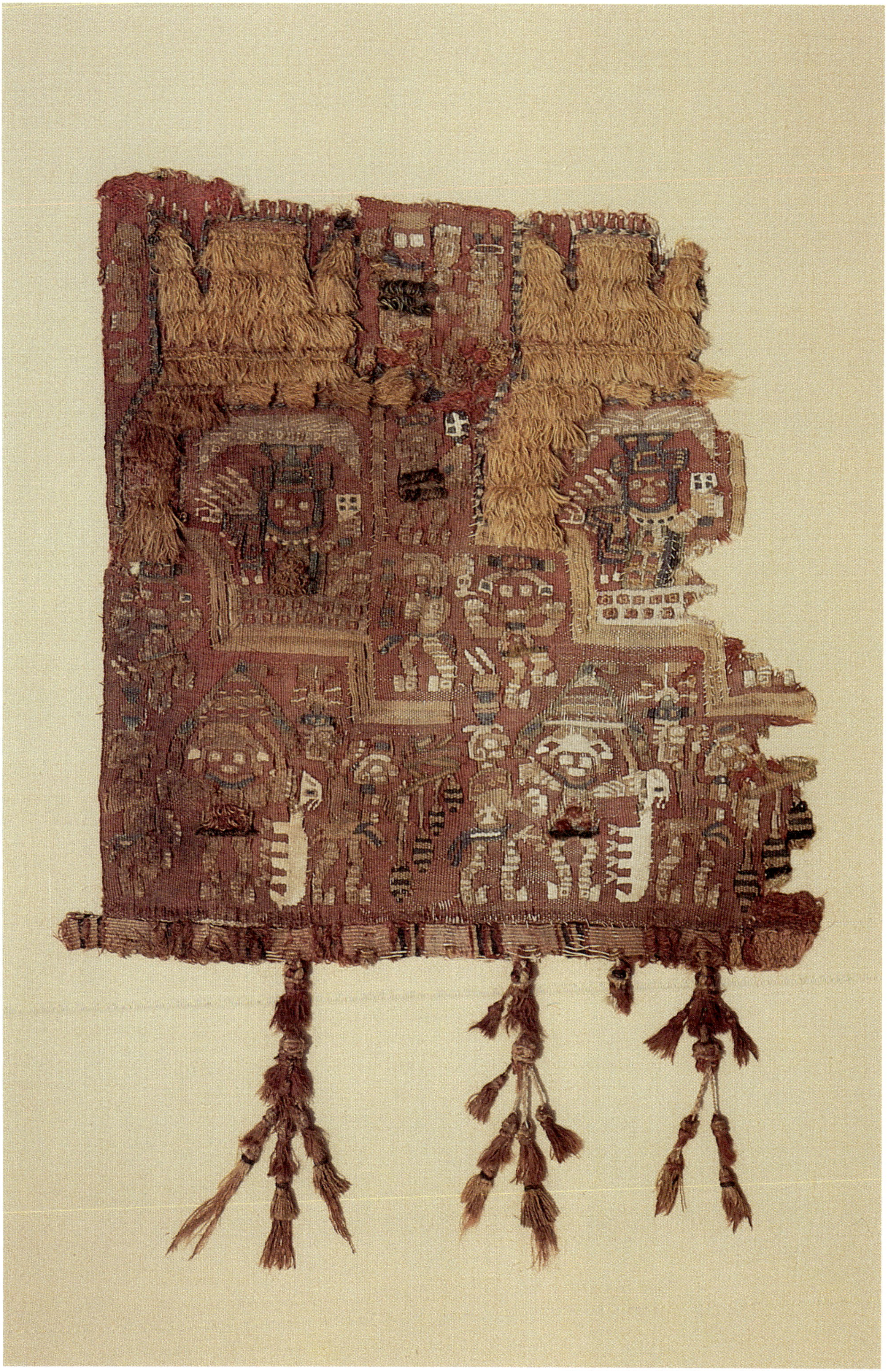

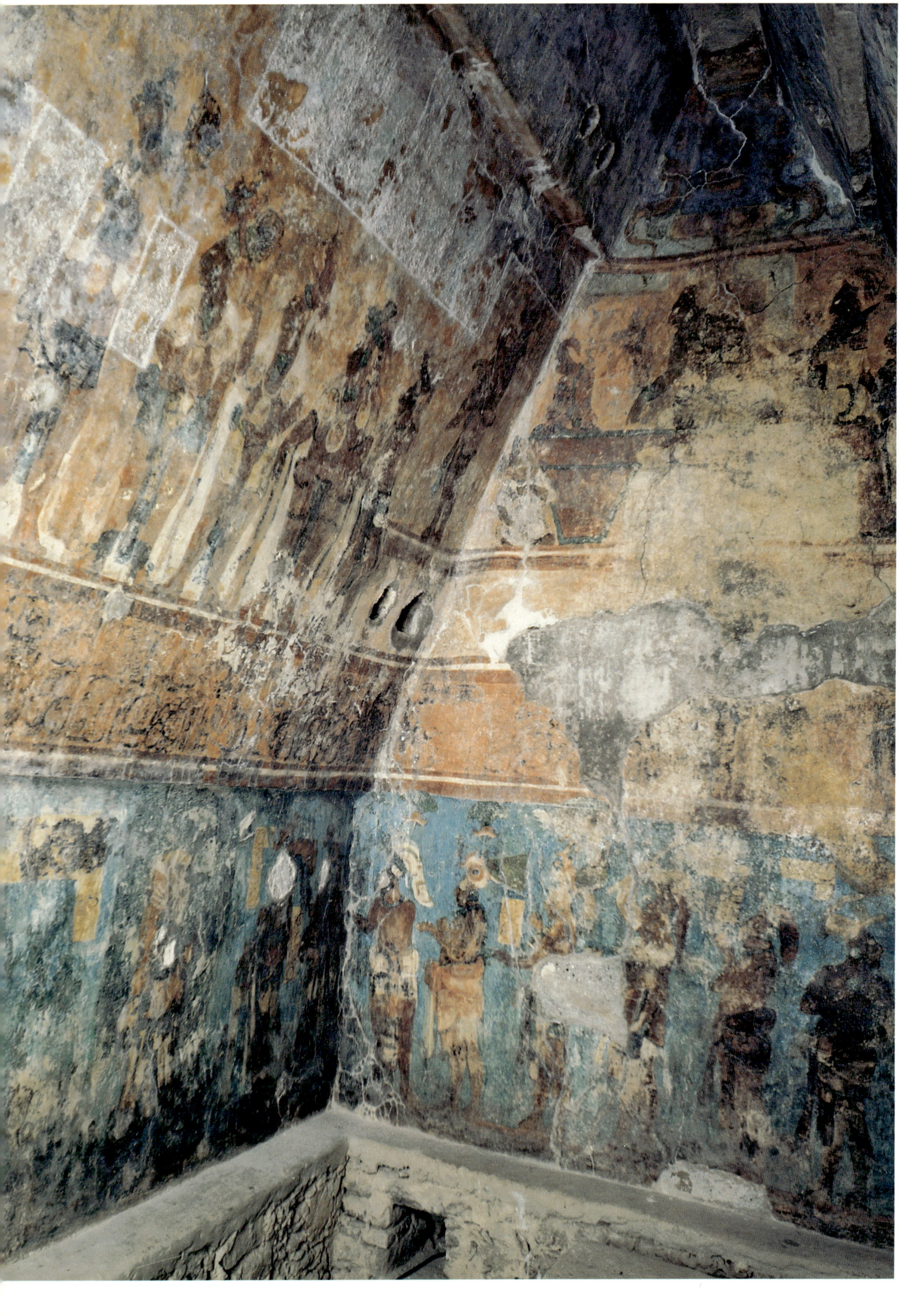

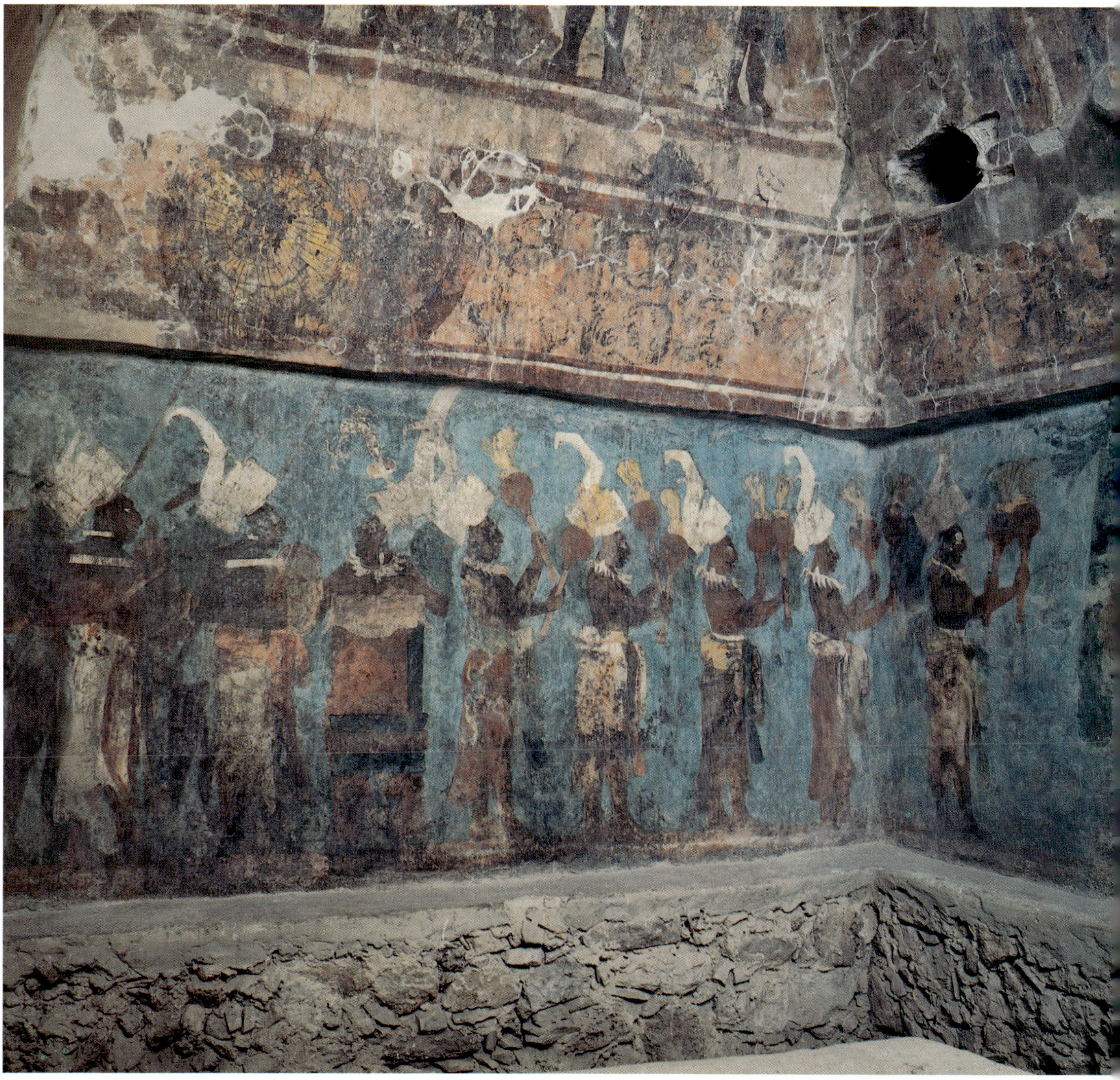

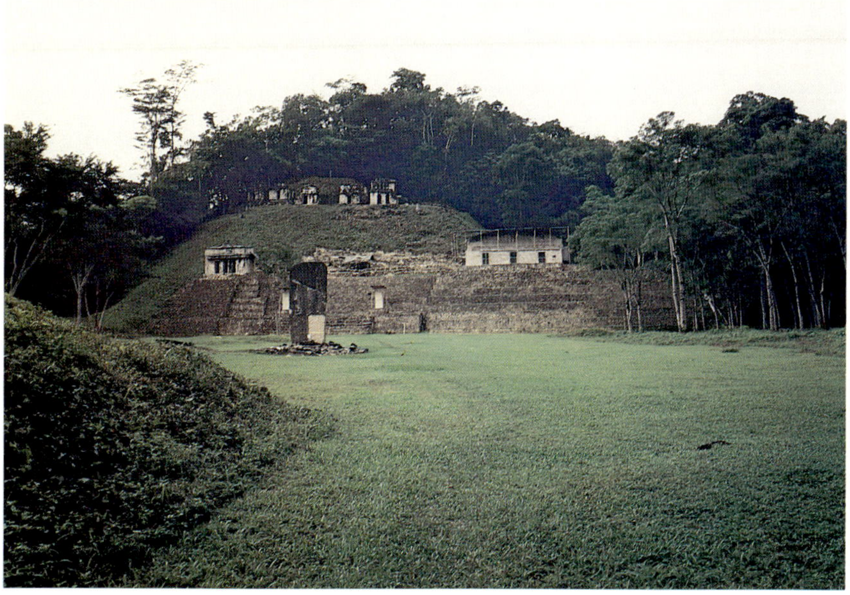

Plate 11 (opposite). Wall painting in the Temple of the Paintings. Bonampak, Chiapas, Mexico.

Plate 12 (above). Wall painting in the Temple of the Paintings. Bonampak, Chiapas, Mexico.

Plate 13 (left). Temple of the Paintings. Bonampak, Chiapas, Mexico.

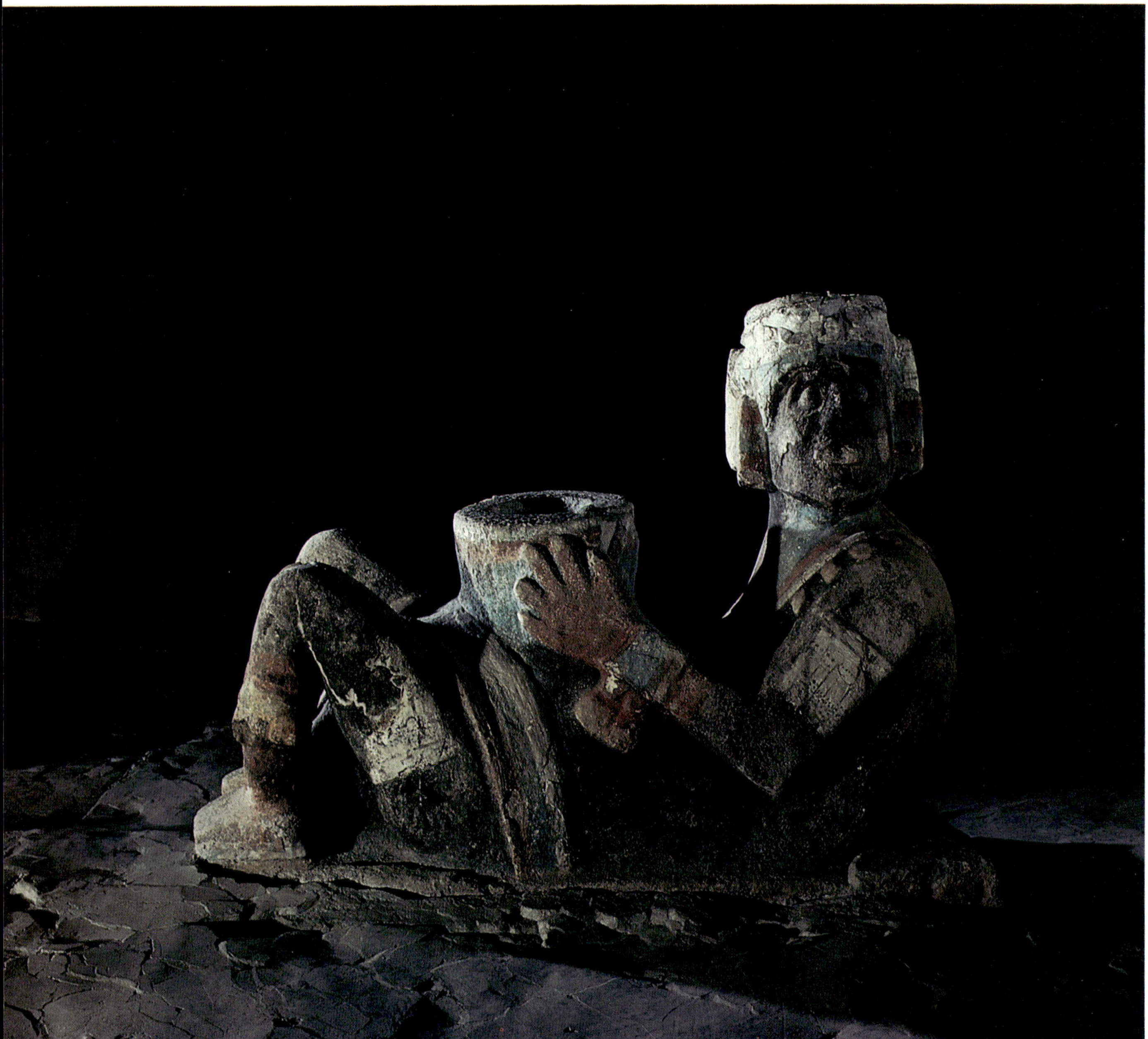

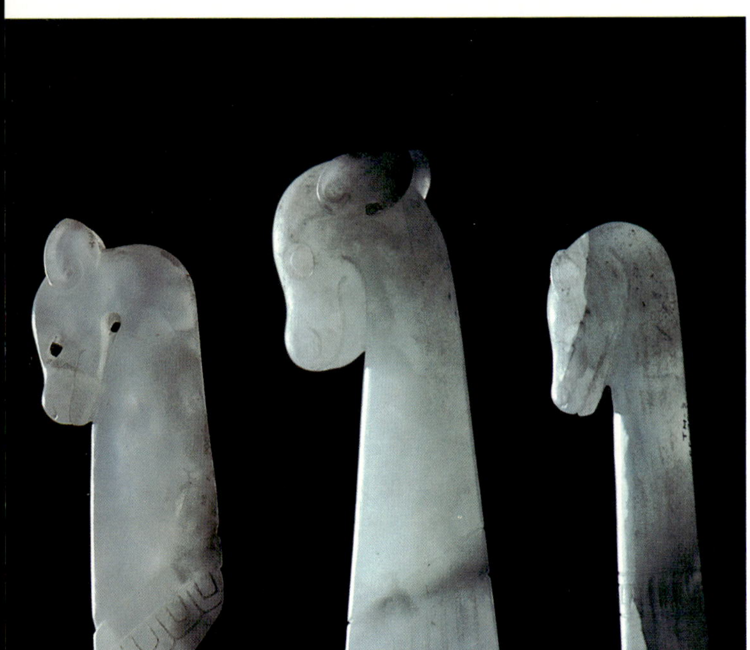

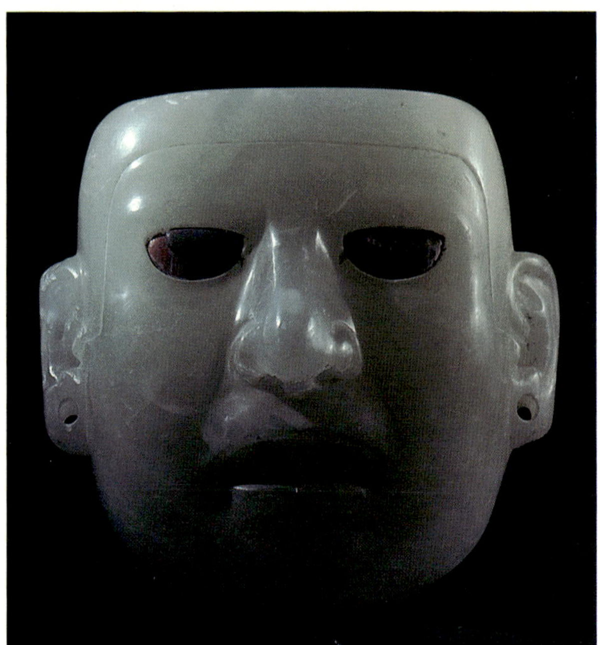

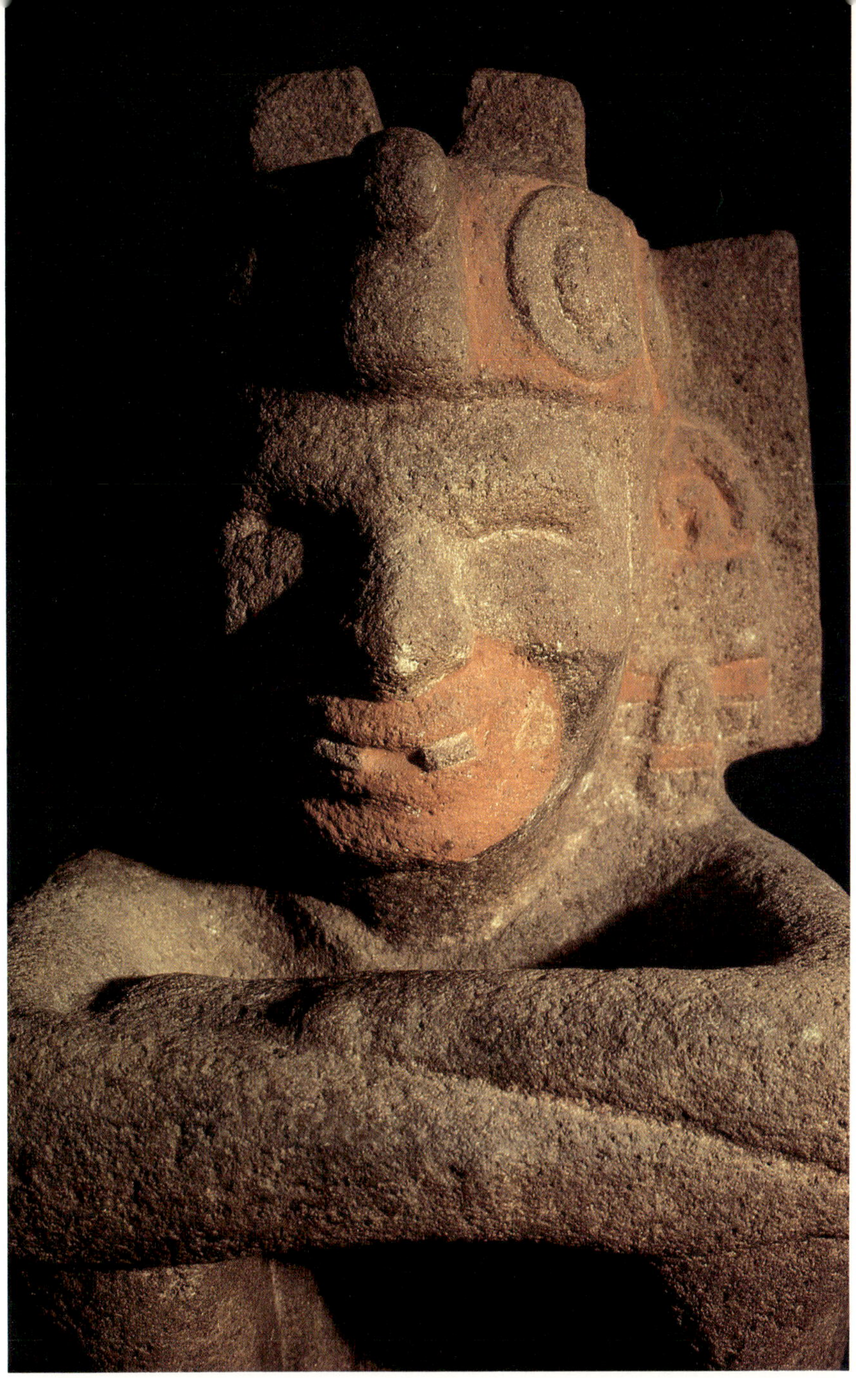

Plate 14 (opposite, above). Chac-mool, god of rain and fertility. Templo Mayor, Mexico.

Plate 15 (opposite, below left). Alabaster deer. Templo Mayor, Mexico.

Plate 16 (opposite, below right). Mexica alabaster mask with encrustations. Templo Mayor, Mexico.

Plate 17 (left). Xiuhtecutli, god of wisdom. Templo Mayor, Mexico.

Plate 18 (below). Stone carving of feathered serpent. Templo Mayor, Mexico.

Plate 19 (overleaf). Stone bifid serpent. Templo Mayor, Mexico.

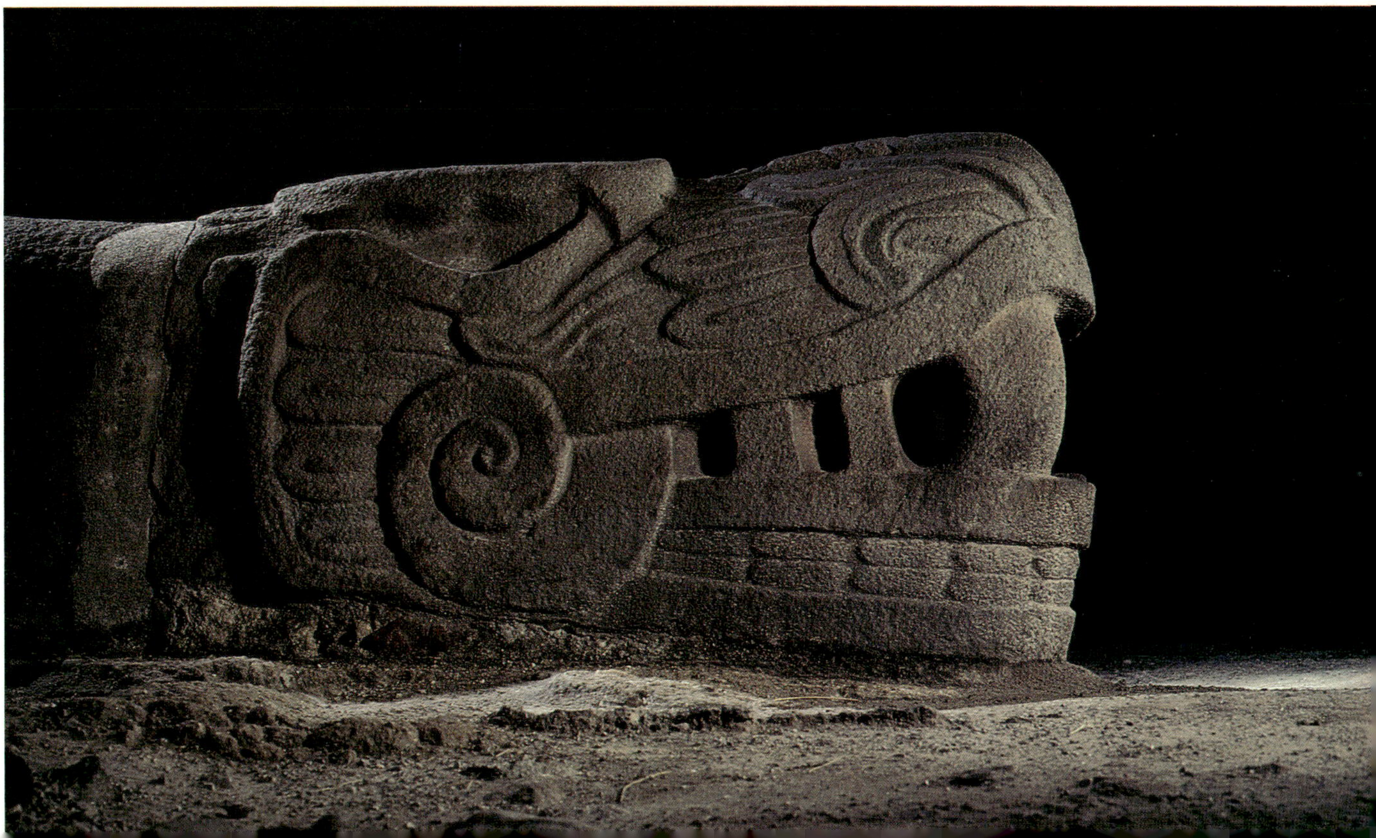

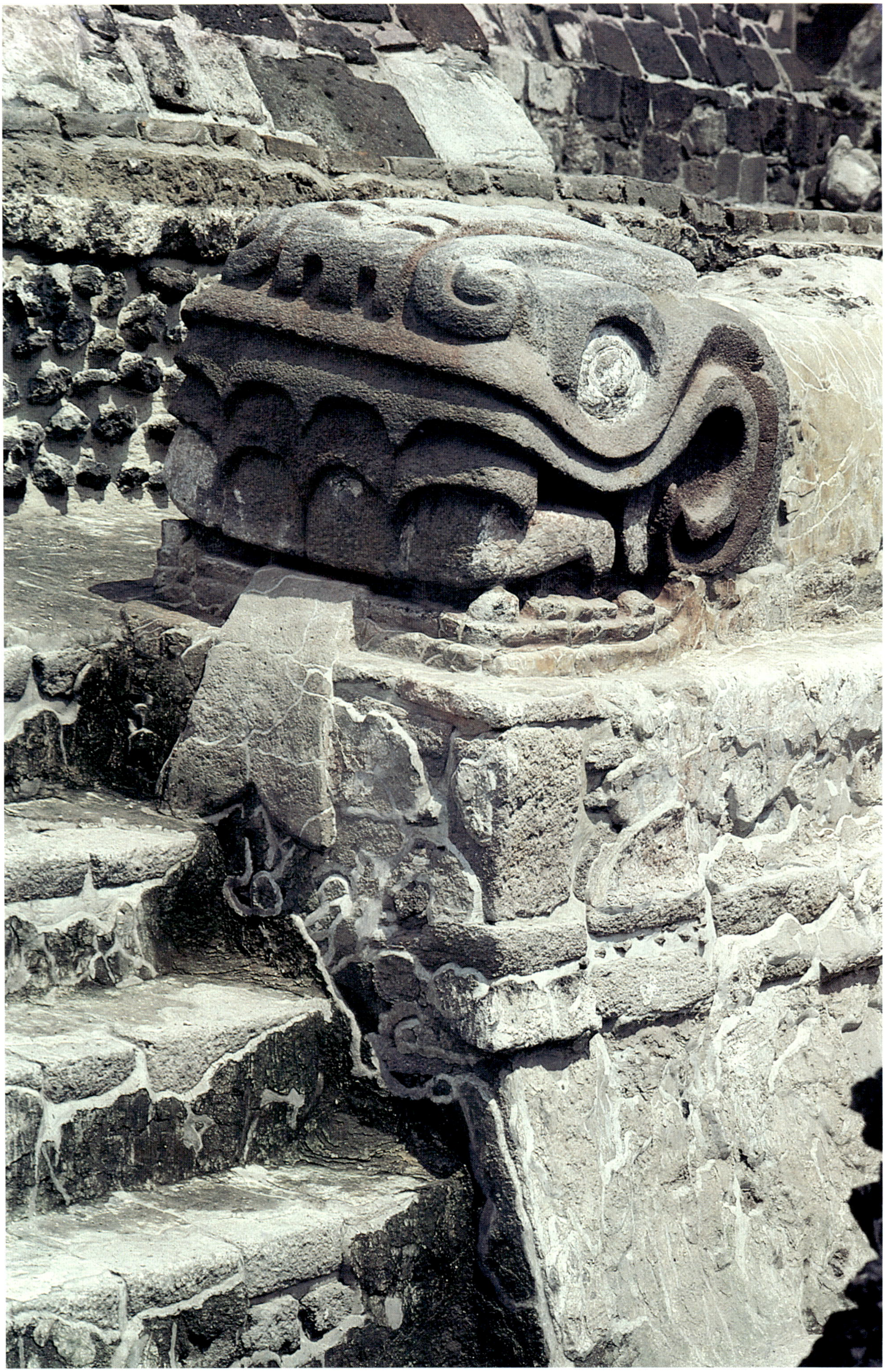

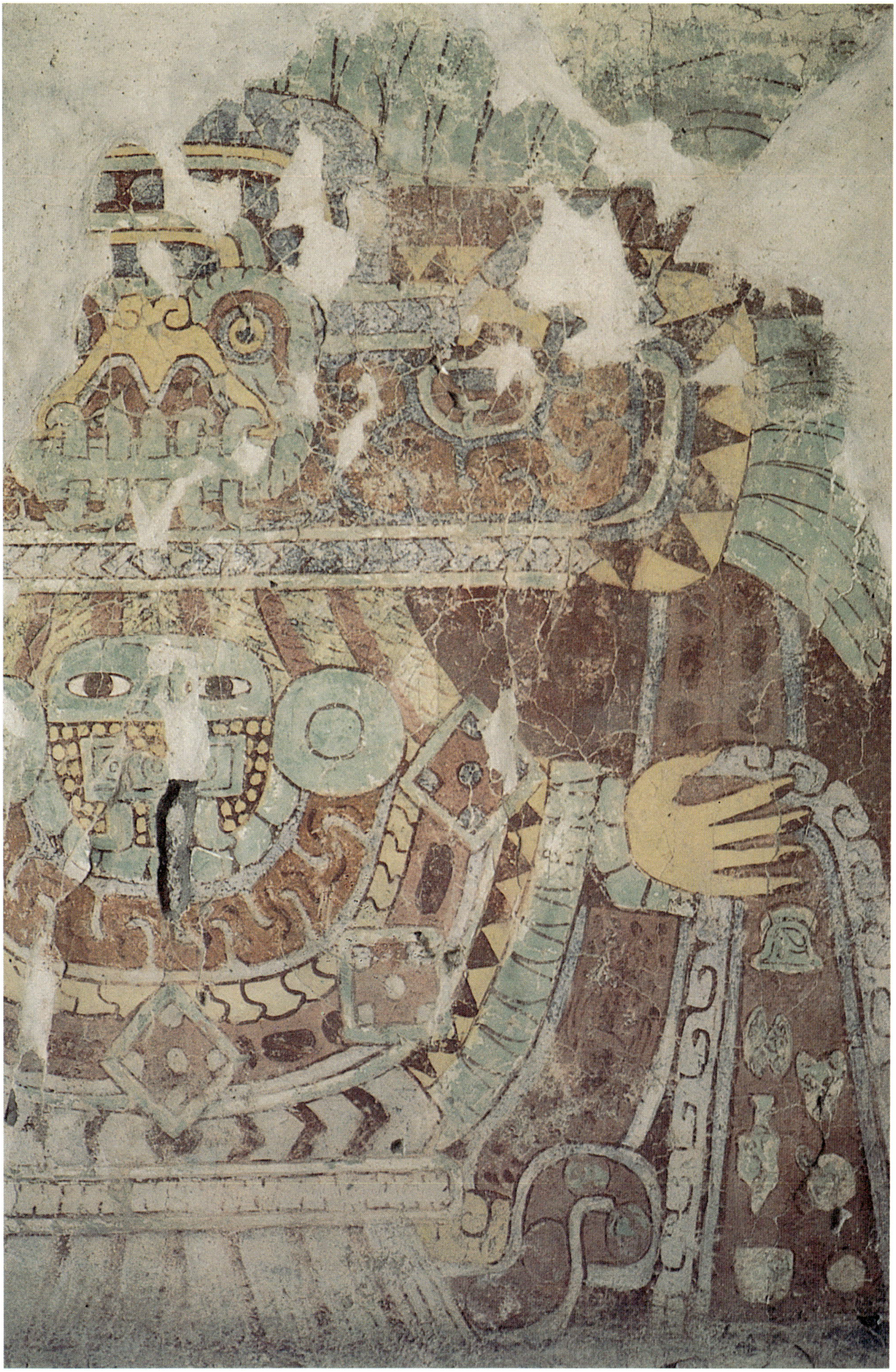

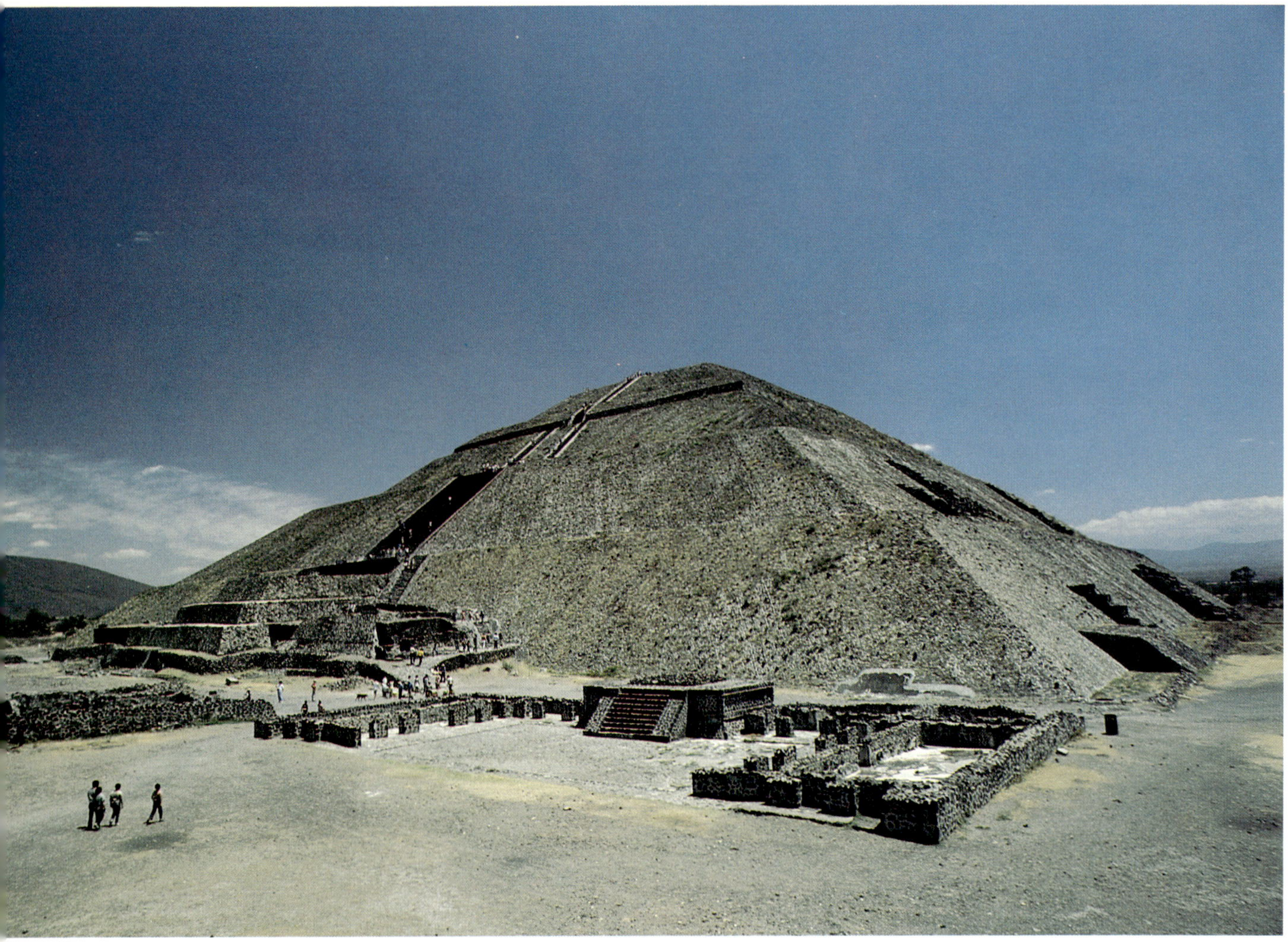

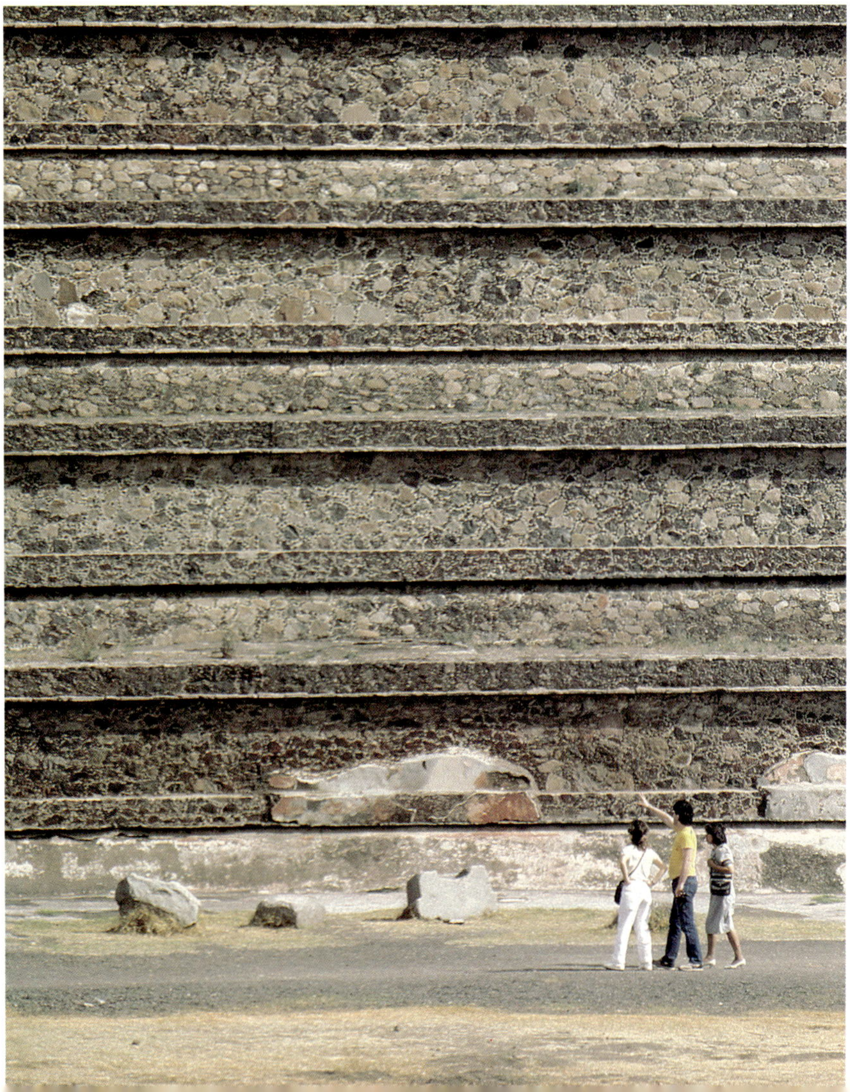

Plate 20 (previous page). Painted mural. Tetitla, Teotihuacan, Mexico.

Plate 21 (above). Pyramid of the Sun. Teotihuacan, Mexico.

Plate 22 (right). Platform. Teotihuacan, Mexico.

Plate 23 (opposite). Stairs ascending the Pyramid of the Sun. Teotihuacan, Mexico.

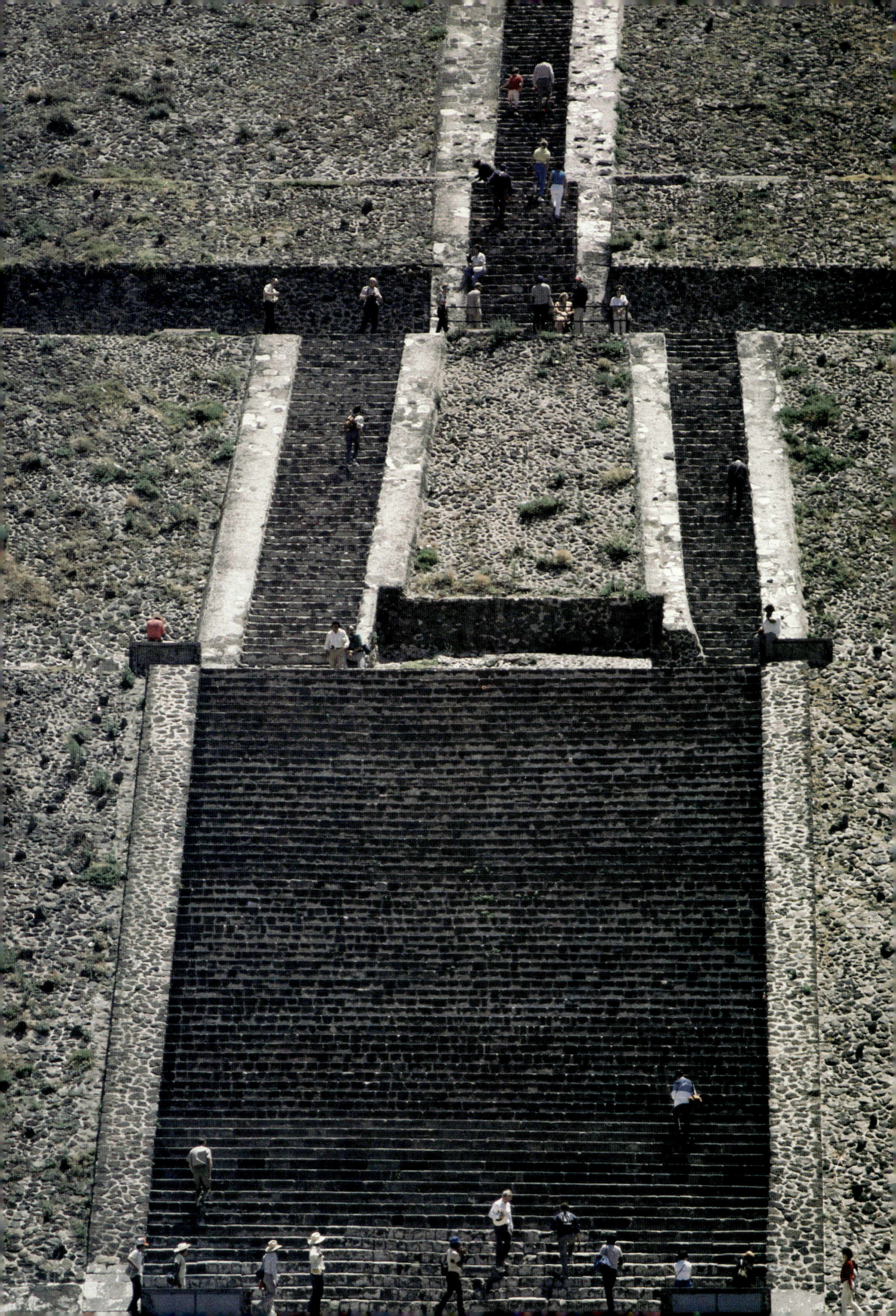

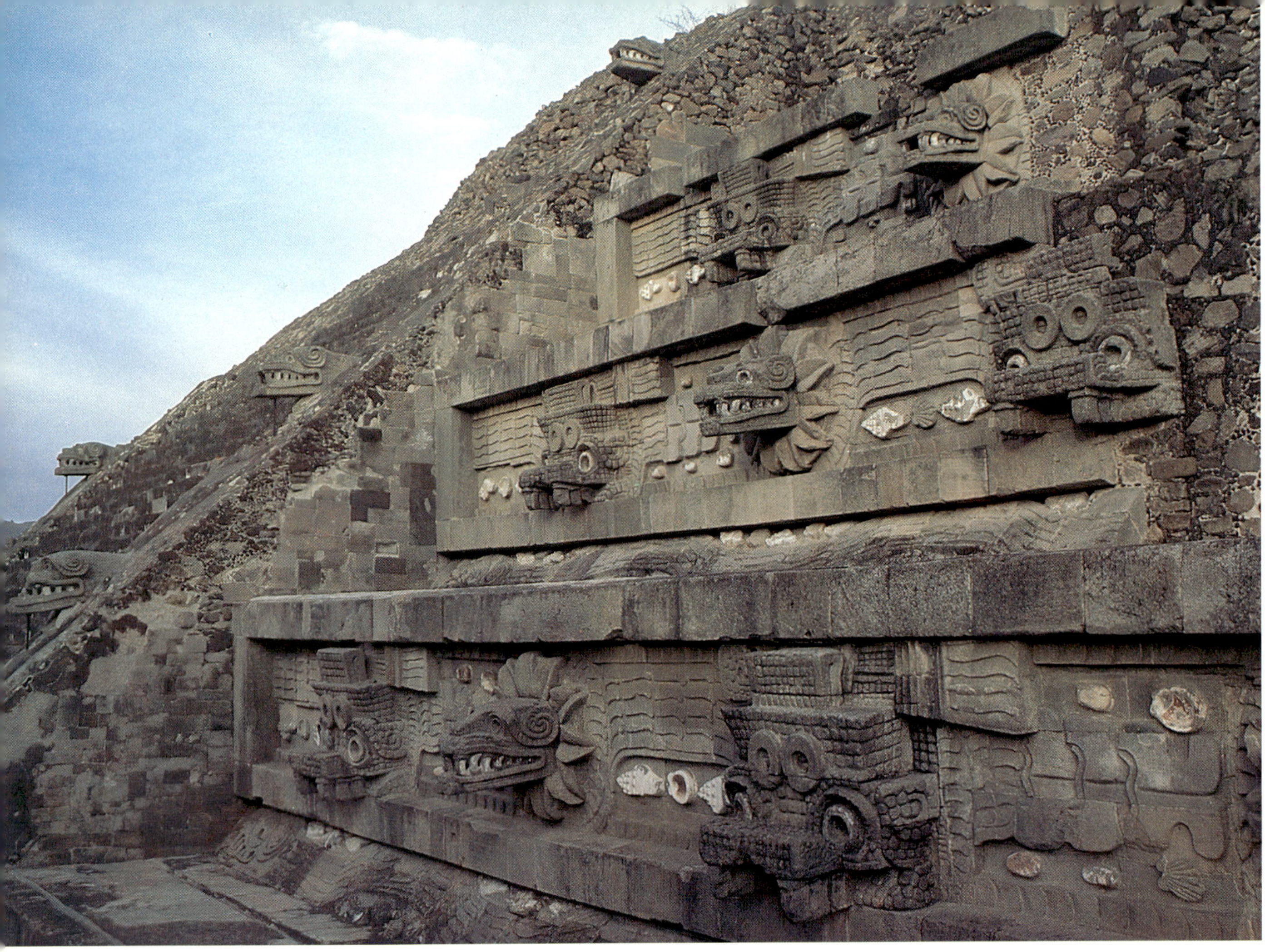

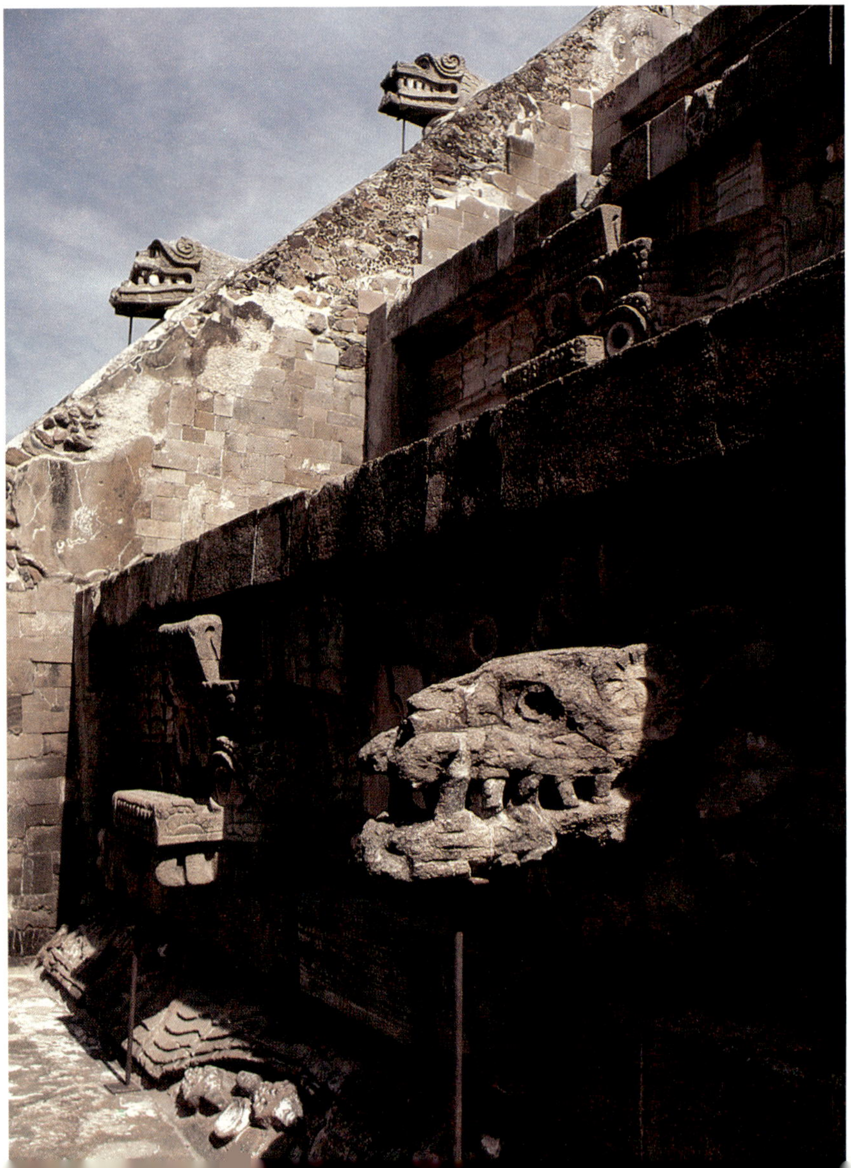

Plate 24 (above). Facade of the Temple of Quetzalcoatl. Teotihuacan, Mexico.

Plate 25 (right). Plumed serpent from the Temple of Quetzalcoatl. Teotihuacan, Mexico.

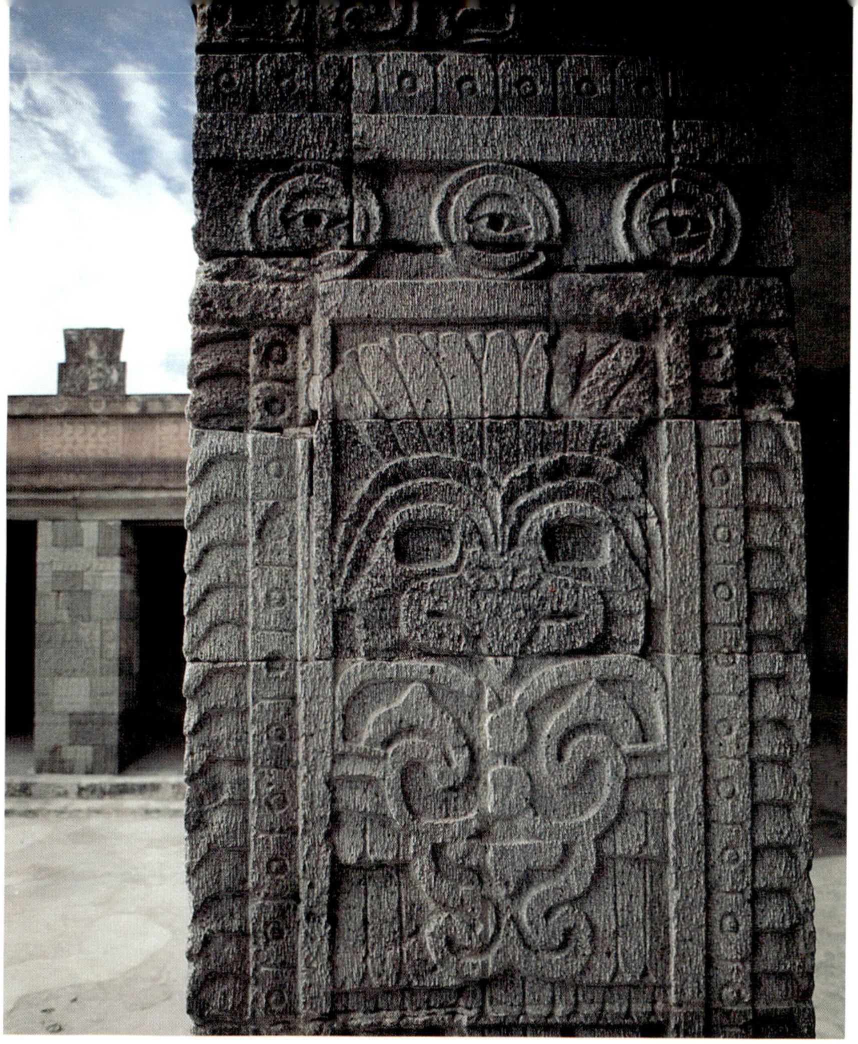

Plate 26 (left). Carved column, Palace of the Quetzal Butterfly. Teotihuacan, Mexico.

Plate 27 (below). Carved relief, Palace of the Quetzal Butterfly. Teotihuacan, Mexico.

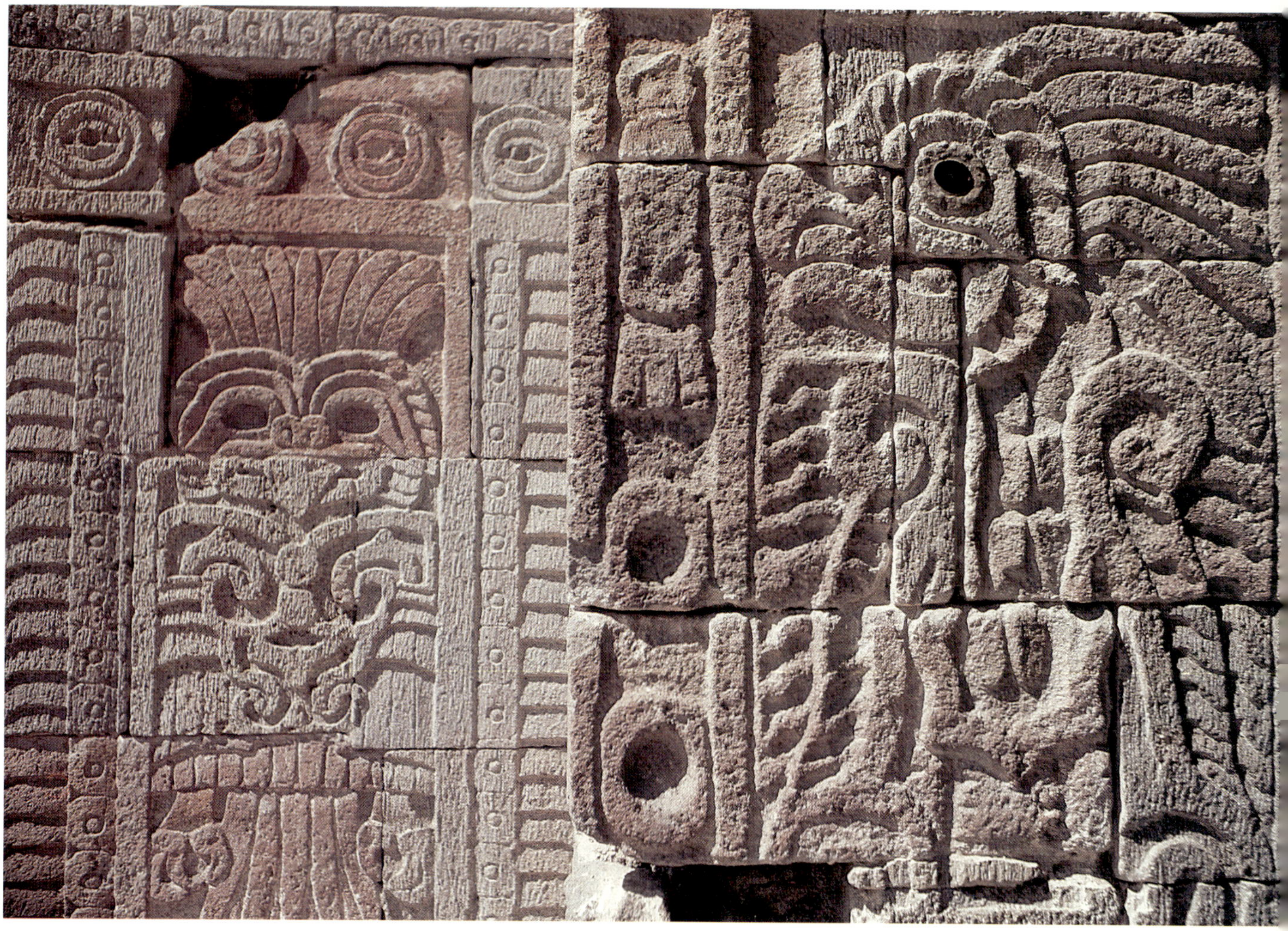

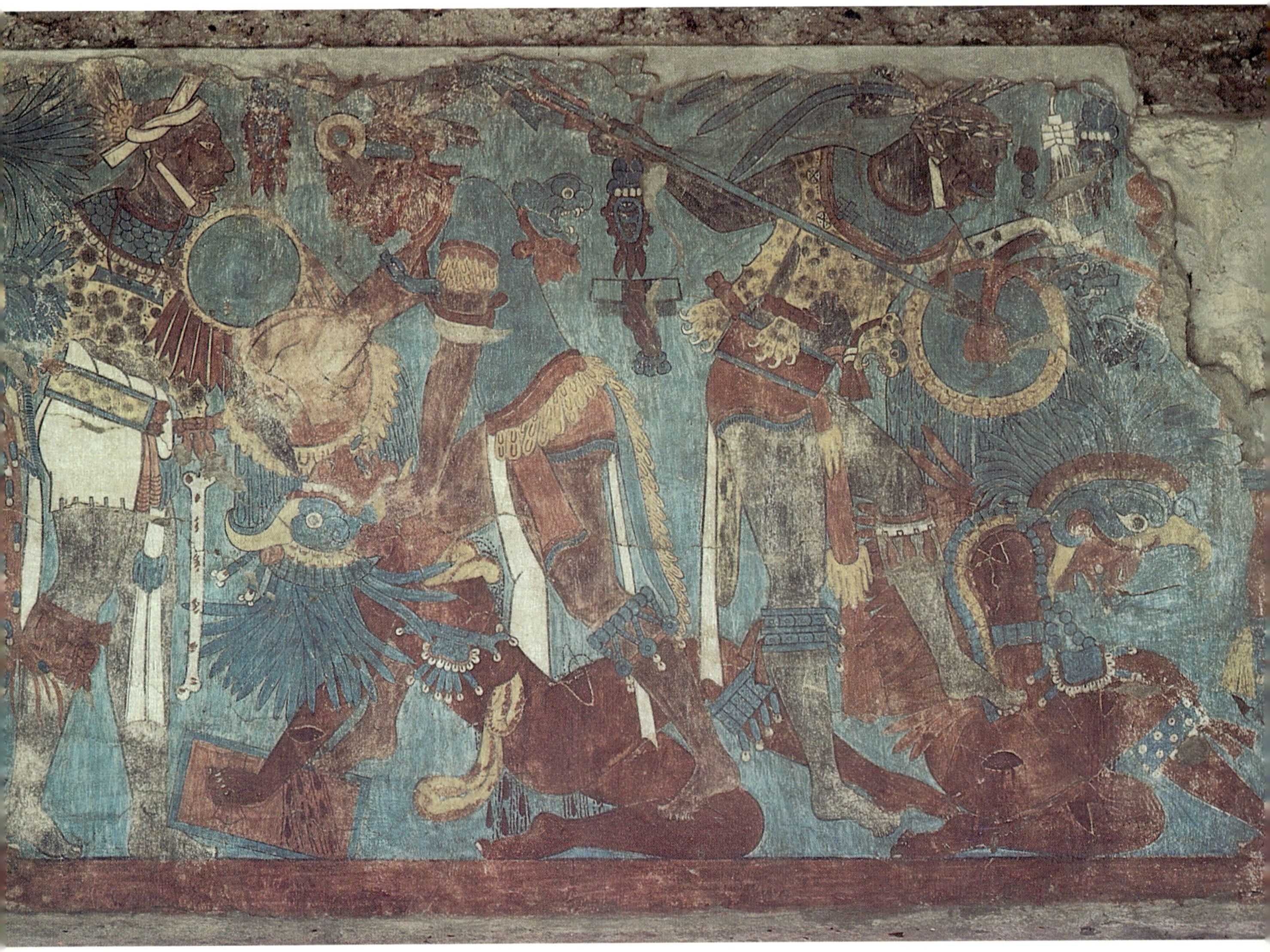

Plate 28. Painted mural. Cacaxtla, Tlaxcala, Mexico.

Brian V. Arthur

FOREWORD

Time, which antiquates antiquities,
and hath an art to make dust of all things,
hath yet spared these minor monuments.
—Sir Thomas Browne, *Hydriotaphia*

Brian V. Arthur trained in archaeological conservation at the Ashmolean Museum, Oxford and the Institute of Archaeology, University of London. He is presently Senior Scientist, National Museum of Man, Canada and is a Fellow of the International Institute for Conservation.

The primary aim of this meeting was to provide a forum for archaeologists and conservators to discuss the problems associated with the conservation of archaeological materials, either before or during their removal from the site, with special attention paid to the conservation of the entire site and to those objects that remain in situ. At the same time, the papers presented provided an opportunity to examine the current state of knowledge in this highly specialized branch of artifact conservation.

Because, as is generally recognized, the roles and aims of archaeologists and on-site conservators are not always complementary, the organizers of this meeting sought to enable the two groups to exchange ideas and information on how both could operate in harmony to recover the maximum amount of information, while also ensuring the preservation of artifacts found on the site.

It is most fortunate that this meeting, hosted by the National Institute of Anthropology and History (Instituto Nacional de Antropología e Historia, INAH) and the Getty Conservation Institute (GCI), was held in Mexico. We enjoyed a unique opportunity to not only hear about the internationally famous excavations of the Templo Mayor, but to visit it and other important sites where in situ conservation plays such a critical role.

The papers from this seminar do not merely review the present state of knowledge of in situ conservation, but also constitute a handbook for practicing field archaeologists. Many areas needing further research have been identified and these may be investigated by the Getty Conservation Institute in the future.

Archaeological conservation has a much longer history than is generally realized. Initial references date to the Roman period. In 1658, Sir Thomas Browne noted that buried iron found surrounded with the bone ash of burnt bodies had only a thin layer of stable corrosion—the ash acted as a preservative. By 1856, Chevreul had described the corrosion process in some detail and had succeeded in restoring a small bronze containing chlorides by using reduction in a stream of hydrogen. And in 1865, von Fellenberg proposed that the corrosion of bronze might be classified according to the material in which it was buried; he listed three types of burial conditions and the forms of corrosion associated with them.

In the field of iron corrosion, Salzer, in 1887, suggested the removal of soluble salts (ferrous and ferric chloride) from iron artifacts by steeping them in a warm alcohol/water bath. That same year, Hartwich was reducing iron objects in a current of hydrogen. And by 1902 in England, Jackson had made photomicrographs of three varieties of bacteria involved in iron corrosion.

Regarding stone, Flinders Petrie suggested, in the *Archaeological Journal* (1888), that stone objects could be strengthened by impregnating them with rice or tapioca water. Rathgen, in 1905, dealt at some length with the problem of extracting soluble salts from fragile stone and clay artifacts. He was, however, overly fond of refiring clay artifacts that would not survive prolonged soaking in water.

Unfortunately, inadequate technical knowledge and the use of outdated methods frequently do more harm than good. Some of the conservation measures used at the turn of the century led to greater damage of cultural property than would have occurred had the property been left alone and simply sheltered. The following is a somewhat humorous example from "Cleaning Copper Coins by Melted Lead" (a conservation publication from the Grand Duchy of Luxembourg) in which the author suggests that, although the results obtained by this method are less satisfactory than with others, it has the advantage of simplicity.

> Using a pair of tongs, dip the coins one by one into melted lead until the crackling, which begins at once, has ceased: this occurs in from 3 to 10 seconds. The coin is then thrown into cold water, cleaned, and placed until the next day in hot milk. It may be necessary to repeat the process when the coin has become cold. By this method an olive colour is imparted to the coin which many antiquaries prefer to dark brown.

In the museum where I was initially trained some thirty-odd years ago, I was perplexed to see that most of the Saxon iron artifacts had, what appeared to be, ocher-colored sawdust adhering to the surface corrosion layer. I wondered if Saxon burials had been sprinkled with colored sawdust or woodchips in the way prehistoric burials were sometimes covered with a layer of ocher. In actuality, the iron artifacts had been thrown into a vat of boiling wax colored with ocher. Those artifacts that survived were then tossed into a cauldron of heated sawdust to absorb the surplus wax.

If we accept the premise that the practice of conservation began with the study of the underlying causes of deterioration, then it was in the 1930s and early 1940s that we saw the establishment of the first conservation laboratories, as well as the appearance of the pioneers of scientific conservation. Many of these early conservators were also talented communicators and teachers; they greatly influenced those who followed, and, indeed, their influence is still felt today. Moreover, their contributions assured the establishment of the profession of artifact conservation.

What then are the threats to the survival of our sites and artifacts? The first to come to mind is the effect of the natural environment. Most changes that take place are, to a great extent, the result of environmental processes. Indeed, precise information on a site environment generally gives a reliable indication of what may have survived underground. The slow and protracted effects of environmental weathering are part of the endless cycle of tearing down and building up found here on earth, as are the more violent and dramatic phenomena of floods, droughts, earthquakes, hurricanes, and volcanic eruptions. But however serious the threats of nature, they pale into insignificance beside the destruction directly provoked by man.

War, vandalism, and the very growth of human civilization—plowing; quarrying; and the building of dams, highways, airports, and pipelines, with the accompanying urban renewal—can, and usually do, affect adversely cultural property of all kinds. The very act of archaeological excavation may itself be added to this list, and thus there is a great need for archaeologists and conservators to be aware of one another's concerns: Excavation without conservation is, indeed, destruction and is morally unacceptable.

In all organizations and institutions involved in the excavation of archaeological sites, the presence of adequate and trained conservators is an essential prerequisite for a successful program. Intellectual isolation is no excuse today when so many avenues exist for the exchange of information: IIC, ICCROM, ICOM, ICOMOS, and now, the Conservation Information Network being developed by the Getty Conservation Institute. However, if one looks at a sampling of excavations around the world, one may conclude that many archaeologists have never heard of conservation. It is also true that many artifact conservators, for their part, have taken very little interest in archaeology or in the conservation problems that face archaeologists on site. It is essential that both sides take a more considered approach and that they spend more time and energy communicating with one another.

I am well aware of another problem facing our colleagues in Central and South America: Little conservation literature is available in Spanish. It is hoped that this meeting will stimulate efforts to correct this deficiency.

Plans for conservation must begin at the same time as plans for excavation; they must be considered together. And it goes without saying that funds must be available for both activities. This is an absolute necessity, since conservation and scientific data that appear after the publication of the excavation report is seldom considered relevant by most archaeologists. The responsibility of the conservator does not end with the close of the excavation—it is an ongoing, never-ending commitment.

The threat to sites from the above-mentioned factors is enormous. Few national or governmental services or institutions possess adequate resources to carry out vital conservation programs on their own. Public interest and support are, therefore, indispensable to the preservation of our cultural heritage. I firmly believe that the only hope for sufficient funds to meet this emergency lies in harnessing the great growth industry of the late twentieth century—tourism. If we do not ride upon its back, this giant will surely swallow us up.

Henry W.M. Hodges

PREFACE

Henry W.M. Hodges is presently Professor of Artifact and Archaeological Conservation at Queen's University, Canada and was previously Lecturer in Archaeological Technology at the Institute of Archaeology, University of London. The author of standard works on the history of technology, he is a Council member of the International Institute for Conservation.

These papers are the product of a conference held in Mexico City in April 1986. The primary aim of the conference was to bring archaeologists and conservators together in order to discuss problems of common interest. From the outset it was evident that it would be impossible to include all the areas of concern that might be voiced by these two groups of specialists, and that we would have to be circumspect in the selection of topics if the conference were to achieve its goal.

We decided at the outset that it would be more profitable to concentrate on site archaeology rather than on activities within a museum. There were two good reasons for making this decision. First, the whole area of stabilizing and reconstructing artifacts in the museum laboratory is amply covered in the current literature, whereas coverage of monuments and artifacts in the field is scant. Second, in a museum, the archaeologist and conservator may consult regarding the best line of action to take, and thereafter the conservator is left very much to his own devices; this contrasts with the state of affairs during an excavation when the two must work together very closely, and coordinated effort is absolutely essential to forestall the development of friction.

The choice of the title for the conference—*In Situ Archaeological Conservation*—may, alas, be somewhat misleading. In retrospect, we might have chosen a different way of expressing the intention of the meeting. By the term "in situ," we had in mind the original sense in which geologists first used the phrase—to describe rocks found in their original or natural context, that is to say, not transported away from the site of formation, as is the case with erratics (i.e., boulders carried some distance from their site of origin by such agencies as glaciers). Unfortunately, the term has also come to mean, "in the precise and exact position in which discovered," and the implications of this interpretation also had to be explored at the conference.

But the confusion with defining "in situ" does not end there. It must be recognized that not every object discovered during an excavation can be retained where it is found. This is especially true of those artifacts whose chances of survival would be imperiled if left exposed to the elements. The preparation for transporting these objects frequently occupies much of the conservator's time on a site, and it was felt that the logistics of this kind of operation should be discussed during the meeting. In the sense that these actions must be taken on the site they, too, are seen as in situ conservation.

Readers will notice that site conservation in one important geographical zone—the humid tropics—was not considered at this meeting. Initially this omission was an oversight; it became clear, however, that the subject is extensive and its attendant problems quite different from those encountered in colder or less humid climes. Instead of endeavoring to include this area in the already crowded agenda, we decided to consider the possibility of a separate meeting devoted exclusively to this topic.

The task of determining what should be discussed at the conference was to some extent made easier by the fact that it was to follow on the heels of another meeting, *The ICCROM/University of Ghent Conference on Site Protection* (1985), a report of which, prepared by Nicholas Stanley Price, is included among these papers. Even so, it was felt that site protection was an area that could not be totally ignored in the present conference, especially since it remains a major concern of both archaeologists and conservators everywhere.

These considerations gave rise to the concept that the conference should be organized along the lines of two complementary sections. The first of these would deal with the archaeological site in its broad context, but without duplicating the work of the ICCROM/University of Ghent meeting. We thus restricted the contributions to topics that were only lightly touched on at the Ghent conference: in particular, how best to preserve sites that are so remote that their protection, especially from vandalism, becomes exceptionally difficult; and how to preserve sites where other activities in the region are so altering the local environment that the site becomes endangered. The second section was designed to deal with materials found on excavations that cannot be left in situ.

It would be hard to think of archaeological sites more removed from centers of urbanization than the many such monuments in Algeria, and we were indeed fortunate to hear the reflections of Malika Hachid on how gentle coercion may be brought to bear on those who live near these sites to act as their guardians. But the damage inflicted upon archaeological sites is often insidious and less deliberate than direct vandalism. The more dramatic examples of such factors that cause destruction—hydroelectric dams, the ill-chosen siting of factories and power stations, and urban development in general—are frequently reported by the news media. Curiously, they are usually described in a style that might lead one to believe that these events are uncommon. Perhaps this stems in part from the fact that until now, there has been no general overview of the many ways in which various human activities can knowingly or unwittingly destroy archaeological sites. The paper presented by John Coles appears to be the first attempt to list all these potentially damaging influences. Indeed, any activity that changes the environment surrounding an archaeological site may surreptitiously cause its ultimate demise. Even such seemingly benign work as, for example, the drainage of marshland to create more ground for agriculture may so lower the water table that buried sites may be irreparably harmed. The steps taken to overcome this kind of problem touch on a subject—cooperation between archaeologists and nature conservationists—for which, unfortunately, there was no opportunity for fuller discussion during this meeting.

It was also recognized during the planning of the conference that archaeologists and conservators do not always, to put it plainly, see eye-to-eye. This, of course, is a state of affairs that both professions can ill afford, and the conference organizers felt that it would be highly salutary to examine both sides of the coin: to hear from archaeologists who throughout many seasons of excavation have worked closely with conservators, and from conservators who have worked closely with archaeologists.

In his excavations of the Basque whaling station at Red Bay in Labrador, James Tuck has worked closely with conservators for the past decade, particularly with Judy Logan, whose name appears with his as co-author in this volume. He generously allows that the operation would have been almost impossible without the cooperation of conservators, the more so since virtually every item discovered had to receive some conservation intervention. His analysis of the logistics of this excavation may well serve as a model for others who find themselves in a similar situation.

As a conservator, Charles Hett is also greatly concerned with logistics, largely because the transportation costs to and from the sites he describes in arctic Canada are so high. Not only does this mean that he may import nothing that is too bulky or too heavy, but also that everything transported must be appropriate. In short, the conservator must understand fully the environment in which he or she will be working before embarking on such an endeavor, and this is true no matter where an excavation takes place.

It is generally axiomatic that the further an archaeological site stands from civilization, the greater the probability that the archaeologist will find it expedient to include a conservator in the team. It is equally true that the further a site departs from the norm, the greater the chances that a conservator will be invited to join the excavation crew. It would be difficult to imagine sites that are more divergent than those in the Arctic Circle and those in Peru, described here by Sharon Gordon Donnan. Not only do their climates differ greatly, but both are far removed from what one might consider to be the archaeological norm. Yet these two sites have more in common than immediately meets the eye: Both are in regions of climatic extremes; both are sufficiently remote to create logistical problems for the excavator; and both have produced delicate materials that, unless properly handled from the outset, could easily be damaged irreparably. Finally, the manner in which materials have been handled on these sites emphasizes the need to appraise critically the local situation and to use wisely whatever resources are available.

Two other topics, although by no means unique to Mexico, are of particular concern to the Mexican archaeologist: the conservation of mudbrick (adobe) and stucco, and the preservation of mural paintings. All of these materials pose numerous conservation problems, more particularly because adobe and stucco have perforce to be preserved in situ, while mural paintings are certain to lose a great part of their impact if removed from their original positions. The preservation of existing structures in adobe is a long-standing problem that demands the attention of archaeologists, conservators, and physical scientists, whose cooperative efforts have demonstrated that what on its face might seem a reasonably simple undertaking is in fact a highly complex matter. Pamela French, who throughout her career as a conservator has been involved with the preservation of mudbrick structures in Turkey, is in an excellent position to survey the state of thinking on this topic today, to describe those procedures that have so far proved to be most successful, and to discuss possible future lines of investigation that appear to offer the greatest potential.

The need to request the cooperation of archaeologists, conservators, and physical scientists is reemphasized by Agustín Espinosa in his description of the work undertaken to preserve and restore the wall paintings in Bonampak, certainly one of the most important such sites in Mexico. Quite apart from any other considerations, the difficulty of preserving these paintings was compounded by the fact that they were supported by adobe walls that themselves required consolidation and reinforcement. The concerns engendered by stuccowork, used in Mexico not only to cover walls and floors but also as a medium for modeling sculptures, are very similar to those of adobe and mural painting. The work done in this field is described by Luciano Cedillo Alvarez, who adds yet a further dimension to an already difficult area: the sheer quantity of material that stands in need of preservation. His paper describes current procedures adopted by the Directorate of Restoration of Cultural Heritage, Mexico, INAH, and some proposals for alternative solutions. An identical set of problems is encountered by those attempting to preserve mural paintings. As Sergio Arturo Montero points out, if such decorations are to be left in situ, some form of protection from the elements—usually a roof—becomes essential. This solution, however, does not apply to all situations, and where several paintings overlay one another, in situ preservation becomes impossible and other means must be sought for their conservation.

The papers described so far concentrate largely either on the preservation of the whole site, or on materials that must—or ideally should—be left in situ. There remains a wide range of materials, usually described under the heading of small finds, whose fate needs to be considered. One needs to ask which, if any, may be safely left in situ. Furthermore, one needs to know what the potential hazards are of leaving some of these materials in their original positions, which of them have scant hope of survival if so left, and how best to cope with those materials that require particularly sympathetic treatment in their removal and transportation. Some of these questions have already been answered, as can be surmised from the papers in this volume—for example, methods of handling textiles (Sharon Gordon Donnan) or organic materials that have become waterlogged (James Tuck) or frozen (Charles Hett).

The remaining organic materials fall broadly into the categories of wood, skin products, or bone and bonelike substances. Since wood is the most likely of these materials to be used structurally, and indeed may constitute the entire structure on a site, one would likely wish to leave it in situ. However, wood is highly vulnerable to attack by biological agencies such as molds and wood-destroying insects. Apart from the use of biocides and water repellents, as James Spriggs points out in his review of case histories, there is a traditional means of preserving wood structures: the repainting or even replacement of elements when they become enfeebled. This approach may not please the purists. Nevertheless, the procedure does preserve the structure in form at least, and it is a traditional method. Therefore it should not, Spriggs argues, be lightly dismissed as being improper or unethical.

On archaeological sites of considerable antiquity, it is unusual for wood artifacts to survive; more commonly the material totally disintegrates leaving little but a stain in the matrix to mark where it once stood. This was the phenomenon encountered during the excavation of the Sutton Hoo ship burial described by Peter Van Geersdaele. It was decided to make a mold of the inner surface of the ribs and planking of the ship, which were perceptible only as a discoloration in the sandy soil of the burial mound. This large undertaking would have been challenging enough had the conservators been able to work alone and unhampered on the site. But more excavation remained to be done after the mold was made, demonstrating once again the need for close cooperation between archaeologists and conservators.

Skin and skin products, such as leather, can never safely be left in situ. As Elizabeth Peacock points out, decayed skin may even pass unrecognized during archaeological excavation and, unless the conditions for preservation are most unusual, the recovery of skin and its products from a site normally present serious problems. Not the least of the difficulties entailed by the recovery of skin stem from the fact that so little research of value has been done in this area. There are, it seems, few valid criteria for dealing with skin either during or after excavation and virtually no established methodology exists. Even with leather, a far more stable product, there are divergent views on how it should best be recovered and later stabilized. To say that the recovery of skin is an area that demands more research is an understatement. This is a field that has been so little studied that our methods are hopelessly empirical.

By contrast, the recovery of bone and other animal hard tissues has unquestionably received greater consideration. Even so, the correct handling of bone in the field, as Lars-Uno Johansson points out, does require some understanding of future preservative treatments, especially as improper handling of the material during excavation may prejudice later conservation methods. The subject is made more complex by the fact that excavated bone may be required to provide two quite distinct types of evidence. On the one hand, for archeologists, the primary significance of bone artifacts lies in the cultural and stylistic evidence they may provide. On the other hand, unworked bones, often the residue of meals, may be used by osteologists in a number of different ways—for example, to determine the nature of the diet of the occupants of the site, to ascertain whether the animals consumed were wild or domesticated, or even to identify the diseases to which the creatures were liable. Thus, while the osteologist is likely to prefer bone materials more or less in the state in which they were discovered, and is not particularly concerned with their longevity, the archaeologist seeks to receive bone artifacts that have been returned to a stabilized condition. The treatment of bone objects during excavation, therefore, must be compatible with their subsequent use.

It is often believed, quite incorrectly, that inorganic materials—metals, ceramics, and glass—being more durable than those of organic origin, may be recovered with far less attention to their well-being. While this may well apply in many cases, it is certainly not a universal truth, a point underscored by Louise Bacon in her discussion of the nonferrous metals—gold, silver, and copper. Although conservators have in recent decades achieved a reasonable measure of success in the stabilization of copper and its alloys, it remains far from proven that the treatments they advocate will provide genuine long-term protection from further corrosion. Fortunately, it is seldom necessary for artifacts of copper to be returned to an in situ position, and even less so in the case of gold and silver objects, if only because of the high risk of theft.

Time-honored methods commonly employed to clean ceramics in the field are, as I try to point out, often inefficient or, worse, potentially damaging. There is a need for fundamental research even in those areas where our methodology seems reasonably satisfactory. This plea for further research is made, directly or indirectly, by the majority of contributors to this volume, many of whom find themselves in an unenviable "Catch-22" position. Engaged to provide a service, whether on an excavation or in a museum, even the specialist conservator has little time, equipment, or encouragement to engage in research. Even the larger national institutions that are able to support a research team tend to restrict their endeavors to immediate issues, if only because they are ultimately controlled by those unable to see beyond the next governmental election. The same restrictions hamper those responsible for the protection and maintenance of sites. The measures they are able to take are essentially empirical, designed more or less ad hoc to meet a specific set of circumstances. The information we do have on this subject results from the success or failure of steps taken to cope with an immediate need. The concept of a research institute devoted solely to the preservation of archaeological sites does not seem to have captivated any national government and less so those who stand to profit most from it—the tourist industry—a point already made by Brian Arthur in his Foreword.

When arranging a conference of this nature there can be no way of predicting the direction of the discussions following each paper. One cannot anticipate what will fire the imagination of the delegates most strongly. Recognizing that these debates would in all probability raise many questions that were not addressed directly by the speakers, or to which they could have no ready answers, we decided to plan a final session wherein those topics that had appeared to generate the greatest concern could be discussed. Obviously not every question raised could be accommodated in this discussion, and we apologize to those who raised topics that time would not allow us to deliberate.

In addition to a visit to the Churubusco Restoration Center of INAH, delegates were able to enjoy guided tours to three important prehistoric sites: The Templo Mayor in Mexico City; the vast group of religious buildings at Teotihuacan; and the more remote religious center at Cacaxtla, parts of which are open to the public while much is still under excavation. These site visits were made even more enjoyable by pre-tour briefings on the conservation methods and the problems associated with the structures.

The visit to the Templo Mayor enabled the participants to experience an environmental phenomenon that must be virtually unique to Mexico City. The pyramid that served as a platform for the temple had been razed in immediate post-Conquest times, leaving the stump of the pyramid covered by a heavy overburden. The truncated pyramid proved to have been faced and refaced many times. The subsoil of drained marshland, behaved rather like a sponge, and as the overburden was removed, the buildings began to rise. This caused some dislocation of the walling of which they were constructed, thus compounding the difficulties of consolidation. Comparing the subsoil to a sponge is made even more apposite by the fact that the subsoil actually holds a high proportion of water. Where the excavations were carried to a considerable depth, as indeed they had to be, there was the problem of rising damp and its attendant deposition of salts on the exposed surfaces of masonry, stucco, and painted plaster.

Rising damp with deep excavation was also a major concern at the two other sites visited. At Teotihuacan, where the structures cover a valley, this was not too great a surprise. It was a considerably greater shock at Cacaxtla, where the buildings crowned a steep ridge, and where one might have expected better drainage. At the moment there seems to be no immediate solution to these problems, especially as the infested structures are for various reasons not susceptible to the usual treatment—the provision of a damp-proof course to isolate them from the source of moisture. It may well be that if a satisfactory answer is not found shortly, the prudent course of action would be to backfill these sites until a suitable treatment emerges.

The potential for a problem of quite a different nature could be discerned at Teotihuacan. There the tallest temple mound—whose apex offers a spectacular view of the entire site—is accessible to the public. Naturally, those who possess sufficient energy scale this monument for the sake of the dramatic panorama. A small number of visitors, suitably shod and moving with care, would presumably do very little damage to the structure. During recent years, however, the number of visitors to this complex has grown steadily, and the foot traffic shows no sign of abating. Ultimately some damage to the structure seems unavoidable if this situation continues. The deterioration of sites because of overvisitation, while recognized as a problem, does not appear to have been given much consideration, save in individual cases, Stonehenge in England and the Acropolis in Athens being classic examples. Obviously we are all proud of our various national heritages, and only with reluctance would we wish to restrict visitors unduly. But there has to be a time when we say, "Enough!" Unhappily, this too often occurs after the damage has been done. It should not be so.

On a far more positive note, the tours of these three sites allowed those of us who were visitors to Mexico to see a small part of the excellent work being undertaken by the members of INAH, who face a truly formidable task when one considers the huge number of sites that stand in need of protection. A major concern has been the provision of canopies to save the less durable artifacts, usually mural paintings, from the elements. At first sight these may seem disarmingly simple structures, but when one considers the corollary requirements—that they should not physically damage the site and detract as little as possible from the overall ambience—it becomes evident that they are far more complex than would appear, demanding the utmost from archaeologists and architects alike.

On almost any archaeological site involving buildings, some reconstruction is inevitable, either to consolidate structures that might otherwise collapse or in order to render the site intelligible to the visitor. The method adopted universally by INAH to distinguish new work from the original is to insert a line of small pebbles in the mortar at the juncture of the two. This system certainly makes it very simple for the visitor to "read" the site and to appreciate the degree of restoration that has been accomplished. In the past, reconstructions were often extensive, pushing the limits of what could be inferred from the remains. The philosophy today is far more conservative, and reconstruction is not even contemplated unless it helps preserve the site or makes it easier to understand. In the latter case, reconstruction is severely limited, the emphasis always being on preservation.

The whole issue of reconstruction, and its ethical implications, was quite deliberately omitted from the agenda of this conference. As far as archaeological sites are concerned, reconstruction rarely demands the expertise of the conservator. Such work is, indeed, usually the domain of the architect or building restorer in conjunction with the archaeologist. Moreover, it can be a highly contentious topic and deserves to be the focus of another conference. In the meantime, I believe, all archaeologists and conservators would agree with the modern thinking of INAH, which seems to say: "First things first; let us primarily preserve what we discover."

I know I speak for all the delegates to the conference when I thank our hosts, the Getty Conservation Institute and the Instituto Nacional de Antropología e Historia, for providing this excellent forum for the exchange of ideas. I would also like to thank our Mexican hosts not only for providing excellent formal papers and discussions, but also for their very great kindness, for the concern they showed for our well-being, and for the care they took in acting as guides and mentors on the sites we visited.

N.P. Stanley Price

PREVENTIVE MEASURES DURING EXCAVATION AND SITE PROTECTION: A REVIEW OF THE ICCROM/UNIVERSITY OF GHENT CONFERENCE, NOVEMBER, 1985

The term "archaeological conservation" conveys different ideas to different specialists. To an objects conservator it means one thing, to an architect it means another, and to an archaeologist perhaps quite another.

"In situ archaeological conservation," however, refers unequivocably to the site that is the subject of archaeological investigation. This, then, can be the common reference point against which all conservation actions are evaluated and the different specialties coordinated.

That said, it must be recognized that archaeological conservation calls on a wide variety of disciplines, and the coordination of them is no easy task. From the safe retrieval during excavation of microscopic organic remains to the preservation of a monument in its contemporary urban or rural landscape is a broad spectrum—too broad, some would say, to make a unitary concept of archaeological conservation at all credible.

Whether or not this be fair, it is undeniable that many instances of archaeological site or object conservation in the past have suffered from a lack of coordination between specialists. Any initiative, such as the present conference, that aims to promote communication is therefore to be welcomed.

The conference organized by ICCROM (International Centre for the Study of the Preservation and the Restoration of Cultural Property) and the University of Ghent in November 1985 also had that aim (1986). Before the Ghent Conference is reviewed, it is worth explaining its background.

Nicholas Stanley Price studied prehistoric archaeology at Oxford University, England and wrote a doctoral thesis on the early settlement of Cyprus. He has been Assistant Director and Acting Director of the British School of Archaeology in Jerusalem and Assistant Archaeological Adviser in the Sultanate of Oman. He was Assistant Coordinator of ICCROM's training course in Scientific Principles of Conservation until joining the Getty Conservation Institute in 1987 as Training Program Officer.

Background to the Ghent Conference

ICCROM has always included the concern for archaeological material within its broad commitment to the preservation of all forms of cultural property. Recent initiatives stem from the decision to organize a conference on conservation on archaeological excavations. This took place in Cyprus in August 1983, with the active collaboration of the Department of Antiquities of Cyprus.

The two aims of the Cyprus meeting were to review basic principles of archaeological conservation in the field and to examine the legislative framework within which they are carried out. The conference was by invitation only; the invitations went principally to a number of specialists, who were commissioned to write a review of a given topic, and to the Directors of Antiquities of the Mediterranean countries, who were asked to explain the legislative context of archaeological conservation in their countries.

The ten invited papers were intended to cover aspects of both object and site conservation, with the aim of bringing together these two areas of expertise that are often considered separately. The papers of the conference—revised, expanded, and edited—have been published by ICCROM (Stanley Price 1984). They have been translated into Italian and a Spanish version is in preparation.

The Cyprus Conference identified a number of areas in which further activity was desirable. One of these was the protection of sites by shelters and roofs (this was to be one of the themes of the Ghent Conference). It also became apparent that while there existed considerable experience with site conservation techniques, there were few means of disseminating that experience outside geographically or professionally restricted areas. Therefore, ICCROM circulated a brief questionnaire in the summer of 1984 to assess the level of interest in establishing a newsletter. A questionnaire sent to archaeologists, architects, conservators, and others potentially interested in the subject enabled the target audience to define itself.

The response to the questionnaire was also used in designing the program for the Ghent Conference and for publicizing it. The Cyprus meeting had had a Mediterranean emphasis; the Ghent one was envisaged as a regional meeting for northern Europe, but with the hope that there would be contributions from a wider area. This hope was fulfilled amply by the response to requests for papers.

The following themes were proposed for the Ghent Conference:

1. The creation of microclimates for excavation
2. The transfer of stratigraphic sections and features
3. Site stabilization and consolidation
4. Methods and materials for backfilling
5. The design of protective roofs and shelters
6. The protection of underwater sites

Based upon the papers that were submitted, the conference was organized around two main themes: preventive measures during excavation for land and marine sites, and protection of sites on display. The final section considers briefly some of the issues that did not receive much attention in Ghent. (A list of the papers presented at the conference appears at the end of this paper.)

Preventive measures before excavation

The papers collected under the theme, "Preventive Measures during Excavation," in fact dealt with a variety of topics ranging from pre-excavation planning to the transport of finds to museums. Most of the papers were by conservators accustomed to working in field conditions, but three were by archaeologists. A recurrent concern of the conference was the relationship between these specialists. If one accepts as ideal the inclusion of a conservator as part of the excavation team, to what extent should an excavator carry out field conservation when no conservator is available?

This question was asked by Melucco in the opening paper. She observed that in Italy instruction in conservation is not included in an archaeologist's training. (One need hardly add that this situation is not unique to Italy.) She noted that existing excavation manuals were also deficient in this respect and that laboratory conservators needed experience in field conditions. She then described a scheme in Italy that utilizes current excavations to train both excavators and restorers in in situ conservation techniques. Particular attention is paid to cases where conservation requirements conflict with stratigraphic investigation so that archaeologists and restorers can gain experience working together.

Several other papers illustrated projects in which they *did* work together. For example, the excavations by the Western Australian Maritime Museum, described by Mills Reid, McCarthy, and MacLeod, include a conservator as a member of the team. For marine sites, conservators are trained as divers so they can assist in the excavation and recovery of material, rather than only receiving it on land in urgent need of attention. On a long-term project in the Dakhleh Oasis in Egypt, the state of conservation of each site is assessed before any excavation is allowed.

On an urban site in Rome, reported by Nardi and Costanzi Cobau, excavation and conservation proceed simultaneously, with conservation requirements sometimes determining excavation strategy. For instance, for the excavation of a wall with surviving plaster on it, conservation experience suggested that, rather than exposing the plaster as excavation proceeded, a thin layer of earth should be left unexcavated against the plaster. This represents a sacrificial surface at which evaporation can take place and the soluble salts can crystallize without damage to the plaster. Once the whole height of the wall has been cleared, then the entire sacrificial layer can be removed. The risks of differential adjustment by the plaster to its new environment are thereby much reduced. Thus cooperation between conservator and excavator has led to better recovery of evidence.

Planning before the excavation to improve data recovery was also featured in other papers. Carver summarized current work at Sutton Hoo in England, where a number of techniques are being tested for maximizing information from the site while also preserving it. It has been found that careful excavation can retrieve details of Anglo-Saxon graves and burials from evidence consisting only of slight differences in soil color and texture. This has led to a project to study the chemistry of decay in sand and other soils and to develop methods of soil color enhancement to detect fugitive traces of decay products. For example, bone residues in graves tend to fluoresce in ultraviolet light and so are distinguished from coffin remains.

The project is also looking at the effects of modern agrichemical farming on the underlying archaeological deposit. On the basis of analysis so far, the site has been divided into six different zones that reflect differences in current survival of evidence. This zoning can be used predictively to guide future excavation work.

Planning before excavation was also referred to by McCarthy in describing the predisturbance analysis of marine sites carried out by the Western Australian Maritime Museum. The data selected for study include temperature, salinity, pH and dissolved oxygen content, water movement and purity, bottom analysis, corrosion products, and marine concretions. The specialists involved, like the conservators, are trained to dive in order to collect these data themselves before any excavation takes place.

Preventive measures during excavation

Regarding preventive measures *during* excavation, Barker reviewed methods of creating microenvironments during work on open sites. From the excavator's point of view, the destructive effects of the weather must be minimized, but not so much as to eliminate those variations in humidity and light that often aid in the interpretation of soil differences. Erecting a temporary shelter over the working area should achieve this end while also allowing excavators to work in greater comfort—and therefore more effectively.

The shelters may be large plastic covers of the type used for temporary outdoor exhibitions, which are kept inflated by maintaining a slightly positive interior air pressure; or extendable tunnels of the kind used in horticulture; or corrugated plastic sheeting on metal scaffolding.

The ability to control lighting and soil conditions under cover depends very much on the type of shelter chosen. In fact, there may be cases in which the disadvantages of installing a shelter outweigh the advantages.

Schofield referred to his experience of excavating inside standing buildings in an urban context. The benefit of a relatively stable climate around the excavation is largely nullified by the difficulties imposed by working in a restricted space. This often dictates the excavation strategy. Similar difficulties are created by the need to shore up the sides of deep trenches for safety reasons—an aspect of site organization that is often seriously neglected. Schofield compared different methods of piling for deep trenches on urban sites and also illustrated a purpose-built protective roof and public viewing-gallery for the Billingsgate excavation in the City of London. The high cost of these measures has to be seen in the light of the total costs of property development in prime locations and of the need to persuade developers to cooperate with and pay for archaeological investigation.

Under these urban rescue conditions, conservation of structures in situ constitutes the exception rather than the rule. In Rome, however, on the extensive urban site of the Crypta Balbi, structures are conserved in situ as excavation continues for future presentation to the public. Nardi and Costanzi Cobau reported on procedures for conserving structures, movable finds, and wall plaster (referred to earlier).

A notable intervention undertaken in the Crypta Balbi was the protection and consolidation of a mudbrick structure, a rare survival in the Rome climate. As with all important finds on the site, a rigid roof of corrugated plastic material on metal supports was erected to divert rainfall, with additional bamboo screening at the sides to reduce the insolation and drying out of the mudbrick. The structure was kept damp by an automatic water-drip system while excavation proceeded. It was then consolidated by injecting cracks and fissures with a solution of clay with a small proportion of hydraulic lime. Static properties were reestablished by infilling collapsed areas with modern half-baked bricks of similar clay. The use of clay-based materials exclusively is designed to consolidate the structure without changing its character until methods of long-term protection are agreed upon.

A different approach formed the subject of a paper on work in Soviet Uzbekistan by Abdurazakov. He had used in situ polymerization of isocyanate resin to consolidate mudbrick remains. His preliminary results after ten years of laboratory and field testing were encouraging and the costs were comparable to other conservation treatments.

For the lifting and removal of small structures or features that cannot be preserved in situ, two papers reviewed experience in Canada. From Quebec, Bergeron described the block-lifting of sections of a heavily mineralized prehistoric occupation floor, using PVA to consolidate and a plaster mold for lifting. He also reported on the conservation and molding of a waterlogged boat landing-stage, using polysulfide rubber. The relative merits of different techniques and materials for soil-section removal, molding, and casting were analyzed in the paper given by Hett. He emphasized how useful these techniques were both for archaeological recording and for museum display purposes. The topic is to be discussed at this Conference and so no further summary is given here.

The Canadian conservators had to contend with adverse cold and wet conditions. Problems in hot and dry climates for the recovery of excavated objects were described by Mills Reid with reference to Western Australia. Excavations on land and underwater are commonly in remote areas with very high temperatures and low relative humidities. Since fresh water is often in short supply, on-site conservation is generally limited to maintaining stable storage conditions and providing safe transport over rough roads. For logistical reasons, locally available materials such as seaweed and sand are preferred for packing marine artifacts for transport. To maintain a relative humidity of 100% during transport, the water-retention capacity of sand is augmented by treating it with polymeric soil-stabilizing chemicals such as Agrisoke and Erosel. Storage solutions used on site are often, of necessity, of salt water, with a biocide added to check the rapid biological growth promoted by high ambient temperatures. Large artifacts are stored wet in pits in the sand lined with black polyethylene. A shade cover has nevertheless proved necessary to reduce evaporation rates.

Preventive measures after excavation

Underwater sites also featured in discussion of site protection between seasons or after fieldwork is over. The work of the Western Australian Maritime Museum provided good examples of various approaches. In general, organic materials are only preserved when buried under coral debris or ballast from the wreck, which has encouraged a build-up of sand over the site. Accordingly, wooden wrecks are protected following excavation by being re-covered with rocks or ballast bricks so as to form a sand trap and thus reestablish an anaerobic environment.

In practice, winter storms may periodically uncover the wreck, leading to renewed marine borer attack. To prevent this, experiments were made using hessian sacking impregnated with a chemical antifouling agent, tributyl tin oxide (TBTO). While apparently successful, the tests were discontinued because of health risks to divers despite the use of protective clothing.

Measures to reduce the scouring of the seabed around wooden wrecks have included the dropping of seabed sand from barges and the construction of artificial reefs using tires, sandbags, or cement-filled bags. A different approach to protecting an unexcavated wooden wreck was reported by Meucci: Silicone-rubber molding compound was used to encase the keel and hull in order to prevent their disintegration through current action.

The lack of storage facilities in museums has sometimes led to the discarding of waterlogged timbers despite the information they provide. A project of the Danish National Museum involves reburying excavated timbers in carefully selected sediments. Jespersen (1985) reported these experiments at the Grenoble Conference and brought them up to date at Ghent. Different materials, including archaeological and fresh woods, paper, and cellulose gauze, have been buried in marine sediments on land but below the local groundwater level. Problems are posed by the weight of large timbers, and also by the tight schedule against which reburial operations have to work. If there is no compelling reason to lift a wreck, then in situ protection with sandbags, for instance, is to be preferred.

The paper by MacLeod et al. described methods to conserve the well-preserved wreck of an iron steamship. Predisturbance surveys of the marine biology and the electrochemical and physical environment of the site were carried out. Prior to raising the ship's engine, one of particular historical importance, a sacrificial cathodic protection system was installed on the wreck to prevent further corrosion while conservation facilities were being built. The protection system proved to be less effective than hoped because of changes in the seabed configuration during the experiment, which resulted in the anodes becoming covered with sand. Nevertheless, the method appears promising for the treatment in situ of large metal structures underwater.

Protection of sites on display

The second principal theme of the Ghent Conference was the protection of sites on display. By "sites on display," I mean those for which there is a deliberate policy to make them accessible to visitors rather than to backfill them or allow them to be destroyed.

The papers in these sessions revolved around three main topics: (1) the conservation and restoration of excavated remains left exposed, (2) the design of protective roofs and shelters, and (3) the complete enclosure of sites to create museumlike conditions. Many of the earlier papers had already touched on these topics; indeed, it is unrealistic to separate the study of preventive measures during excavation from those taken following it.

Two case studies were presented of simple conservation measures where local facilities were limited. Frashëri described procedures of wall consolidation and the installation of a small site museum at the medieval castle of Cruja in Albania. Zielinski reported on work on two Roman buildings in the Western Desert of Egypt. There, architectural conservation forms an integral part of a multidisciplinary research project. One notable aspect was the concern for pre-excavation planning before investigating an important Roman temple buried to its roof in a sand dune. In view of the deteriorated state of the exposed stone of the roof, a protective windbreak using indigenous plant species has been planted upwind of the temple, with the aim of reducing wind erosion and helping to stabilize the dune. A phased sequence of operations has been planned so that excavation and conservation can proceed simultaneously.

The topic of protective roofs and shelters was introduced by Stevens in a wide-ranging review of protective structures in use around the world. Many of his examples were of sites with mudbrick remains, on which temporary shelters in local materials may be combined with various capping or consolidation techniques. At the other extreme, Stevens surveyed contemporary solutions to wide-span roofing for spaces such as exhibition halls and sports stadia. Speaking as an architect, he advocated more adventurous approaches to the roofing of archaeological sites than are presently used.

Case studies presented by other speakers emphasized the need, first and foremost, for roofs that are effective and affordable in terms of local resources. Waane explained the difficulties of conserving well-known open-air Palaeolithic sites in Tanzania such as Olduvai and Isimila. Excavated many years ago, and provided at selected spots with shelters of local materials, the sites suffer from a lack of maintenance, due to the fact that they are inaccessible for several months of the year; during this time they are subject to severe erosion.

The question of whether sites, however important in archaeological terms, should be preserved visibly if they are in remote areas receiving few visitors, was posed more sharply by Agnew and Wade. They summarized ten years of excavation and conservation of the Lark Quarry dinosaur trackway site in central Queensland. Protective measures at this very remote site have included the consolidation of the trackway and the construction of a roof and walkway for visitors, which have in turn created further conservation problems. The lack of a conservation plan at the outset has resulted over the years in a series of errors and extremely high costs that are unacceptable, given the low number of visitors to the site.

The conflict between technical aims and aesthetic values was brought out by De Silva in a review of roofs erected over monuments in Sri Lanka. He recalled that there has been a local history of erecting roofs over existing monuments since the second century B.C. Aesthetic and historic factors have a particular importance in the design of protective roofs for monuments that are part of a living religious tradition.

Three papers dealt with the complete enclosure of sites for protection and viewing by visitors. In this context, though, mention should first be made of the contribution by Vouvé and Brunet, who illustrated their paper with examples of naturally enclosed sites, namely the decorated caves of southern France. Their paper outlined an overall approach to the protection of archaeological sites in the broad sense. They made a distinction between "active protection," which consists of urgent intervention against natural deterioriation factors; and "direct protection," which is a continuing program of measures against damage caused mainly by man.

"Active protection" includes the important concept of "protection perimeters," which are zones that influence the preservation of a site in different ways. The extent of the zones is carefully defined in terms of factors such as hydrology, vegetation cover, pedology, and geomorphology. Three perimeters of varying importance are delineated according to the degree of risk to the site that they represent (immediate, near, and distant perimeters). The result is a "map of vulnerability" for each site in its environment.

Both for naturally enclosed sites, such as caves, and those enclosed by man, a key factor is the maintenance of an appropriate interior environment. In the case of the Roman baths at Heerlen in The Netherlands, the site was excavated in the 1940s and then backfilled for thirty years, when a museum was constructed around the ruins in 1977. The enclosure of the site in a building immediately caused problems, including cracks in the subsoil, salt efflorescence on the Roman bricks, and condensation on the museum walls.

Analyses carried out five years later, reported by Boekwijt, indicated that by then the enclosed remains had almost reached a new environmental equilibrium and little further damage could be expected. The problem caused by salt efflorescence was now mainly an aesthetic one, requiring the development of appropriate cleaning methods.

Two other studies demonstrated how such difficulties could be foreseen and avoided by installing air-conditioning systems for remains in situ that are completely enclosed. De Witte explained the system designed for exhibition of medieval painted tombs under the floor of a church in Bruges. The initial high cost and present running costs of this project are now more than offset by visitor admission fees.

A more ambitious project was reported by Scichilone, who has extended concepts used in museum showcase design to the problems of protection of excavated remains in situ. Excavated Roman structures in a central public square were incorporated into a series of four showcases. A constant interior environment is maintained by a computer-controlled air circulation system that also excludes dust and biological contaminants. The showcases are made of specially developed glass that is condensation-free and resistant to visitor impact. This project represents perhaps one extreme of technological sophistication in this field. Scichilone would maintain, however, that long-term in situ protection of archaeological remains is not an easy task. Any proposal is likely to be expensive initially, if it is to be equal to the task.

Site protection and security

The final paper in the Ghent program dealt with the security of sites on display, an aspect of site protection that is often neglected. Biasiotti described the installation of alarm systems at a fenced archaeological zone in the Rome suburbs. He also demonstrated, by example, the inadequacy of several popular security devices.

The security of a site during its excavation must also be considered, and several speakers alluded to this problem. Carver mentioned security measures taken at Sutton Hoo, where the publicity about renewed fieldwork on a such a well-known site might have put it at risk.

Underwater sites are notoriously prone to looting while an authorized excavation is in progress. One response is to create "decoy sites" in the immediate vicinity of a wreck to divert looters. A more positive approach, adopted by the Western Australian Maritime Museum, is to encourage amateur diving on selected archaeological sites while minimizing publicity about others. To this end, underwater trails have been established that lead divers to wreck sites provided with an information plaque on the sea bottom. This idea of the "underwater display case" represents a promising approach to in situ conservation and display of marine sites.

Commentary

To close this review of preventive measures and site protection techniques discussed at the Ghent Conference, it is worth mentioning some topics that received little attention at the meeting.

One of these is the backfilling of excavated sites. If there is no explicit policy to conserve and maintain a site visibly following excavation, then it must somehow be re-covered so as to remain accessible for future investigation. This is not uncommon in practice, but rarely do excavators mention the techniques they have used.

At the Ghent Conference, both Barker and Melucco described methods for covering the excavation surface overnight and between seasons. Barker, the archaeologist, refers to the beneficial role of black polyethylene sheets in promoting the formation of dew overnight, which helps to keep the excavation surface damp. Melucco, the conservator, emphasizes how dew must be prevented from forming on objects left in situ overnight. Obviously, each point of view is tenable, depending on the particular conditions of the excavation, but dialogue between the archaeologist and the conservator is necessary to define what those conditions are.

Backfilling was also mentioned by MacLeod with reference to underwater sites, and by Boekwijt referring to the thirty-year lapse between the excavation and the enclosure of the Roman baths at Heerlen. Still lacking, however, are long-term studies of the effects on excavated remains of different backfilling materials and techniques. This is all the more necessary now that new materials, such as expanded clay and polystyrene sheeting, are being used (Mora 1984:97–107).

Little mention was made of the role of vegetation, as either a positive or a negative element on a site's potential for preservation. The choice of appropriate plant species is important for stabilizing backfilled trenches and for helping to consolidate actively eroding sites in desert and beach environments. There are few studies (see Zallar et al. 1979, and Hamel and Jones 1984 for exceptions) devoted to use of vegetation on archaeological sites to this end.

The consolidation and capping of walls received more attention (e.g., in papers by Nardi, Abdurazakov, Frashëri, and Zielinski), than did, for example, the foundations of walls and drainage. Several sites with a long history of conservation provide the opportunity for comparative studies of different consolidation techniques. In fact, very few follow-up studies have been made of site conservation or restoration carried out in the past.

If effective site protection appears to depend heavily on technical conservation expertise, there are two areas at least in which it also owes much to the discipline of field archaeology. Preventive measures during excavation amount, in the end, to good excavation and recording techniques. In general, the poorer the excavation technique, the more conservation problems are created.

The other area in which field archaeology has much to contribute is experimental archaeology. The information accumulated from such projects as experimental earthworks designed for academic research purposes is also potentially relevant to site deterioration and conservation studies.

Techniques of site protection are, of course, only a means to an end. They need always to be subordinated to the question "Is this site worth preserving visibly or not?" The answer will depend on those values that we ascribe to the site, whether aesthetic, economic, symbolic, or historic. Necessary technical measures may then be decided upon accordingly.

The Ghent Conference provided a forum for comparing solutions that have been adopted in a wide variety of cases. Though far from being complete as a "state-of-the-art" review, it gave grounds for believing that archaeologists and conservators are working toward a better mutual understanding.

If the site itself is taken as the shared point of departure, and if excavation and conservation are planned together, this trend must lead to a better preservation and understanding of the past.

References

Hamel, G. and K. Jones
1984 *Manual of vegetation management on New Zealand archaeological sites.* 2nd ed. New Zealand Historic Places Trust 17. Wellington.

ICCROM
1986 *Preventive Measures during Excavation and Site Protection.* Rome.

Jespersen, K.
1985 Extended storage of waterlogged wood in nature. In *Waterlogged Wood: Study and Conservation.* Proceedings of 2nd ICOM Waterlogged Wood Working Group Conference, Grenoble, 1984.

Mora, P.
1984 Conservation of excavated intonaco, stucco and mosaics. In *Conservation on archaeological excavations* N.P. Stanley Price (ed.). Rome.

Stanley Price, N.P. (ed.)
1984 *Conservation on archaeological excavations.* Rome.

Zallar, S.A., K. Siow, and P.J.F. Coutts
1979 *Stabilisation of coastal archaeological sites in Victoria: A pilot study.* Joint publication by the Soil Conservation Authority and the Victoria Archaeological Survey. Victoria.

ICCROM and the University of Ghent Conference on PREVENTIVE MEASURES DURING EXCAVATION AND SITE PROTECTION
Ghent, 6–8 November 1985

List of papers

A. Melucco	Archéologie de fouille et conservation: limites et domaine d'intervention, respectivement de l'archéologue et du restaurateur	*Session 1*
N. Mills Reid	On-site conservation of organic materials in Western Australia	
M. Carver	Sutton Hoo: management and intervention on acid sandy sites	
P. Barker	Temporary shelter and site protection	
J. Schofield	Practical problems in the excavation of Roman and medieval London	
J. Vouvé, J. Brunet	De la protection des sites archéologiques (S.L.), gisements, fouilles et monuments préhistoriques	

A. Bergeron	Conservation et archéologie au Québec	*Session 2*
C.E.S. Hett	Soil section transfers, moulding and casting	
A. Abdurazakov	Conservation experience of clay construction in Uzbekistan	
R. Nardi	Conservation on the excavation: the Crypt of Balbus in Rome	
A. Costanzi Cobau	Excavated wall plasters—conservation problems	
J. Podany	Observations from the Forum of Trajan	

I. MacLeod et al.	The excavation, analysis and conservation of shipwreck sites	*Session 3*
M. McCarthy	The protection of Australia's underwater sites	
K. Jespersen	Extended storage of waterlogged wood, when excavated and in situ	
C. Meucci	Le bateau arabe-normand de Marsala: la protection in situ	

A. Stevens	Structures nouvelles de protection des sites archéologiques du Tiers-Monde	*Session 4*
G. Scichilone	The site of the Cathedral at Atri: a case-study of in situ conservation of archaeological remains	
H. De Witte	Protection of archaeological sites: the Bruges situation and the example of the Church of our Lady	

S. Waane	The design of protective roofs and shelters: the Tanzanian problem	*Session 5*
N. Agnew, M. Wade	A case study of a palaeontological site—the need for planning and protection	
T.K.N. De Silva	Roof over a monument—Sri Lankan experience	
W.O. Boekwijt	Salt and crack problems in the Roman bath at Heerlen and the therapy	

G. Frashëri	Archaeological conservation in the castle of Cruja	*Session 6*
A. Zielinski	Conservation unit of the Dakhleh Oasis Project—an approach to on-site conservation and preservation in the Dakhleh Oasis, Egypt	
A. Biasiotti	Theft protection of archaeological excavations: a test case study	

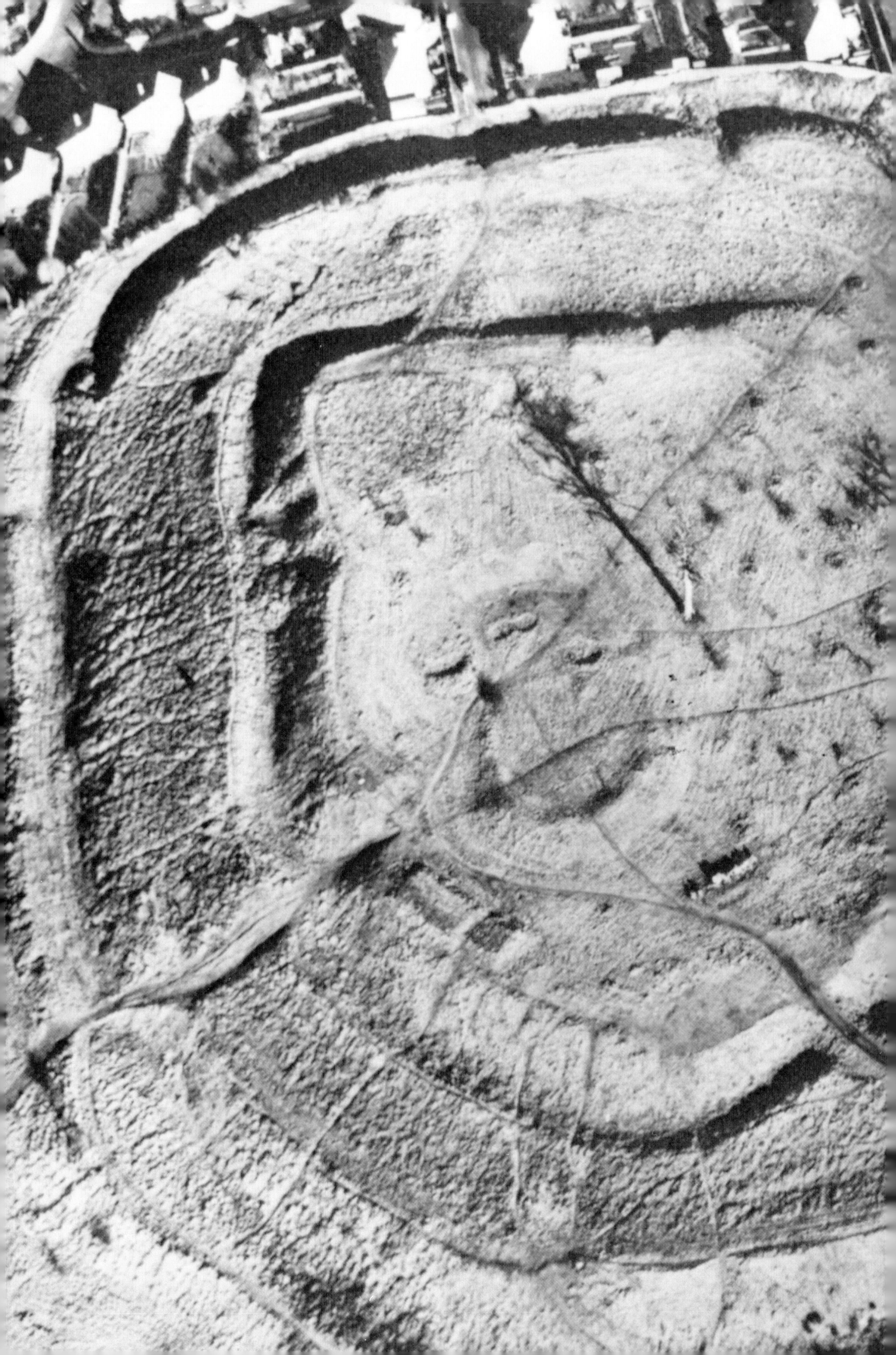

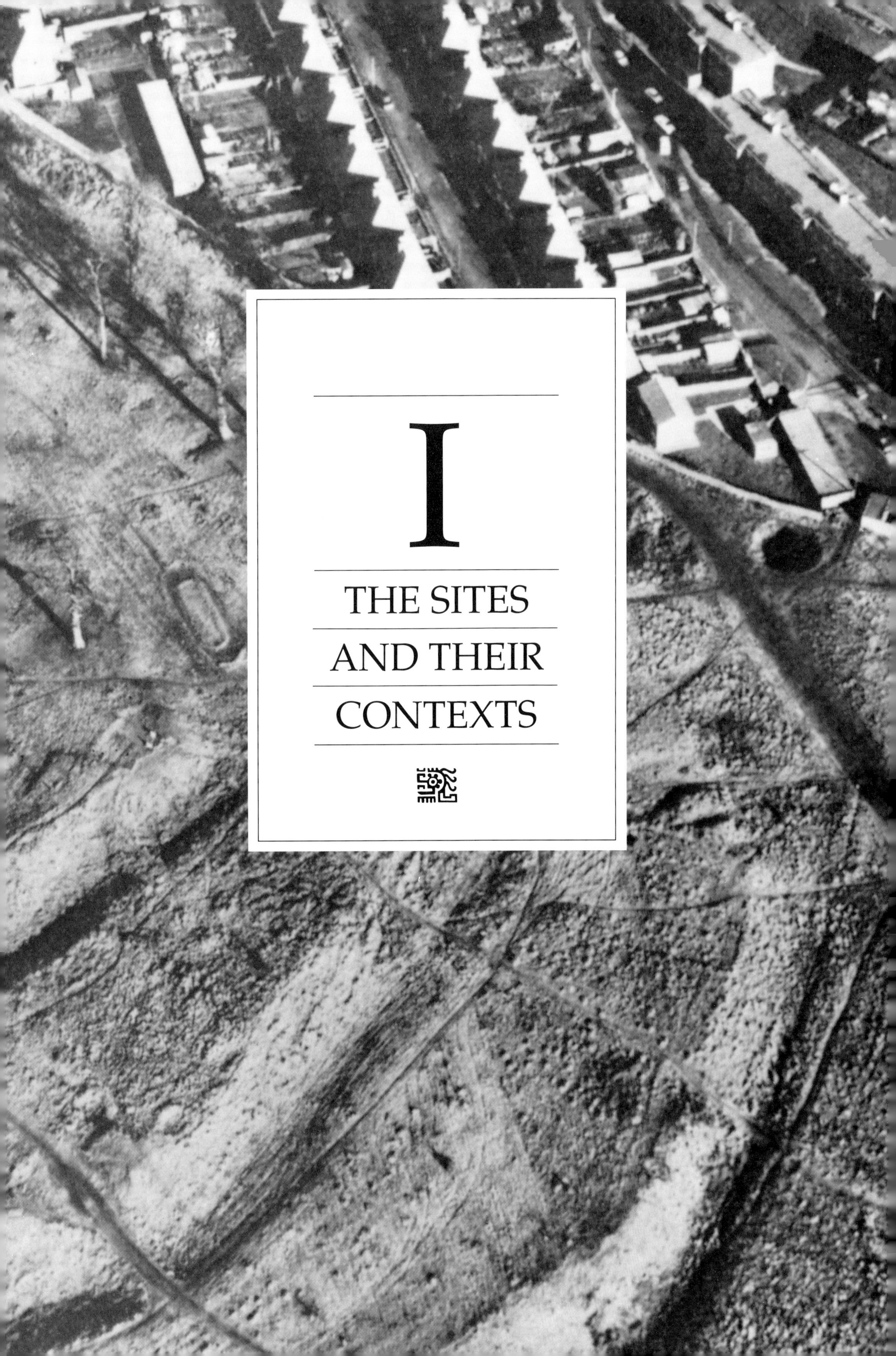

I

THE SITES AND THEIR CONTEXTS

Malika Hachid

IN SITU ARCHAEOLOGICAL HERITAGE IN ALGERIA: CHALLENGES AND RESOURCES

Malika Hachid is a prehistorian, specializing in the study of North African and Saharan rock art. She is a Researcher at the National Center of Historic Studies and Senior Assistant at the Institute of Archaeology in Algiers. She is the author of many articles which have appeared in national and foreign publications.

When Algeria achieved independence, it, like all developing countries facing pressing problems of economic and social development, at once saw the need for a policy that would protect its cultural and historical assets. At such a dramatic juncture, a nation's archaeological patrimony represents an excellent instrument for the reestablishment of its historical and cultural identity.

Despite valiant efforts made by the younger, often poor, countries, enormous difficulties still remain in managing the multifaceted tasks of restoration, conservation, and development of the national heritage. Algeria, for one, faces the most profound challenges: Its immense territory contains a multitude of significant archaeological remains, but these sites are difficult to protect from the continuous threat of destructive human encroachment. Like most Third World countries, and many developed countries, Algeria must contend with limited human, technical, and financial resources.

What follows is an analysis of the major problems encountered in the conservation and protection of archaeological sites in Algeria. Some short-term solutions available to archaeologists and the authorities are suggested, ones which can be implemented prior to a more substantial course of action. Algeria's case is not unique: Many countries in Africa and some in Europe face similar problems.

Rock art sites

The prehistoric rock art site illustrates the importance and complexity of in situ archaeological conservation and touches on all issues germane to open-air archaeological sites (pl. 6). It is also the type of site on which we have conducted the greatest number of protection and conservation experiments over the years. Of all the archaeological remains scattered across the country, rupestrian sites are the most vulnerable. Graffiti and lichen may alter a Roman wall to some extent, but they will destroy a painting or engraving on rock. Immovable by definition, rock art has meaning only in its original context. It presents all the problems inherent in the preservation of archaeological objects in a natural and historically significant environment. Exposed to all sources of natural alteration, its study must be closely linked to its conservation.

The first challenge Algeria must confront is the sheer immensity of its territory, some 2,381,741 square kilometers. In Africa, it is second only to Sudan in size, and a major portion of the country comprises Saharan zones with moderate to substantial archaeological content. For thousands of years, it has seen numerous civilizations succeed one another, from prehistory to the Ottoman period, embracing the Punic, Numid, Roman, Vandal, Byzantine, and Islamic epochs.

In the Sahara and the sub-Saharan regions, archaeological sites are so extensive that adequate control and protection are difficult to achieve. Some rock art sites, such as the prehistoric funerary monuments grouped in necropoles, may spread over several hundred square kilometers. Rock engravings on the Saharan Atlas follow exactly the natural contours of the 600-kilometer-long massif; the Tassili n'Ajjer Mountains, abundant in prehistoric remains, stretch across 700 kilometers (fig. 1). Thus, each site remains geographically isolated, exposed to all environmental dangers (Hachid and Maka 1979:107–114).

Such inaccessible, unmanageable terrain poses endless logistical problems. Every action, every type of work, is affected by the geography and climate. Travel, particularly to the central Sahara, requires specially equipped vehicles; but often walking or riding animals is the only possible form of transportation. Moreover, the scarcity of reference points, such as oases, makes the establishment of a base of operations very difficult. In such areas, interventions turn into expeditions, sometimes requiring acts of courage. To undertake research work or site conservation under such conditions means accepting considerable hardship and risk. For example, on the Tassili n'Ajjer Plateau, a standard meteorological shelter has been installed at an experimental station as part of a rock art conservation program. Thermohydrographs linked to recorders and powered by solar energy are placed near the painted rock shelters in order to obtain detailed microclimatic data. The control and maintenance of this experimental station, located in the northwest part of the Plateau, is very difficult. The closest village, Djanet, lies at the foot of the Plateau, 700 meters below the station—a walk of several hours (Brunet et al. 1985).

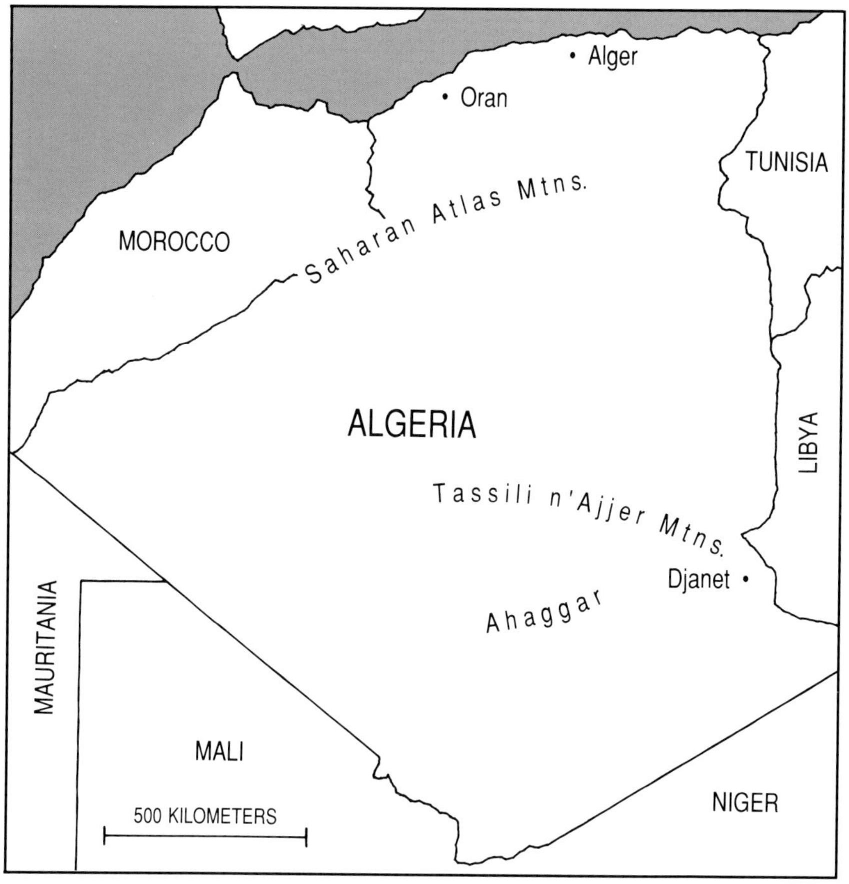

Figure 1. Map of Algeria.

Obviously, such an abundance of scattered archaeological sites raises the question of how they can be mapped and recorded. Locating a site and assessing its importance must necessarily precede the development and implementation of a policy of protection, as well as any research, conservation, or restoration programs.

While almost a majority of European nations possess archaeological maps, inventories in developing countries are still being compiled. In many African nations, this work increasingly takes precedence over other cultural policy needs. Since 1969, Sudan has been involved in a large-scale archaeological survey in collaboration with France (Raimbault 1983). Niger is in the process of completing an inventory stressing emergency zones where rescue archaeology is needed and compiling a list that will ultimately serve as a basis for a national research policy (Poncet 1983). Other nations—Gabon, Morocco, Mauritania, Mali, Senegal, Mozambique, and Lesotho—have undertaken this job with varying degrees of commitment.

In Algeria, where such a rock art inventory by multidisciplinary teams is urgently needed, the lack of human and financial resources has stood in the way of its completion. Nevertheless, a small research group of archaeologists and conservation specialists has devised a method of taking inventory (Hachid and Maka in press). It is based on a system of standard cards, all of them illustrated, that list the salient administrative, geographical, physical, and archaeological facts for each site. A "state-of-conservation" card accompanies each archaeological card, facilitating the establishment of conservation programs. In high-risk areas, this inventory is attached to a list of particularly sensitive sites, and a "red" list covering the most seriously threatened sites is given to local authorities.

Unfortunately, this inventory campaign, which represents the first attempt to come to terms with the vast size and singular nature of the archaeology of the Sahara, can only serve as a guide to action over the long term. The capacity for action is hampered by a lack of human resources, as is usually the case in countries that have neither enough trained scientists nor enough technical management personnel. Rock-art-rich Lesotho is so short of manpower that to gather information on just one of its 1,700 sites takes more than two years (Anonymous 1983:70). In Algeria it takes two months to record a painted wall, as thousands of others await attention. And in Mali, there were only two archaeologists capable of taking charge of activities, while a number of others complete their training (Raimbault 1983).

The impact of socioeconomic change

The second pressing problem facing Algeria and others is how to reconcile site protection with socioeconomic development. Not only does the passage of time adversely affect archaeological sites, but so does man and his activities.

Algeria is a vast territory now in rapid development. It is acquiring an infrastructure and increasing its industrial capabilities. Its population is growing, and urbanization is proceeding apace. New areas are being opened to agriculture, and the transportation network is expanding. But all of these activities pose an obvious threat to the environment and to archaeological sites in particular; progress goes hand in hand with the intensification and acceleration of destructive—and usually irreversible—processes. With few exceptions, the field of archaeology has fallen victim to degradation and destruction from these activities all over the world. Archaeology competes with profit-generating sectors of the economy, which take precedence in matters of development. Historically, the advance of industry takes a disastrous toll on the earth. In France, for instance, it was recently noted with alarm that for more than two decades the interactions between archaeologists and national development agencies constituted a chronicle of appalling scandals, resulting in a piecemeal policy of archaeological rescue in response to catastrophic events (Demoule 1982). The challenge thus becomes one of integrated development, whereby the closely linked economic, social, and cultural factors together contribute to progress.

In 1981, Oxford University hosted a colloquium on development and the historic environment. The consensus was that archaeology professionals throughout Europe have not yet addressed the problems of the preservation and presentation of cultural resources (Hughes 1982). However, North American countries, by requiring preliminary impact studies, are achieving the preservation of many important archaeological sites. For example the Canadian province of Quebec is a model of thoughtful policy (Pradeau-Moisson 1982).

But in Africa, the situation is acute. The building of two dams in Mali, in an area rich in archaeological remains, compelled the authorities to deal with the matter of inventory on a national scale (Sidibe 1983). In Niger, an emergency program was set up in response to the threat of the exploitation of mineral resources (Poucet 1983). And in Algeria, the national research agenda, especially regarding archaeology, is constantly being delayed by the numerous emergency rescue excavations.

The protection of rock art in the Saharan Atlas illustrates this problem perfectly. Because of geographical location and easy accessibility, these sites are now under severe threat (pls. 2–5). Located in the northern part of the country, this archaeologically rich zone represents an area where socioeconomic development, urbanization, and increasing road density subject archaeological sites to constant danger. With two conflicting demands for space—as an archaeological site and as a habitat for sedentary and nomadic peoples—it is vital that this region be transformed into an "eco-museum." Hundreds of rock art sites are permanently exposed to destruction by both public and private enterprises, including building sites and gravel production for roads. In addition, pointless defacements (e.g., graffiti and daubing with paint) are perpetrated by individuals.

The local population as site guardians

In situ archaeology is a class in living history that must become integrated into the social and cultural fabric of the country as well as its economic life. Clearly the protection of archaeological sites will not be assured unless an effort is made to integrate them into the modern environment. Ignorance of the archaeological and historical significance of a site most often explains the damage done to cultural property by the local populace. Some inhabitants of long-colonized countries share such a profound lack of cultural consciousness that some nations have felt the need to launch promotional campaigns through their national education programs. Between 1978 and 1982, the Mozambique government, for instance, conducted a "national campaign for the preservation and appreciation of the cultural patrimony" to encourage people to study and critically examine their own past and culture. Mature preservation agents led this investigation and received assistance from "interested circles," such as school children and their teachers, who collaborated in drawing up inventories of historical and cultural sites in their neighborhoods. People responded *en masse* to the campaign, thus proving that public awareness of the cultural value of the sites constitutes the first stage of protection (Anonymous 1983:48). And in Switzerland, the national program of research on methods of conservation of cultural property views the sensitization of the population to the issues involved as the first and foremost method of conservation (Lapaire 1985).

In the Saharan Atlas, one of the means of action, modest but efficient, similarly consists of educating the general population and the local authorities. Recently, the latter took the initiative of hiring paid guardians, thus relieving higher officials of some of the burden. This example proves that heightening awareness, especially in high-risk areas, can trigger a movement from awareness to reflection, and from reflection to action.

Conservation measures at threatened and damaged sites

But educational campaigns to heighten public and official awareness, a tactic that should not be overlooked, have only medium- and long-term effects. In the short term, namely in the case of imminent danger when emergency intervention is necessary, it appears that the only possible way to make a record of an imperiled site is to make a rapid reproduction of it. In such cases, we have chosen a three-dimensional molding process— casting with elastomer silicones (RTV)— which gives permanent impressions (Garcia and Hachid in press). The casting of rock art today has evolved into a scientific technique applied with rigor and caution, with archaeologists and conservators engaged in a joint effort to save the object. They also collaborate closely when they clean and restore damaged, decorated rock faces in an effort to come as close as possible to the original appearance. Suitable products and techniques can remove charcoal, paint, and tar smears, or fill scratches left by graffiti. An experiment of this kind was conducted on a heavily damaged Tassili n'Ajjer site. The clean-up process succeeded because conservation specialists joined the Tassili National Park guides and the personnel of the municipal services of Djanet (Brunet and Vidal 1982).

Impact of inadequate human and financial resources

No doubt the availability of financial resources ranks as the most important factor in the preservation of natural sites in almost all countries. Moreover, developing and developed countries alike suffer from a shortage of technical and human resources. The French minister of culture, for instance, does not hesitate to point out that "the inadequacy of means available to the services in charge of protecting, managing, and developing our rich archaeological heritage is patently obvious" (Lang 1982).

Algeria has not been spared this problem. Inadequate means and the centralization of those that do exist place the authorities in the difficult position of not being able to fully discharge their responsibilities. In such a vast country, protection and conservation are either impossible or difficult to carry out efficiently without help from regional administrators close to the specific problems. After all, it is they who are responsible for carrying out the policy set up by the central organization (regional centers, research departments, conservation and restoration laboratories, etc.). Moreover, this basic control is itself difficult to establish without scientific and technical personnel and equipment. Soon after winning independence in 1965, Algeria was forced to confront this vast and complex task with only three archaeologists. In addition, there was no tradition in conservation and restoration. The number of skilled personnel was barely sufficient to carry out all the administrative, much less the scientific, duties entrusted to them. The most important rescue and restoration operations were handled by experts with technical equipment provided by UNESCO.

With only four architects qualified to inventory, study, and restore sites and historical monuments, Algeria established a training program for restoration technicians and architects. Another technical training program with the aid of the United Nations Development Program is currently under way. While the number of archaeologists trained for field work has multiplied, there is still a shortage of specialized technical personnel. This shortage is critical in developing nations, and money alone cannot offset it; competent and sufficient staff is indispensable.

Faced with budget constraints, developing countries are forced to turn to foreign and international organizations for help. Their needs vary from assistance with site inventory and support in the study of protective measures, to the excavation of sites under threat, the building of museums and laboratories, the training of personnel, and the preparation of publications. Algeria is carrying out the conservation of the Tassili n'Ajjer site with financial, technical, and scientific help from UNESCO. Mali's five-year inventory program depends on French financial support, as does Niger's emergency archaeological project.

However, while people from these countries are able to acquire good training abroad, they are usually unable to apply their expertise without access to proper equipment and materials. For instance, in order to properly record rock art, certain supplies, such as polyethylene, special types of felt, elastomer silicone, and photographic paper and film, are necessary. Since these supplies are sporadically or consistently unavailable on the national market, researchers must order them from foreign suppliers. But these purchases are often impeded by bureaucratic delays, disrupting the continuity of the work. The specialist, therefore, may have to compromise by using substandard materials or go without completely. As a result, scientific research drops to mediocre levels. In India, rock art specialists with a yearly budget of only $500 must resign themselves to tracing frescoes onto plastic sheets (Anonymous 1983:73). And in Algeria, the installation of a fence can deplete the entire maintenance budget of a site.

The impact of tourism on archaeological sites

Developing countries rely greatly on visitors as a solution to their national financial problems, but tourism may cause equally acute site protection problems. While only 230,000 tourists visited Algeria in 1983 (compared to 41 million in Spain), much of the damage and looting of sites may be attributed to them. In the Sahara, artifacts are collected from open-air sites and clandestinely transported across the borders. In Mali, tourism ranks second as a factor contributing to the degradation of the archaeological environment; the indigenous population has primary responsibility for plundering ethnographic material. Argentina spends more money recovering stolen goods than it spends on educating the public. And in India, excavated artifacts are stolen routinely, often finding their way into European and American museums (Anonymous 1983:84).[1]

Unauthorized scientific investigation often goes hand in hand with archaeological theft. Each year, completely illegal publications on the Saharan archaeological patrimony appear in Europe without the permission or knowledge of the exploited nations.[2]

To strike a blow at such offenders, Algeria has passed strict laws. Broadly phrased and extensive regulations place all cultural property of historical, artistic, and archaeological national interest under legal protection. But a good many countries still lack a legal framework, and even when one is in place, it may not be fully operational. Ignorance of the law by the population and the lack of coordination and information can render its application ineffective. As a result, offenders are rarely apprehended or prosecuted. Nevertheless, this legislation remains the best support upon which to base decisions to protect sites under threat. Frequently, the application of the law depends on whether or not the relevant agencies, and especially the public, have been made aware of the importance of protecting artifacts. Here, too,

1. Today, European and North American museums require proof of the antecedents of an artifact. Questionable or nonexistent documentation usually precludes acquisition. Unfortunately, the same cannot be said of private collectors, who constitute the majority of purchasers of illicit goods. [Ed.]

2. This raises interesting, and so far unresolved, issues of international copyright law. For example, to what extent does copyright protect against the unauthorized use of photographs of a nation's antiquities? What may be regarded as illicit by one nation may be deemed as legal by another. Even so, there is no excuse for such blatent disrespect. [Ed.]

education plays an obvious role, sometimes leading to the spontaneous emergence of a body of volunteers. The amateur tradition is a valuable resource. Local people can be the most reliable guardians of their own heritage. For instance, in some regions of Algeria where the archaeological stock is under severe strain, an informal network of monitors—composed of teachers, high school students, official tourist bureaus, and "white fathers"—has been increasingly successful in gathering information and taking action. When well managed and properly channeled, the help this body of volunteers can give is not to be overlooked; it can develop into an instrument of first-rate guardianship. These monitors are presently collaborating in the drawing up of inventories for the Atlas region; in the central Sahara, they have thwarted numerous looting attempts.

In conclusion, drawing upon the experience of my own country, I have attempted to describe briefly the main problems, the very real challenges that confront a young nation: the abundance and the vastness of archaeological sites, the development of the territory with its harmful effects on the environment, vandalism, the lack of sufficient resources, and the administrative inadequacies. I have also described the limited resources and courses of action that must address urgent and diverse problems: the education of officials and the public, the need to recruit volunteers, the potential and limitation of legislative actions, and the difficulty of ensuring the necessary cooperation at all levels.

While the scientific community agrees that the archaeological and cultural heritage must be preserved, only governmental commitment can ensure it. The situation calls for organizational and budgetary support; without it, the authorities in charge of safekeeping and managing archaeological property cannot effectively carry out their mission. In developing nations, the archaeological site must assume an important social and cultural role. Above and beyond its historical value, it shapes the present and the future by contributing to the affirmation and enrichment of a people.

References

Anonymous
1983 "Préservation et mise en valeur de l'art rupestre, 1981–82." *Etudes et documents sur le patrimoine culturel* 1. UNESCO.

Brunet, J. and P. Vidal
1982 *Conservation des figures rupestres du Tassili n'Ajjer. Mission UNESCO en Algérie du 3 au 24 avril 1982.* Ministry of Culture and Communications. Paris.

Brunet, J., P. Vidal, and J. Vouve
1985 "Conservation de l'art rupestre." *Etudes et documents sur le patrimoine culturel* 7. UNESCO.

Demoule, J.P.
1982 "Archéologie et aménagement." *Nouvelles de l'archéologie* 9. Paris.

Garcia, M. and M. Hachid
In press "Les gravures rupestres de l'Atlas saharien." *Etudes et documents sur le patrimoine culturel.* UNESCO.

Hachid, M. and M.F. Maka
In press "Une méthode d'inventaire de l'art et des sites préhistoriques en Algérie." *Actes de la réunion internationale sur la conservation et la documentation de l'art rupestre de la région méditerranéenne.* Barcelona.
1979 "Prospection et conservation des oeuvres rupestres du Tassili n'Ajjer. Sahara algérien." *Caesaraugusta.* Zaragoza.

Hughes, M.
1982 "L'exemple britannique: Aménagement et environment historique dans les années 1980." *Nouvelles de l'archéologie* 9. Paris.

Lang, J.
1982 Correspondence in *Nouvelles de l'archéologie* 10. Paris.

Lapaire, C.
1985 "Méthodes de conservation des biens culturels." *Revue suisse d'art et d'archéologie* 42.

Poncet, Y.
1983 "La région d'In Gall-Tegiddan Tesemt (Niger). Programme archéologie d'urgence, 1977-81." *Etudes nigériennes* 7. Niamey.

Pradeau-Moisson, F.
1982 "Archéologie et grands travaux au Québec." *Nouvelles de l'archéologie* 9. Paris.

Raimbault, M.
1983 "Pour une plus grand efficacité de l'inventaire des sites archéologiques du Mali." *ler Symposium international sur l'archéologie africaine et les sciences de la nature appliquées à l'archéologie.* Phototype. Bordeaux.

Sidibe, S.
1983 "Pour l'urgence d'une archéologie de sauvetage." *ler Symposium international sur l'archéologie africaine et les sciences de la nature appliquées à l'archéologie.* Phototype. Bordeaux.

J.M. Coles

THE PRESERVATION OF ARCHAEOLOGICAL SITES BY ENVIRONMENTAL INTERVENTION

The nation behaves well if it treats the natural resources as assets which it must turn over to the next generation increased and not impaired in value. . . . Conservation means development as much as it does protection.

—Theodore Roosevelt

John M. Coles is presently Professor of European Prehistory in the Department of Archaeology, Cambridge University, England. In addition to excavating important waterlogged sites on the Somerset Levels, he has been a pioneer in the preservation of archaeological sites by environmental intervention.

With these words, the American naturalist and statesman Theodore Roosevelt formulated a principle of conservation that has often been neglected and sometimes willfully ignored. A nation's natural resources include its cultural heritage—those surviving traces of past human endeavor that mark the individual character of a nation and its people. Such visible signs of identity provide cohesion and help establish relationships that may otherwise be difficult to recognize in the makeup of society. Archaeologists have a primary obligation to identify, explain, and conserve the heritage of the past and to present it to the public not merely as a collection of antiquarian curiosities, but as a valuable and important resource for education and for the development of an awareness of human existence, struggle, and achievement. It is in the patterns of the past that we see our present state, and our aspirations for the future are conditioned by such patterns.

One of the more obvious ways in which our archaeological purposes and obligations may be met is through the presentation of important sites and monuments to society as a whole. It could be argued that such presentations, accompanied by explanations of the significance of the monuments, constitute the particular aspect of archaeological endeavor that justifies the continued existence of our discipline. All our other work could be seen as mere internal reorganization of the data base, satisfying only to those few who have the time to participate in the endless assembling and sorting of evidence. Whether or not this is true, almost every nation that has developed a policy for its heritage has emphasized the fundamental importance of preserving those monuments that have been identified as significant to society as a whole. In 1985, ICCROM (1985:315–316) agreed to principles that concur with such aims, recognizing the need for the development of strategies to identify and protect monuments, for adequate controls on the restoration of sites, and for suitable conservation measures both during and subsequent to excavation. For too long, passivity and neglect have led to the gradual deterioration of important and unique sites.

The examples of environmental control in this paper are drawn from Europe, where there are particular problems perhaps not encountered in other regions of the world. These problems derive from the dense settlement of Europe over a very long period of time. Today, large areas are being intensively cultivated or otherwise worked, which makes the need for environmental management extremely urgent. In northern and western Europe in particular, damp conditions lead to special problems affecting vegetation growth, land fertility, and (for archaeology) wet-site conservation. This is not to say that other regions of the world do not have other, even more severe, problems. Isolation of sites, lack of documentation and monitoring, vandalism, pollution, land erosion, and the like are among these major threats. The European examples may help to throw some of these into perspective and may suggest possible action.

All my information indicates that sites and monuments can be preserved successfully only under the following conditions: (1) legal intervention—protective orders, often incapable of application and imperfectly framed; (2) cultural intervention—management of human activities, including public education; and (3) environmental intervention—the establishment of suitable survival conditions for sites based on natural features of the landscape.

Sites and landscapes

No archaeological monument has ever existed divorced from its environment, no site was ever established in the past without a landscape setting, and no society has ever contemplated placing a structure at random. Every site we deal with has, or had, a landscape that was a part of the monument, that provided not only a base for the structure but also an environment in which it functioned. In most cases, each site was designed to suit its contemporary environment; the occasional cracked and slipped castle provides ample proof of the results for those who neglected to assess the situation. Lakeshore settlements set on deeply driven piles, massive stone monuments set on living rock or on wide flat rafts of stone or wood, structures at high altitudes strongly built to withstand snow and frost, fortress earthworks laced with timber and engineered against landslip, and burial platforms and pathways made of wood and pegged into marsh or swamp all demonstrate the ancient acknowledgment of the intimate relationship of site and setting. But the monuments of ancient societies—constructed to suit a particular environment and a specialized regime of human behavior—have often survived beyond those humans and now face conditions that threaten their very existence. Many monuments that were lost perished through the ignorant or willful actions of people who had not acknowledged the importance of the cultural heritage. The loss of thousands of burial mounds by plowing in Denmark and Poland, the demolition of stone monuments for building materials in Ireland and France, and the destruction of settlements by quarrying in Germany and Britain are examples of such cultural annihilation.

The pressures on archaeological sites are very great, yet measures can be taken to prevent or redress their destruction. Some of the pressures exerted on archaeological monuments in recent years are listed in Figure 1, and those for which environmental intervention is possible are discussed below. A few comments on cultural intervention are also given, although this is a separate subject, much concerned with education.

Farming	Clearance, cultivation, biocides, drainage, grazing
Forestry and vegetation	Planting, harvesting, drainage, scrubland
Quarrying	Rock, sand, gravel; on land and underwater
Water erosion	Lake levels, wave action, rivers
Land erosion	Temperature, precipitation, leaching, landslip
Tourism and urbanism	Erosion, housing, industry, pollution, transport
Wetland drainage	Desiccation, erosion, biochemical degradation

Figure 1. Pressures on archaeological environments.

Farming

Probably the greatest threat to archaeological monuments is posed by farming, that fundamental activity that has dominated the landscapes of many parts of the world for at least 6,000 years. It is only in the past few decades, however, that farming has changed from a relatively stable operation with noncumulative effects to an aggressive and relentlessly extractive process systematically damaging many traditional aspects of the land (Lowe et al. 1986).

The clearance of hitherto untouched woodland, including hedgerows and bands of vegetation, has caused erosion and deteriorated a stable environment. Deep plowing and panbreaking has removed the traces of untold numbers of ancient sites already reduced to mere shadows by natural processes of decay (Hinchliffe and Schadla-Hall 1980). The negative impact of agricultural activity on monuments has increased substantially with the advent of national or international agencies charged with the mandate of producing food at any cost. The European Economic Community (EEC) has a particularly bad record here, with millions of tons of excess grain in storage awaiting disposal either through natural means (deterioration beyond use) or uneconomic means (by sale below cost of production).

The plowing of earthen monuments gradually levels sites until their significance, in terms of heritage, is gone. There are almost too many examples of this phenomenon to choose from, but instructive aerial photographs of the deserted medieval village earthworks of Bureth Tupholme, England, show the immaculate precision of identifiable streets, houses, and small fields in 1963, and only their flattened, distorted images in 1976 (Baker 1983). The fact that earthen monuments, particularly burial mounds and hillforts, are often protected by law does not necessarily mean that agricultural pursuits already in operation can be stopped (fig. 2), nor does it deter erosion or the effect of animals. This seems to be the case in many countries, such as Britain, France, Switzerland, Germany, and The Netherlands. Even sites as important as the Wauwilermoos in Germany, where Mesolithic and Neolithic settlements cluster around a gradually shrinking wetland, cannot effectively be protected from activities related to farming, although major disturbances of the underlying layers are prevented (Wyss 1979).

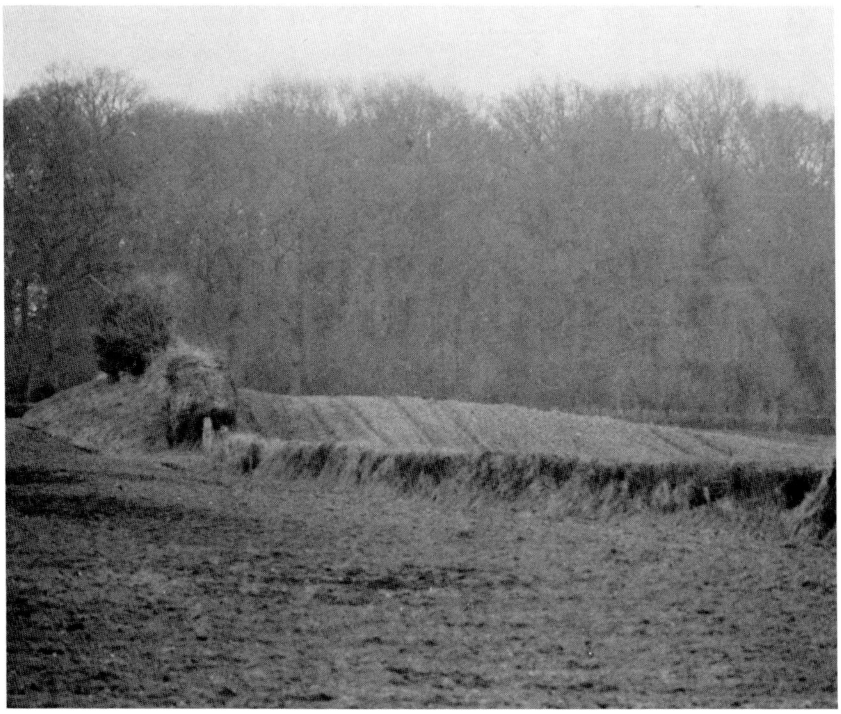

Figure 2. Plowing of a Bronze Age barrow, a long burial mound in Devon, England.

Agricultural damage to sites results from land drainage, which leads to desiccation; the spreading and spraying of fertilizers and biocides directly on the site or by drift from adjacent land; and uncontrolled grazing by livestock. Drainage in certain environments, such as marshlands and moorlands, can have disastrous effects on buried waterlogged monuments (as we will see below), and only firm environmental policies and interventions can control this practice. Although the deteriorating effect of acidic fertilizers on monuments is not fully accepted or understood, biocidal sprays designed to clear unwanted wildlife are known to affect adversely natural agents that aid archaeological conservation (see below). Ancient earthworks, as well as stone-built structures, suffer severely from overgrazing by cattle or sheep (fig. 3); the damage to an earthwork caused by rooting pigs may be compared to the impact of uncontrolled tourists.

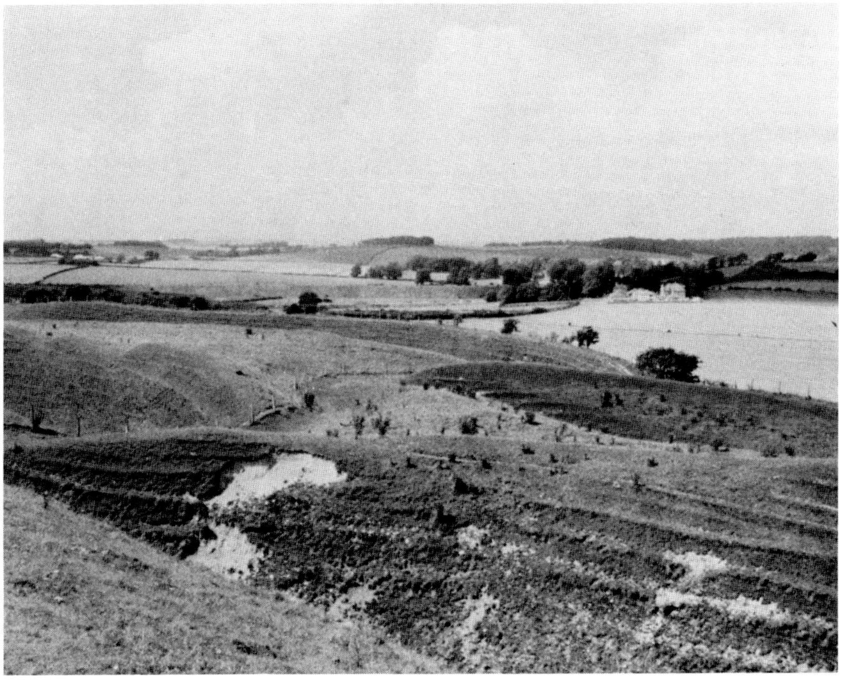

Figure 3. Cattle and tourist erosion on the ramparts of an Iron Age earthwork in Dorset, England. Although it is "protected," this monument has been damaged by overgrazing, weather, and lack of maintenance.

There are, however, hopeful signs amid these scenes of disaster. If an agency with jurisdiction over the land is sympathetic to the conservation of archaeological sites and monuments, opportunities for preservation exist, even if the land is being worked for agricultural purposes. Denmark's Conservation of Nature Act (1978) recognizes that cultural/historical features in the landscape are as important as its natural attributes, and preservation must take into account both scientific and recreational needs (Ministry of the Environment 1978). The Ancient Monument Act of Sweden (1942) protects every monument (about 450,000); approval is required for any action that could damage or alter a site (Hyenstrand 1979). In Britain, agencies, such as English Heritage, CADW (Wales), the Scottish Development Department, and the Department of the Environment for Northern Ireland have statutory requirements for the protection of ancient monuments, but their powers are limited; until adequate surveys are conducted and sites are assessed, the listing of "Scheduled Ancient Monuments" will be incomplete. In England alone, for example, of 635,000 known ancient sites only 12,600 are currently protected in this way. Moreover, this legal protection may have no effect on normal plowing (if it is already practiced on the site); erosion by cattle; rock fall, river, or wave action; scrub clearance or tree growth; or a decline in the water table (Wainwright 1985).

Many other countries have such policies, and, indeed, some embrace the entire landscape. Nonetheless, damage to sites still occurs in every country. Unfortunately, in France the problem is even worse because only formal acquisition of the land can ensure control over agriculture. Official policy statements seem to be universally rejected by French farmers, and while this reaction may be laudable, it rarely benefits archaeology.

The cooperation (or at least acquiescence) of landowners and landusers is crucial to the provision of appropriate environmental conditions for monuments. Such protection can sometimes take the form of environmental intervention. For example, the National Trust in the United Kingdom was originally established to acquire places of historic and natural beauty in order to allow the "preservation of their natural aspect, features and animal and plant life." The Trust is one of the largest landowners in the U.K., and among its holdings are perhaps as many as 20,000 archaeological sites (Thackray and Hearn 1985). In many cases the lands held are worked under tenancy agreements, and conservation interests are negotiated with each successive tenant. On the Isle of Wight, a strip-field system, rich in wildlife and ancient hedgerows, is managed along traditional, nonintensive farming lines in order to preserve both human and wildlife features. In Cornwall at West Penwith, part of a very ancient land-use pattern (probably Iron Age in origin, ca. 500 B.C.–A.D. 100) is preserved by stone fences forming small fields. The stone walls support specialized wildlife forms, while the fields are generally rather sparsely occupied by plant species. The archaeological interests, however, are paramount and the tenancy agreements restrict changes in farming patterns. Broadly comparable stone-lined holdings in other regions (western Ireland, for example) have lost up to 80 percent of their monuments because of changes in traditional farming practices inspired by the expansionist policies of the EEC.

The perpetuation of traditional, noninnovative agricultural practices—by inducement, agreement, or law—represents a form of environmental intervention that often comes close to the aims of other agencies, particularly nature conservancy interests. So much land in Europe has now been taken for agriculture, and the methods are so artificial—broad deep sweeps for the plow and heavy use of fertilizers and biocides—that few areas remain for remnant colonies of traditional wildlife, plant or animal. Earthworks of many kinds provide habitats for these relict survivors, and increasingly there has developed a common ground between archaeologists and nature conservationists. For example, several Bronze Age barrows (ca. 1000 B.C.) in Zealand, Denmark, have been studied as habitats for plant communities that rarely survive elsewhere in the area (Ravnsted-Larsen 1983). Shade and temperature preferences for a wide variety of these plants were studied on the Hagbards Høj, a round barrow, which by definition offers all aspects to the sun. The preservation of such monuments through environmental control is thus desirable for both wildlife and archaeological interests (see also Kelly 1981).

In a few cases, these interests may diverge. Archaeologists will want a light and uniform grazing on large earthen monuments and landscapes of ancient fields, but environmentalists might prefer a varied grazing pattern that allows dense turf to develop, enabling the support of certain insects (e.g., butterflies). Archaeologists are keen to prevent the erosion of earthworks and the exposure of subsoil, but others may see areas of erosion as an ideal environment for boring invertebrates. Burrowing animals in earthworks are another problem for archaeologists; unwelcome, but some are protected by law. Notwithstanding, the association of archaeological and natural conservationists is a logical outcome of their overwhelmingly common aim: the preservation of the landscape.

A proposed program of environmental management of an Iron Age fort at Cherbury, Oxon, England, is an example of the synthesis of the various aspects of such an intervention (Lewis and Miles 1985). This hillfort of eight hectares formerly sustained richly varied and prolific wildlife, particularly that of botanical interest. With farming, the interior of the site was plowed, cattle grazed and eroded the earthwork defenses, scrub developed, herbicides and fertilizers drifted onto the banks and ditches, and rabbits and badgers burrowed into and occupied the banks. Environmental intervention in this case would require the following: a botanical survey and assessment, some excavation to determine the extent of archaeological damage, fencing against humans and animals, careful removal and transport of the protected badgers, gassing of the rabbits, filling and repairing earthwork damage, removal of scrub vegetation, and the reallocation of the interior from arable land to pasture with fenced-in cattle. The cost of the program will be negligible compared to the value of the monument as a surviving example of the cultural heritage of the region.

Forestry and vegetation

In assessing the impact of environmental change on archaeological sites, it is clear that we must deal with landscapes. The historic evidence of human occupation and activity has often survived in tangible form, readily visible and understandable to all. The Roman frontier walls and forts, often restored, represent the earliest of these historic monuments, but medieval field systems, villages, and farms are equally preserved, if less heavily intrusive, in the landscapes of today. Their European predecessors—prehistoric sites and monuments—are less apparent and often survive only as shadow or cropmark sites. They are easily overlooked by both archaeologists and those working the land, and when obscured by heavy plant growth, they are defenseless.

Few surveys have been carried out on the effects of forestry operations on ancient landscapes, but the evidence suggests a very heavy loss of archaeological information. Forestry involves the clearing and uprooting of trees and scrub, machine planting and, sometimes, plowing to depths of nearly 1 meter and widths of almost 2 meters. Environmentally, the effects are severe; there can be an alteration in intensity or direction of drainage, monuments can be truncated, and upper deposits can be weathered and eroded (Lowe et al. 1986). In such cases, archaeological intervention can only consist of attempts to educate forestry operators, the surveying and marking of monuments, and the planning of subsequent action. I know of no examples in this field of an adequate archaeological response, or a successful environmental intervention, although in parts of Scotland there is a clear awareness of the problem and attempts are being made to inform operators and prevent damage (Jackson 1978).

Scrub vegetation is a natural result of the abandonment of land, whether it be farmland or archaeological sites. In Europe, scrub can involve both trees and shrubs, and its encroachment over a site can be serious, masking contours and disturbing underground features. The Iron Age forts of Badbury Rings and Hembury in southern England have been overwhelmed by scrub woodland and bracken; they are now being cleared, mostly by hand, and seeded with grass to permit grazing. Sheep and cattle will have an intervention role here, but only strict management will prevent the recurrence of the problems.

The common practice in some European regions of planting

trees on conspicuous burial mounds (e.g., the Bronze Age barrow at Olhøj in Denmark) had profound effects on the survival of such monuments, particularly where secondary burials were inserted into barrow sides (fig. 4; Thorsen 1982). These were easily disturbed by roots and, in addition, the dense foliage of tall trees, such as elms or oaks, discouraged the regeneration of grasses and herbaceous plants, which would have stabilized the mound surface and provided a habitat for other plants and animals. An intervention program for such monuments would involve the clearing of trees, the planting of grasses mixed with species of wild flowers and perhaps several low shrubs, and an annual mowing and clearing of grass cuttings to develop a firm thin turf.

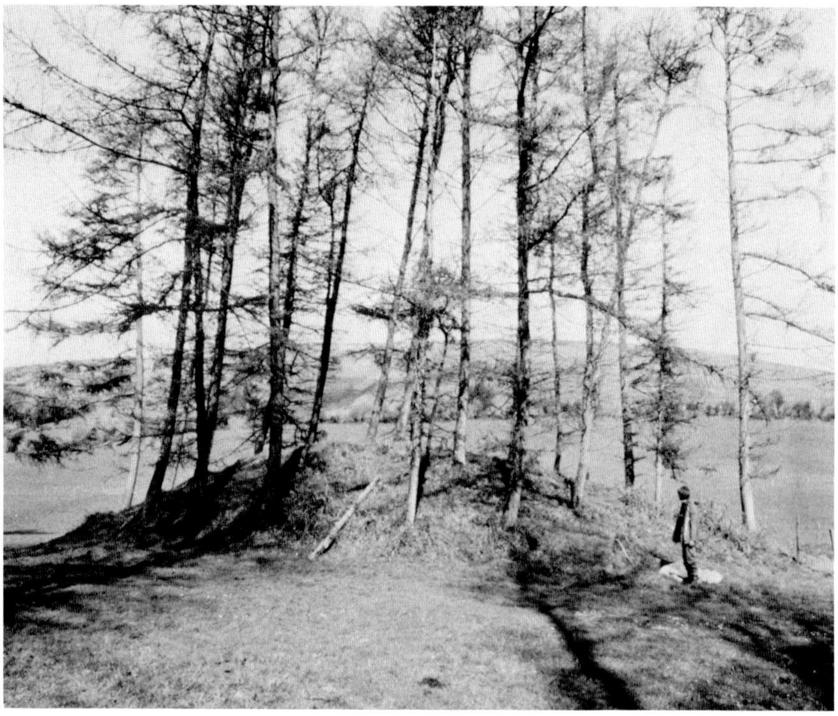

Figure 4. Scrub trees on a Bronze Age barrow, Slievebloom Mountains, Ireland. Whether self-sown or deliberately planted, such trees will affect the mound structure by root growth.

Quarrying

Many modern operations have a more direct and wholly irreparable effect on monuments and landscapes than the pressures of agriculture or forestry. There is an understandable, yet indefensible, attitude on the part of many agencies that, while some sites may have archaeological significance and should be left alone, their immediate surroundings do not deserve consideration and can be razed or otherwise damaged. Environmental decay around a monument will seriously damage the site, in the sense of loss of information and loss of appreciation of the place and role of the site within an ancient world. Perhaps the best example of such a relationship, and of such blatant damage, is the landscape of a hilltop enclosure at Navan in northern Ireland (Mallory n.d.). Navan was the focus of Celtic rituals in 600–100 B.C. and had an elaborate central monument overlooking a ritually important lake and earthworks. The site later became the capital of the Kingdom of Ulster. The limestone rock upon which Emain Macha (its Irish name) sits is now being quarried away in perhaps the greatest act of heritage destruction in Europe. Archaeologists are resisting extensions to the quarry, which now approaches the outer bank of the monument itself in an immense cutting about 25 meters deep (fig. 5). The spoil from the quarry is dropped in the lake (which contains many Iron Age and Celtic artifacts), thereby separating the two ritual sites and truncating the landscape. Only environmental intervention can stop

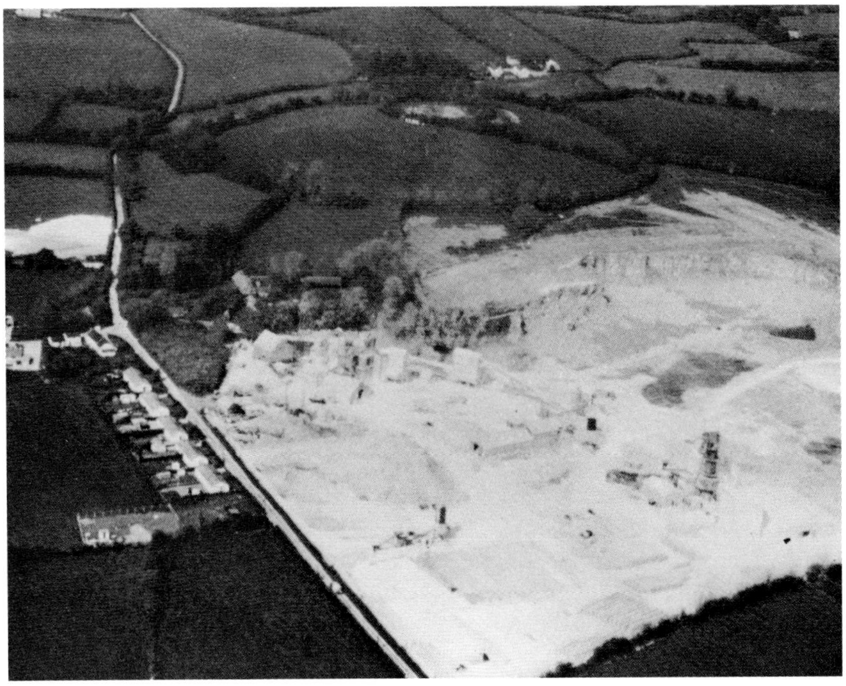

Figure 5. Quarrying for limestone at Navan, Northern Ireland on a Celtic ritual landscape. The circular Navan monument is at the center with the quarry edge approaching the periphery of the site. The Celtic lake lies off the photo to the right.

this total destruction. The enormous hole made by quarrying must be backfilled, the lake and contours of the land must be restored, and the landscape must be turned over to grass with a regimen designed to enhance the archaeological importance of the monument in its setting. The law must sometimes intervene in order that environmental and cultural control can be achieved over such nationally significant sites. A recent application to extend the quarry at Navan even further onto the site has been refused, through the vigorous efforts of a group of archaeologists; but the present damage remains unchecked.

Quarrying is not restricted to rock alone, and many archaeological monuments are at risk because of such operations, both on land and under water. Gravel digging by dredger, combing of fossil oyster beds for fertilizer, and channeling work along the coasts of western Europe (and sometimes in inland lakes, as in Switzerland) have yielded many archaeological finds. Only in Denmark does there seem to be an adequate environmental response. Off the coast of southern Fyn, for example, diving surveys have located more than forty Mesolithic and Neolithic sites, some with good organic preservation (Skaarup 1983:137–161; and see Fischer and Sørensen 1983:104–126 for other surveys). All of these are now protected by legislation, but environmental control over adjacent dredging is necessary to prevent erosion and currents that could damage the sites. Similar procedures have to be imposed to protect the many submerged wrecks known to exist in some of the fjords and harbor approaches in Denmark, such as the scuttled Viking boats in the Roskilde fjord. Although there are agencies for protecting sites by legislation, there are great difficulties in imposing and maintaining particular environmental conditions underwater (see Muckelroy 1980 for notes on sites and protection).

Water erosion

An underwater environment may seem to offer no problems requiring intervention, but erosion and shifting deposits can be severe, difficult to detect, and almost impossible to prevent (fig. 6). Yet, in some circumstances, shallow waters can be controlled. In Finland, rock painting sites along lake shores are usually above the water level, although some northern sites are subject to water damage

Figure 6. Erosion by flood waters creating a gully in sand and gravel deposits in northwest Ireland. Neolithic stone walls (upper right) from ancient field systems are being displaced by such erosion.

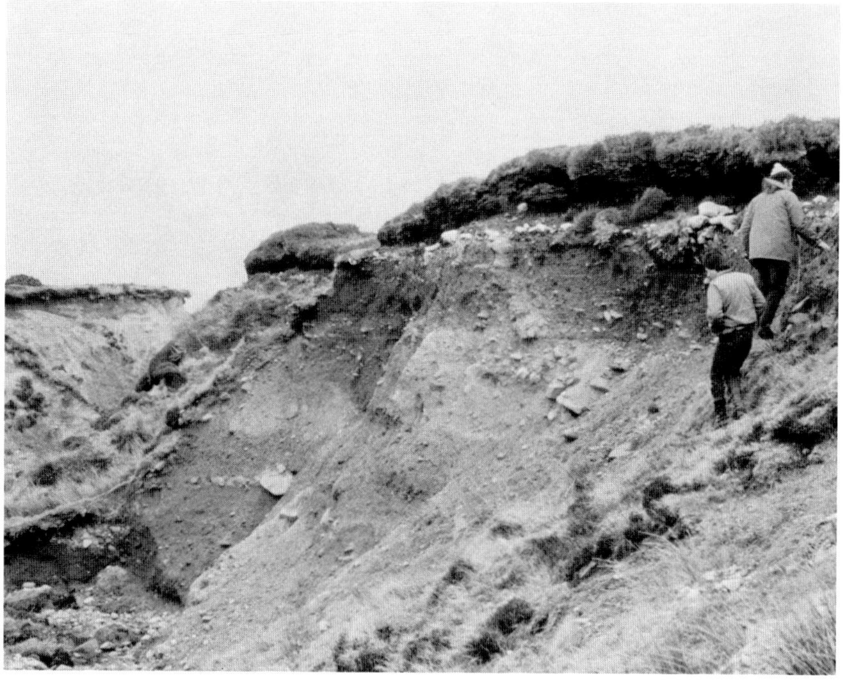

caused in part by storm action, but primarily by changes in lake levels. Erosion may be partially prevented by building rock walls to break the force of the waves. The rocks for the walls are brought to the sites across the frozen lakes in winter, a good example of an environment aiding in its intervention.

Elsewhere, archaeological monuments are more severely threatened by lake waters. In western Switzerland, many Neolithic and Bronze Age settlements were constructed on the shores of lakes Neuchâtel and Bienne. Overwhelmed by rising lake waters, these sites were protected by sands and chalk silts and kept waterlogged, thereby preserving unique evidence about structures, industries, and economies of the period circa 3000–500 B.C. Adjustments to the overall water sources in the Jura in 1880 lowered the lake levels; falls of more than 2 meters exposed the protective sands and silts. Storms and wave action began to erode the deposits, and further adjustments of the Jura waters stabilized the lake waters at the level of the ancient structures. This shoreline erosion led to the exposure of thousands of wood piles from the prehistoric sites, some of which are now destroyed. Others, like Cortaillod-les-Esserts, are "leached" and most of the organic components have been lost. Several attempts to conserve the surviving sites have been made. One popular approach is to bury the site under sand and rocks. The overall effect of this procedure has been to deform and damage the structures and to introduce decay of the organic elements that made these sites so important. Neuchâtel-Bains du Crêt, a very large Bronze Age settlement, has suffered in this way, as have Neolithic Saint-Blaise and the leached site of Cortaillod-les-Esserts. Here, then, a policy amounting to environmental intervention has not succeeded (see Egloff 1982:109–118 for a survey of the problems).

A further development, seemingly of destructive character, has had surprisingly positive results. The planned construction of a motorway along the north shore of Lake Neuchâtel led to an archaeological survey that revealed complete Bronze Age villages submerged in the deeper waters of small bays. An archaeological service was set up to conduct a major excavation program, and one of its tasks was to devise measures to protect sites still buried by sands and silts that lie perilously close to the erosion-line of the lakes. At Corcelettes, a Bronze Age settlement, some piles are already exposed and decaying in the water (Vaud 1983:223–251). Intervention in 1983 consisted of a curtain of wood poles driven into the lake-bed just offshore, with the pole tops level with the average height of the lake. Behind the curtain, a beach of cobblestones has been laid up to the present shore, providing added protection against erosion. The aim is to stabilize the beach with vegetation, maintaining the water-saturation of the ancient deposits beyond, but preventing any further erosion. To date, the intervention appears to be successful.

Where archaeological sites lie near marine waters, added difficulties may well occur, not only through tidal and wave action, but also through the action of salt. At Lothal, for example (a city of the Indus culture, ca. 2500–2000 B.C.; see Sankalia 1974 for a general survey), the mudbrick structures have suffered from salt spray and atmosphere. Experiments are underway to solve the problem internally, as environmental change is hardly possible.

Land erosion

The effects of erosion and weathering by climatic events are similarly difficult to counter. Archaeological sites at high altitudes, subject to severe frost and heavy snow, need special measures to ensure their survival. The Great Moravian sites of Staré Mesto and Mikulcice (Poulík 1980:24–29) in Czechoslovakia have ancient stone walls with upper courses of modern stonework that strengthen and protect the older remains from winter frost. More elaborate intervention at the thirteenth-century monastery of Oujon in Switzerland has involved the excavation and reburial of some walling, with overlying deposits of soil shaped to depict the wall lines and provide protection against frost and snow at the 1000-meter altitude (Weidmann 1985). New forms of concrete and cement for use under these conditions fall more properly into the category of restoration.

Even at lower altitudes, sites may suffer damage through severe climatic conditions. Some Bronze Age engravings on rocks in Sweden have been affected by frost and snow. When snow melts, water covers the rock surfaces and severe frosts may then crack the rock, detaching flakes or small "plates" (fig. 7). The interventionist action here involves

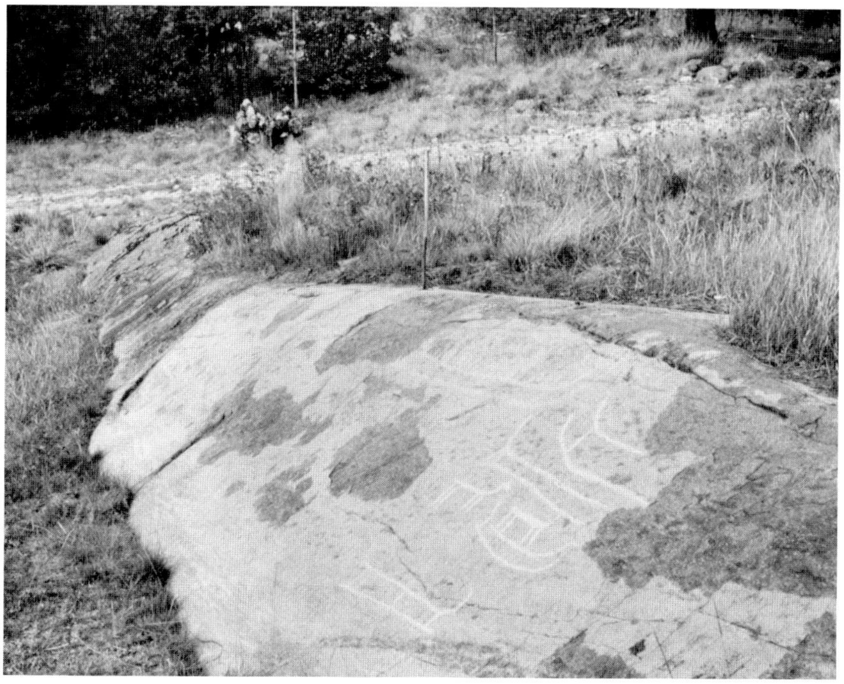

Figure 7. Severe frost damage of a Bronze Age rock carving at Vastervik, Sweden. Burial under earth and turf during winter months can prevent such damage.

burying the sites at risk by piling earth over the engraved areas; this also helps protect the monuments from vandalism. But there are thousands of such rock art sites in Sweden, Norway, and Denmark, and it is impossible to protect them all.

A variant approach to monuments, in a totally different environment, has been adopted in The Netherlands. For centuries, the Zuiderzee was an important body of water for transport and fishing. The shallow waters were usually calm, but mud banks and occasional storms often proved fatal for small and large ships. Since 1930, reclamation of the Ijesselmeer polders has taken place (see van der Heide 1959 for a historical survey), and as these lands have emerged through damming and pumping, hundreds of wrecks have appeared, the oldest so far dating back to the thirteenth century (fig. 8). Many await investigation and their conservation poses problems. Basic, of course, is the loss of the water that constituted their environment both during use (as boats) and after sinking. Drying the deposits would be disastrous to the survival of these sites.

Over the past ten years, about twelve wreck sites have been protected by environmental intervention. While the polder is still damp,

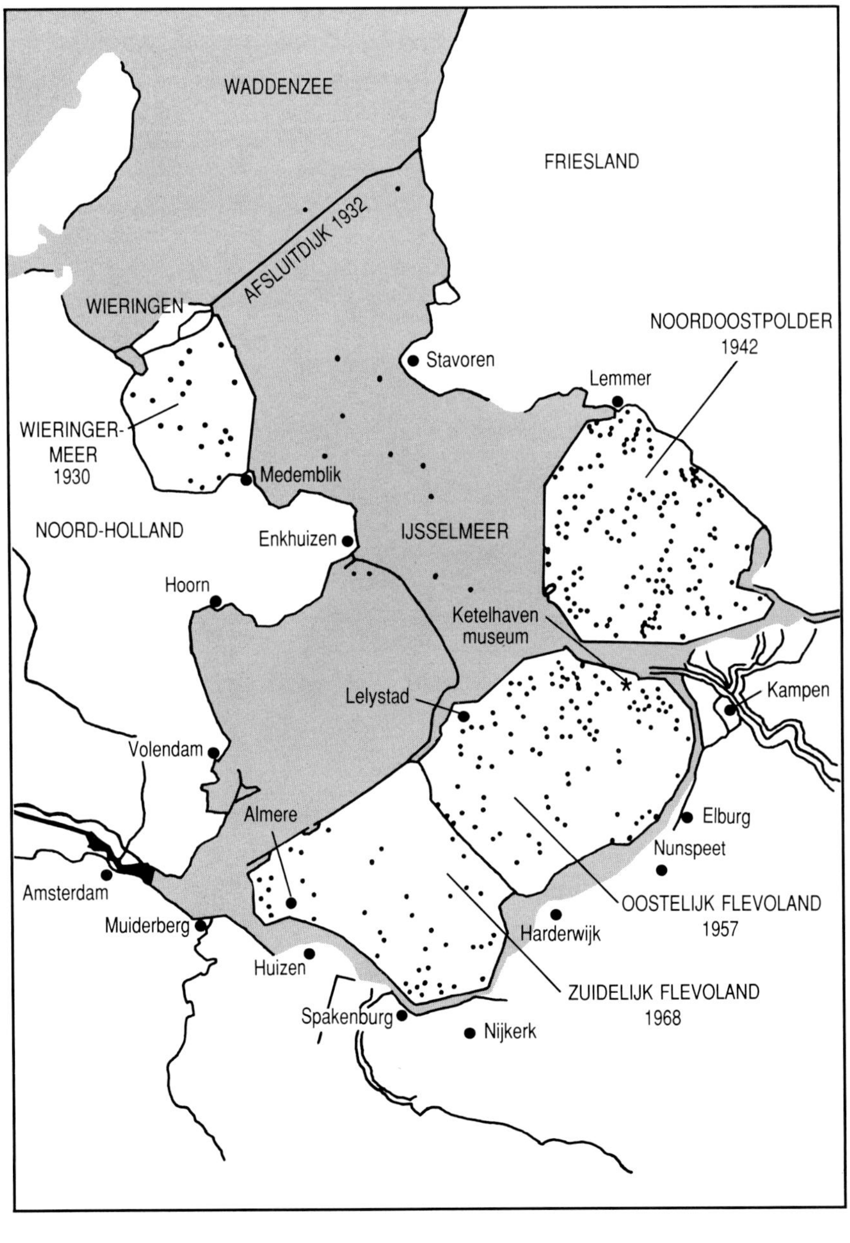

Figure 8. Map of the Ijsselmeer polders reclaimed from the North Sea, 1930–1968. Distribution of shipwrecks in the polders indicated by dots; some of these have been sealed to prevent desiccation.

the site is surrounded by a vertical wall of plastic foil extending from the surface down into the thick clay subsoil. As this base is impermeable, the effect of the wall is to create a "tub" of water, keeping the shipwreck and its enclosing silts and clays waterlogged (fig. 9). Where the wreck lies partially exposed on the surface, deposits of 1 to 2 meters of clay have been placed on top to help seal the site, making it possible to continue to introduce new water from time to time. In this way, the wreck may survive for ten to twenty years.

The reburial of sites of all kinds is increasingly accepted as a form of environmental and cultural intervention. The Neolithic site of Hekelingen near Rotterdam in The Netherlands has been steadily approached by urban expansion over the years (Louwe Kooijmans 1985), and it has recently been "submerged" beneath 1 meter of soil to help retain a measure of the formerly high water table, as well as to deter pillaging. Less obviously interventionist are Dutch restoration projects, such as the repair of eroded and damaged sand barrows, the sealing of the floors of megalithic monuments with concrete, and, of course, the imposition of legal requirements for the management of ancient sites to prevent plow damage, for example. Some of these measures are di-

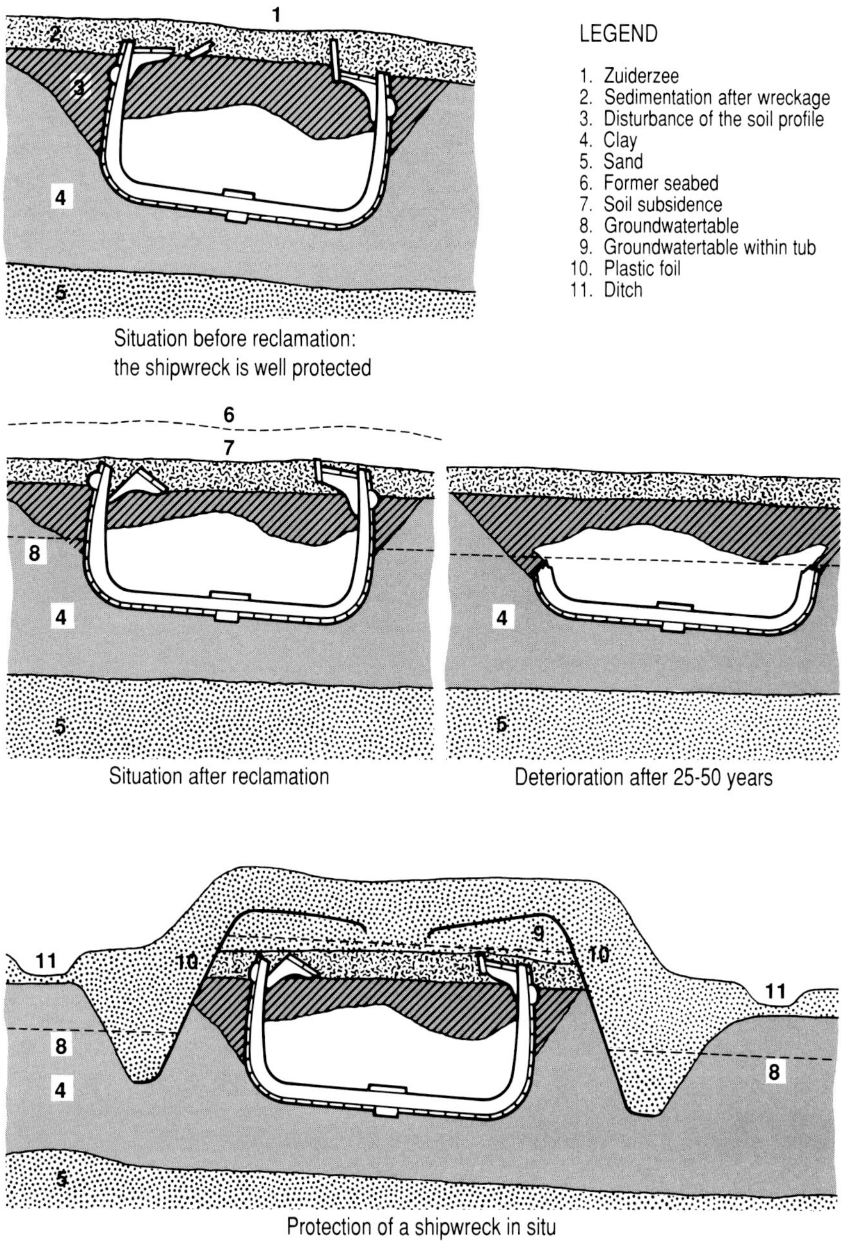

Figure 9. Scheme for intervention to preserve shipwrecks in the polders of The Netherlands.

rected at the public—the supporters (direct or indirect) of archaeology—whose interests must at the same time be encouraged and controlled.

Tourism and urbanism

There are few sites stable and strong enough to withstand an excess of popularity; sometimes environmental control, rather than intervention, is necessary. Take for example Stonehenge, one of the best-known prehistoric sites in the world. The small central megalithic monument (only 30 meters in diameter) received about 500,000 annual visitors until 1978 when entry to the circle of stones was barred. By then, the land surface of the monument had been seriously flattened and eroded (fig. 10). Even with barely any access to the monument's stones, there are now about 750,000 visitors to Stonehenge each year, and the pressures on the land, roads, and facilities are overwhelming (English Heritage n.d.). In addition, cultural intervention on the monument is required because of the erroneous belief that the site was Druidic. Hordes of people holding this and other equally preposterous beliefs descend annually on Stonehenge to go through certain acts, some of a ritual nature. The exclusion of visitors from the monument has led to erosion on and near the site and has transformed the landscape, which is so clearly a part of Stonehenge, into a battleground, with conservationists and archaeologists on one side, and the public on the other. The matter will not be resolved without firm environmental action, which by definition will seriously inconvenience the public. And it cannot be forgotten that it is the public who will be asked to pay for such action. As a bizarre example, the budget for rescue excavations for all of England in 1986 was £450,000; the proposed budget for police action to prevent the 1986 attempted invasion of Stonehenge by those with Druidic and other fanciful ideas was £500,000. Clearly, the costs of cultural intervention are high. One is tempted to suggest that a solution to this problem might be found by adopting one of the features of the Dartmoor National Park in England—it is a military training ground (Greeves 1985:85–88). This has several disadvantages, of course, namely the damage to sites by bombs and practice trenches, but on the other hand it discourages visitors! And wildlife (e.g., frogs) find suitable environmental niches in water-filled shell holes.

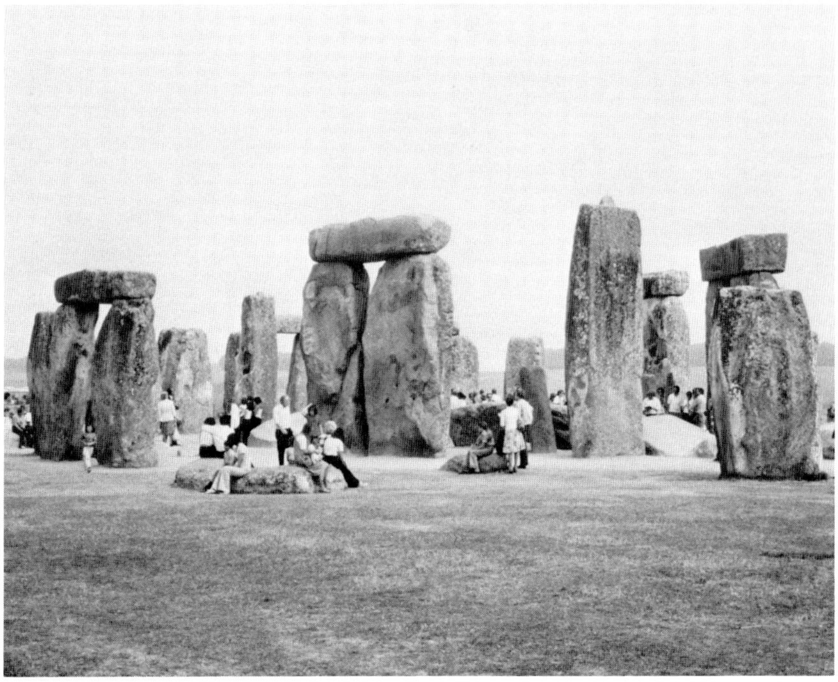

Figure 10. Stonehenge, England before tourists were banned from the central area of the monument. Note the eroded and worn area around the stones.

Human pressures on monuments are not restricted to tourism alone, of course, and whereas farming can sometimes be regulated to permit the survival of sites within their landscapes, this is not the case with urban or industrial development. Such pressures usually lead to serious damage, almost certain loss of the cultural environment of the site (e.g., Tredegar Camp, Wales, is almost totally surrounded by houses; fig. 11), and sometimes the initiation of decay by pollution or subsidence. Even the Taj Mahal, so immaculately and imperiously constructed, has suffered from oil refinery pollution. Its seventeenth-century marble has only recently had a coating of soot removed and its surface protected chemically from further damage.

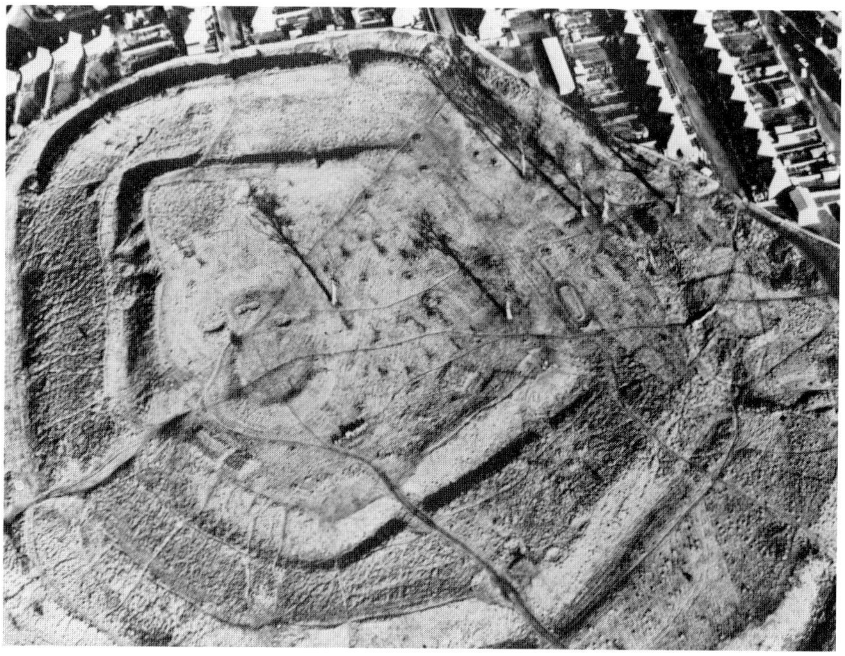

Figure 11. Loss of landscape by urban development around an Iron Age camp in Wales. The environmental setting of the site has been totally destroyed.

The public and conservation interests

It is not appropriate here to consider the wider implications of "public archaeology," but what is important to note is that in the long run it is the public who will determine the fate of all monuments of the past. Education at all levels is crucially important—about the past, about the monuments of the past, and about the duty of all to treat the cultural heritage with respect.

Legislation for the protection of monuments, as enacted in most European countries, rarely allows or makes provision for environmental intervention in the classic sense of those words, but some of the examples above demonstrate that opportunities for protection have been seized by archaeologists in certain circumstances. In Bavaria, for example, although sites and areas of archaeological interest are protected in general (*Grabungschutzgebieten*), each case also requires separate negotiation in order to ensure its survival and appropriate management. The wide public interest in restoration, particularly of Roman forts and towers (Weber 1984; Bayerisches Staattsministerium 1983), as well as of other monuments (e.g., Iron Age ramparts; Abels 1984), show the genuine value of a broadly interpreted monuments protection act (*Denkmalschutzgesetz*; Bayerisches Staatsministerium 1974; Schwarz 1977).

To the north, in Schleswig-Holstein, conditions for the survival of monuments are different, and restrictions on agricultural and other practices for areas of archaeological interest are strict and legally enforced. Recognition of the unique habitats that ancient monuments provide for wildlife has already been noted, and here there is excellent collaboration between nature conservation and the archaeological authorities—to such an extent, in fact, that complete protection of monuments, such as the Danewerk, is made possible (Müller and Neergaard 1908). The Danewerk, or Danevirke—an earthen rampart more than 7 kilometers long, in parts 6 meters high, and 30 meters wide—protected the Danish north from the Saxon south in the eighth century A.D. It was used again in 1864 and in 1945, both times unnecessarily. This massive fortification has required protection by legislation and by the combined efforts of archaeological and nature conservation interests. The environmental regime for the monument has thus been controlled so that both disciplines benefit.

Similarly, the unusual topographical configuration of glacial "tunnel-valleys," in the Ahrensburg region, for example, where late glacial hunters operated a highly sophisticated reindeer strategy (Clark 1975), has also encouraged the establishment of many species of plants and animals. Protection of both cultural and natural habitat interests necessitates strict environmental control and intervention. It has been acknowledged in this area that adherence to rigid management practices can only be guaranteed if the land is owned by sympathetic interests. In northern Germany, it is the nature conservation authorities who are able to acquire land, with archaeological support, rather than the reverse. The Schleswig-Holstein example demonstrates the success of good documentation of resources, of careful use of legislation, and of close collaboration with nature conservation interests. The two subjects are, of course, included in the Conservation of Nature Act in Denmark, embodying an ideal toward which others are striving, and which was foreseen by the true pioneer of collaboration, Gudmund Hatt (Hansen 1984). It may be of some interest to note the comparable ideas for conservation of both sites and wildlife in Denmark and Germany (e.g., Solvang 1983; Glutz, Grewe, and Müller 1984). That the interests of archaeology and nature conservancy are very closely linked is also borne out by a recently published list of priorities for conservation in England (fig. 12).

Figure 12. Priorities in the selection of monuments and nature conservation sites for protection in England.

Archaeology	*Nature Conservation*
Survival/condition (quality)	Survival: quality and extent
Period: all types	Habitat type: all types
Rarity (unique sites)	Rarity value: rare species
Fragility/vulnerability	Fragility or vulnerability
Diversity (multiple or single features)	Diversity of interest (plants and animals)
Documentation: adequacy of data	Documentation: species records
Group value: landscape association	Group value: wildlife reservoirs
Potential (anticipation of importance)	Potential: future value

(Bracketed comments are paraphrased from the original. Based on Wainwright 1985 and Vittery 1985.)

Wetland drainage

While some archaeological monuments and sites exist in advantageous circumstances, there are also certain problems. Many sites are in wetlands—flat expanses of marsh and moor, often with peat deposits laying over rock, clays, and sands, all the result of glacial outwash and ice sheet withdrawal. Archaeological monuments in such environments are often well preserved (e.g., the mid-first millennium A.D. Feddersen Wierde settlement with house walls, sheds, and fences standing 1 meter high in places [Haarnagel 1979], and the still-intact wood roads from the Neolithic and Bronze Age [Hayen 1986]). Such wetlands are rich in wildlife and thus attract nature conservation interests. In Lower Saxony, evidence of settlement since Neolithic times may be found in the peatbogs. Wetland wildlife is abundant, and traditional rural pursuits are still practiced (e.g., peat cutting, moor burning, water-meadow grazing, and special cropping of reeds). All combine in efforts to conserve monuments, wildlife, and human traditions in the same areas (Hayen 1980, 1981).

On the other hand, however, all these interests are brought together by one element—water. High water levels are necessary for organic preservation on sites, for specialized wildlife habitats, and for traditional wetland farming sustenance. Environmental intervention to retain sites is therefore a subsidiary benefit to more widespread interests in the conservation of nature and of traditions. Thus the priorities for the preservation of particular ancient monuments cannot be selected independently.

There is, however, one fundamental difference in the effect of the outcome of any combined nature conservation and archaeological action in a wetland. Wetlands attract specialized wildlife that evolve and change according to environmental dictates. A drying wetland supports certain life forms, but if re-wetted, the land will again attract the specialized forms common to a true wetland. In other words, with nature conservation, wetlands may be restored to previous levels and even enhanced in some cases. This is not the case for waterlogged archaeological remains. Once they begin to dry out, they cannot be returned to their original well-preserved condition. The loss is permanent, and no future wetting of the environment can restore the archaeological elements.

In Denmark, wetland sites are known to have high archaeological potential, and environmental control and intervention are expressed in a direct form. In the Åmosen bog of Zealand, where sites of the Mesolithic period are well preserved, efforts are being made to identify particular sectors for conservation, surveys, test excavations, and coring, as well as to terminate drainage and restore water regimes suitable for wetland conservation (Clark 1975). Similar surveys are underway in the Illerup bog in eastern Jutland, where enormous sacrificial deposits of weapons, circa A.D. 200, have been identified and more material undoubtedly remains (Ilkjaer and Lønstrup 1982). The ritual practices that resulted in these deposits were almost certainly comparable to those described by Caesar in Gaul: War booty was bundled up and deposited in a water meadow or small lake. At Illerup, there are at least three such sacrifices in a drying deposit now being assessed for conservation. A similar analysis is going on at a sacrificial spring or bog pool from the sixth century A.D. at Isokyrö Lev aluhta in Finland (Hackmann 1913), and provisions are being made for a continuing water supply.

Just as wetlands are probably the most threatened archaeological environment, their physical survival as a land form is also in jeopardy (Coles 1987a). Wetlands are repositories for peat (the partial decay of centuries of reeds, sedges, mosses, and other plants) and are a valuable fertilizer for impoverished soils. The quarrying of peat in Ireland, Lower Saxony, and England has removed millions of tons of organic deposits, often containing or protecting waterlogged archaeological remains. Settlements, platforms, trackways, and countless prehistoric and early historic artifacts have been discovered; sometimes they have been examined, but generally they have been destroyed. In the Irish and northern German boglands, it has not been possible to conserve any monuments in situ, either because of the lack of an archaeological strategy or because of the difficulties of water retention (Coles 1987b; Hayen 1980).

In England, comparable problems exist for the huge wetland of the Fenland. Here, centuries of drainage have succeeded in transforming a wet organic peat and silt land into a dry fertile landscape suitable for cultivation. As a result, meters of organic soils have been lost through wind erosion and shrinkage, and the whole landscape has been lowered, except for the embanked watercourses that carry water above the surrounding land. Previously, the land was both wet and dry, with streams, marshes, islands, and peninsulas that supported many populations from Neolithic to recent times. As the land shrunk and eroded, the archaeological monuments that were buried by peats and silts have begun to emerge. Thus, some flat, featureless fields have become richly varied cultural landscapes, distinguished by protruding Bronze Age barrows (Hall 1985, 1986). Such sites may still have waterlogged deposits in their lower sections—uneroded, unexposed, and unexamined; in other words, they offer a virtually intact prehistoric landscape. The great and urgent need is for some measure of environmental intervention to ensure the survival of this unique evidence. How this can be achieved is not at all clear. For individual sites not truncated by drainage ditches, some form of walled enclosure (as used in The Netherlands for shipwrecks) might help to hold waterlogged structures for several decades. In areas with settlement patterns and ritual activities still intact, the procedures would involve the use of existing ditches to bring water to the area, barriers to retain high water tables and prevent shrinkage of the peat body, and grassland regimes to curtail erosion. The costs for such an operation would be high, but the results would justify them.

The same wetland problems have been encountered and overcome in the Somerset Levels in England, where archaeologists have identified hundreds of prehistoric structures built in the swamps and moors of a wide basin (Coles and Coles 1986). The earliest known structure circa 3900 B.C. is a Neolithic trackway, the Sweet Track, built of many varieties of wood (fig. 13). Detailed environmental and dendrochronological analyses, excavations, and experimental work have revealed that this 1800-meter-long track was built in a very short period of time (perhaps only one or two days!) by two groups of Neolithic farmers and gatherers using huge oaks and other trees from different managed woodlands on the hills and islands of the Levels. The many studies of this structure inform us about local and regional environmental conditions, climatic change, the elm decline that often signals the onset of farming practices in western Europe, carpentry and craftsmanship in wood, refinements in tree-ring dating, and sophisticated use wear analysis of flint tools found beside the structure. Soon, we hope to determine the exact year in which almost all of the many thousands of pieces of wood were felled, and even eventually to estimate the exact month when the two Neolithic groups met and decided on their courses of action.

Some of this unique monument has been destroyed by peat-cutting, and parts have been damaged by drainage that has allowed the fragile Neolithic wood to decay beneath the ground. Our overall aim in the Levels has been to conserve whatever sites we can, and for the Sweet Track a collaborative effort was undertaken involving archaeology and nature conservancy. This is yet another example of the importance of combined interests working toward a common goal, as we have seen in northern Germany and Denmark. For the Sweet Track, matters were taken a stage further with the purchase of 16.8 hectares of peatland from the major peat-cutting company in the area at a cost (£95,000) well

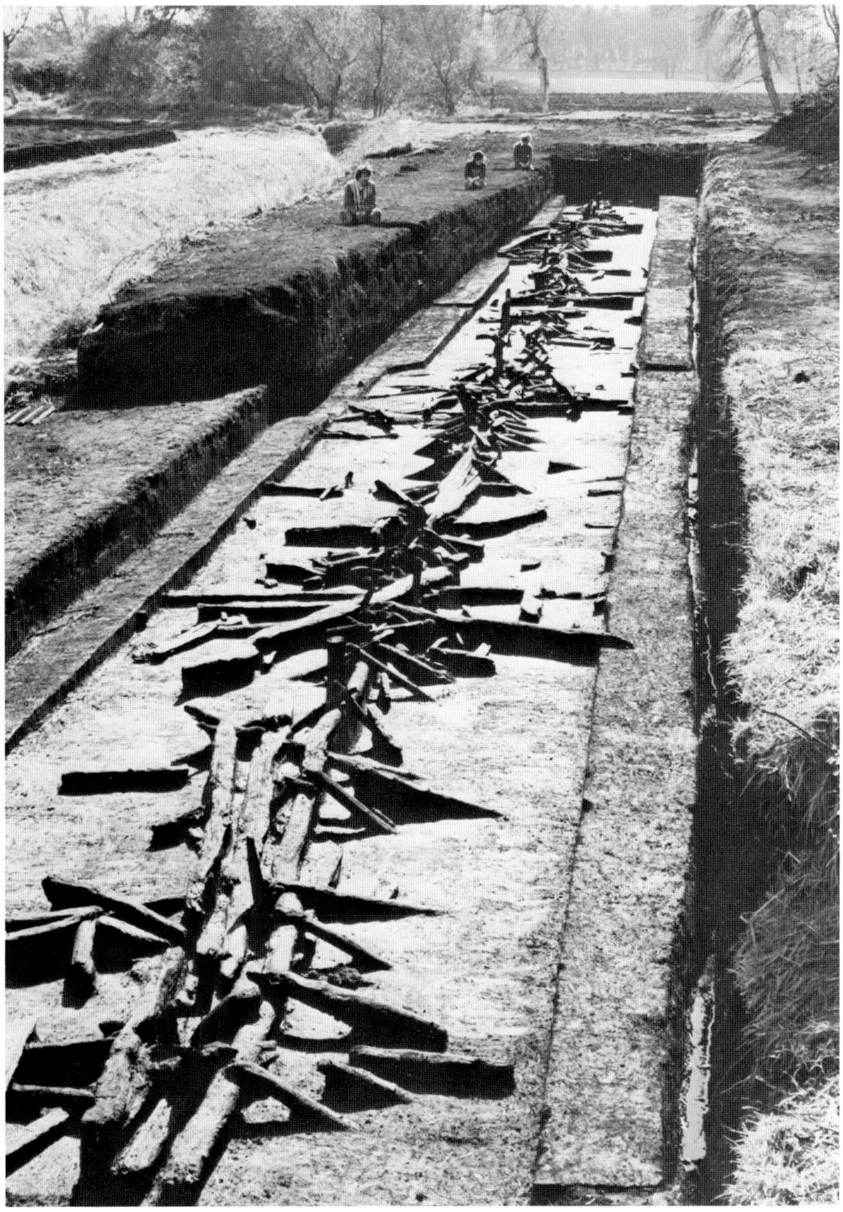

Figure 13. Early Neolithic wood trackway, the Sweet Track, in the Somerset Levels, England. The track was overwhelmed by water and then peat formation, and all its components are immaculately preserved. The entire structure is organic. Peat removal and drainage now threaten survival.

below the commercial price. This company (Fisons plc) had a fifteen-year record of cooperation with archaeologists. The funds for the purchase were obtained from the National Heritage Memorial Fund by a combined approach from the Nature Conservancy Council, English Heritage, and the Somerset Levels Project (the archaeological unit). This action prevented any further peat-cutting and thus preserved about 500 meters of the track (fig. 14). However, drainage around much of this reserve was drying out the peat, affecting the unique wetland wildlife of the reserve (which was of the highest grade of nature reserves in Great Britain), and decaying the prehistoric wood structure (a scheduled ancient monument). It must be emphasized again that legislation protecting a nature reserve or an archaeological monument is ineffective if the adjacent land is not also managed and controlled in sympathy with scientific and cultural interests. Thus, we had an association of interests in organisms as variable as butterfly orchid, royal fern, nightingale, and silver-washed fritillary, and the 6000-year-old wood monument of the Sweet Track, with all its planks, pegs, pottery, seeds, flints and wood artifacts—all requiring protection.

The solution involved environmental intervention on a massive

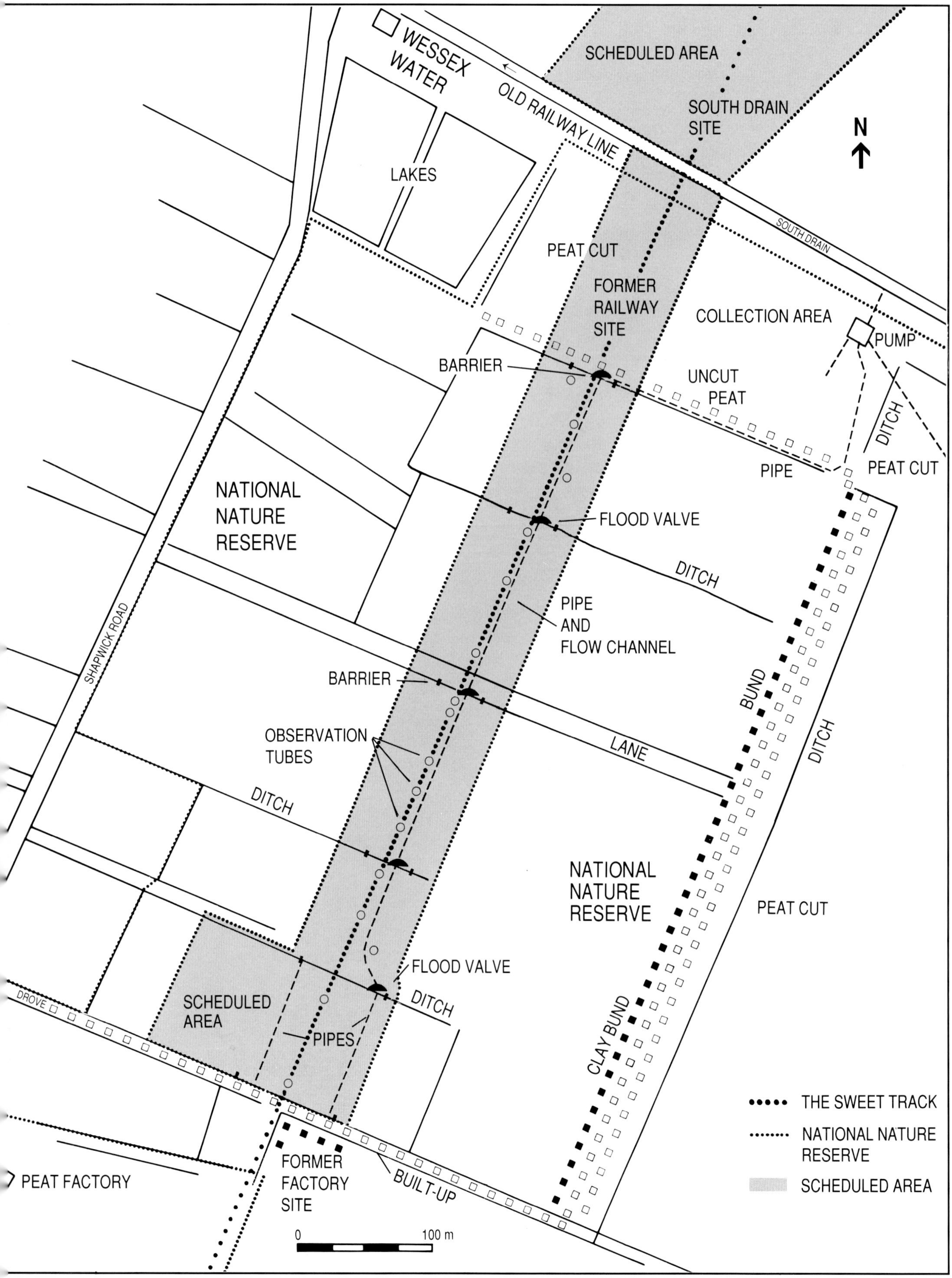
WESSEX WATER
SCHEDULED AREA
SOUTH DRAIN SITE
N
OLD RAILWAY LINE
LAKES
SOUTH DRAIN
PEAT CUT
FORMER RAILWAY SITE
COLLECTION AREA
PUMP
BARRIER
UNCUT PEAT
DITCH
PIPE
PEAT CUT
NATIONAL NATURE RESERVE
FLOOD VALVE
DITCH
PIPE AND FLOW CHANNEL
SHAPWICK ROAD
BARRIER
BUND
OBSERVATION TUBES
LANE
DITCH
DITCH
NATIONAL NATURE RESERVE
PEAT CUT
FLOOD VALVE
DITCH
SCHEDULED AREA
PIPES
CLAY BUND
DROVE
PEAT FACTORY
FORMER FACTORY SITE
BUILT-UP
0
100 m
THE SWEET TRACK
NATIONAL NATURE RESERVE
SCHEDULED AREA

scale. An additional grant (£35,000) was obtained from the same fund and this was used to construct a clay bank (bund) around part of the reserve and to reinforce peat banks on the other sides. These measures prevented drainage onto the adjacent lower lands. Near the line of the monument, which lies buried about 1 meter below the peat surface, narrow flow channels were dug by hand, and, on the surface, a 20-centimeter-diameter pipe with a series of flood valves was laid (fig. 15). (This effort was funded separately by national archaeological and nature conservancy agencies). The flood valves were placed over cross-ditches with barrier walls to retain high levels of water within the central area where the monument lies. Water is regulated by a timing device (with manual override for special conditions). All the ditches and flow channels are kept filled with water, thereby soaking the peat, encouraging the wetland wildlife for the nature reserve, and keeping the monument waterlogged. Small observation tubes have been inserted into the peat beside the track, and measurements are taken at least weekly throughout the year to monitor the water level by direct sighting; any lowering can then by corrected by additional pumping. Since this work was done in 1984, the monument has been maintained in a satisfactory condition, and the wetland plants and animals have been attracted back to the reserve (fig. 16). The environmental intervention here has been entirely successful; it must surely provide another example of the opportunities for archaeologists to combine with other interests in taking positive action to preserve significant parts of our cultural heritage for the future.

Figure 14 (opposite). Sketch map to show part of the alignment of the Sweet Track, a scheduled ancient monument, as it passes through a National Nature Reserve. The irrigation system that has preserved the ancient structure includes water collection areas, pumps, pipe and flood valves, ditches and barriers, and a bund (bank) of clay to retain moisture in the protected area.

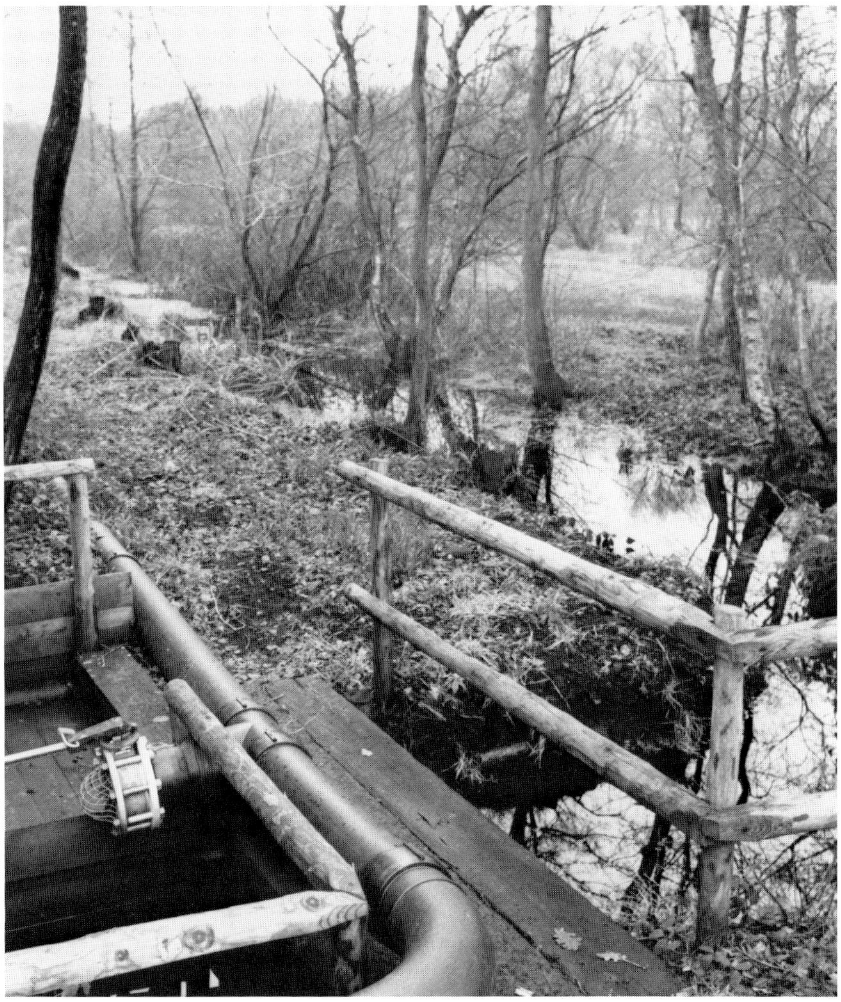

Figure 15 (left). Entry of the pipe, flood valve, cross ditch, and main flow channel in the Reserve, which ensure maintenance of waterlogged conditions for the Neolithic structure.

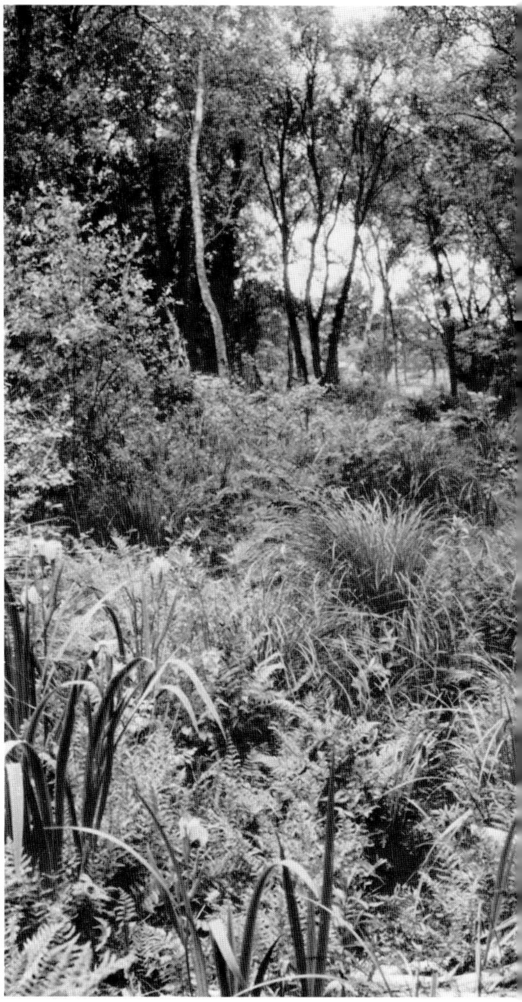

Figure 16 (right). Interior of the Reserve in 1985 with luxuriant wetland vegetation, indicating the success of the irrigation system in the preservation of the wetland habitat and archaeological monument.

Conclusion

For too long, monuments have been neglected, stemming from the belief that legislative orders alone will somehow guarantee their unchanged survival. This virtual abandonment cannot halt, and sometimes cannot even delay, the processes of attrition and decay that threaten the monuments of the past. Such passive attitudes are unconscionable when they affect monuments of national significance. Cultural intervention requires positive management and public control, but often little can be done beyond delaying deterioration. Environmental intervention is a more fundamental and dynamic process that can positively affect the preservation of monuments, at least retaining their condition upon discovery.

No archaeologist today could deny the contributions of the natural sciences to the understanding of ancient sites—their environments and economies—as well as to the conservation of artifacts. Collaborating with natural scientists would yield not only a broader knowledge of ancient environments, but also of present environments; this would aid in the development of appropriate environmental controls to protect the sites from the effects of modern climatic, geographical, and human pressures. The financing for such efforts is often more easily obtained by nature conservation interests than by archaeologists, whose contact with the public in some regions is less well established.

The enormous public interest in nature conservation as it relates to plant life and animal life in nature reserves and national parks, the public enthusiasm for archaeological monuments, and the overwhelming concern with destruction and pollution of all kinds, could be channeled by archaeologists into efforts to establish environmental intervention on sites and monuments.

Education plays a major role in the preservation of the cultural heritage. It is, after all, the public who will make the decisions affecting the survival of the heritage. Without adequate education, particularly for the young, efforts at archaeological preservation cannot succeed. Education can provide the key to solving problems of archaeological destruction, but the public must be in a position, both socially and economically, to receive and accept that education.

Finally, all efforts to conserve the environment must take into account and accept the natural evolution of the world. All monuments and all sites cannot remain untouched. Choices regarding the preservation of monuments and landscapes must be based on adequate documentation of the cultural record and presented to an educated public and an informed government. This will require positive action on the part of archaeologists and conservators in making available our results and in asserting our beliefs.

Acknowledgments

Because the literature on environmental intervention is virtually nonexistent, I have had to rely upon many colleagues for information about current developments in Europe and elsewhere. I am grateful to many archaeologists and conservationists who have responded to enquiries about measures employed and problems encountered. I particularly thank Dr. K. Kristiansen (Denmark); Prof. L. Louwe Kooijmans (The Netherlands); Dr. H. Hayen, Prof. W. Torbrügge, Dr. J. Reichstein (Germany); Dr. D. Weidmann (Switzerland); Dr. J. Bouzek (Czechoslovakia); Dr. J-P. Taavitsainen (Finland); and Prof. Dhavalikar (India) for detailed information and/or illustrations. And I thank the following for information about conservation and site preservation assessments: Dr. G. Wainwright, Dr. D. Thackray (England); Mr. P. Ashmore (Scotland); Dr. A. Hamlin (Northern Ireland); Prof. G. Eogan (Ireland); Prof. P-R. Giot and Dr. M. Ilett (France); Dr. T. Capelle (Germany); Prof. M. Egloff and Dr. J. Bill (Switzerland); and Dr. P. Kjaerum (Denmark).

References

Abels, B.-U.
1984 "Ausgrabungen und rekonstruktion der spätlatènezeitlichen Befestigungsmauer auf dem Staffelberg." *Das archäologische Jahr in Bayern 1983.*

Baker, D.
1983 *Living with the past. The historic environment.* Bedford.

Bayerisches Staatsministerium
1974 *Bayerisches Denkmalschutzgesetz.* Munich.
1983 *Flurbereinigung und Denkmalpflege. Ein Mosaik aus Mittelfranken.* Munich.

Clark, G.
1975 *The earlier Stone Age settlement of Scandinavia.* Cambridge.

Coles, B. and J. Coles
1986 *Sweet Track to Glastonbury.* London.

Coles, J.M.
1984 *The archaeology of wetlands.* Edinburgh.
1987a "Precision, purpose and priorities in wetland archaeology." *Antiquaries Journal* 66. Oxford.
1987b "Irish bogs. The time is now." *North Munster Antiquarian Journal* 28.

Egloff, M.
1982 "Un avenir pour notre passé? L'archéologie Neuchâteloise entre trax et truelle." *Musée Neuchâteloise* 3. Colombier, Switzerland.

English Heritage
n.d. *Stonehenge study group report.* London.

Fischer, A. and S.A. Sørensen
1983 "Stenalder på den danske havbund." *Antikvariske studier* 6.

Glutz, R., K. Grewe and D. Müller
1984 *Zeichenrichtlinien für topographische Pläne der archäologischen Denkmalpflege.* Cologne.

Greeves, T.A.P.
1985 "Archaeology and nature conservation in Dartmoor National Park." In *Archaeology and nature conservation.* G. Lambrick (ed.). Oxford.

Haarnagel, W.
1979 *Die Grabung Feddersen Wierde. Methode, Hausbau, Siedlungs- und Wirtschaftsformen sowie Sozialstruktur.* Wiesbaden.

Hackmann, A.
1913 "Ein Opferfund der Volkerwanderungszeit in Finnland." *Opuscula Arch. O. Montelius dicata.* Stockholm.

Hall, D.N.
1985 "Survey work in eastern England." In *Archaeological field survey in Britain and abroad.* S. Macready and F.H. Thompson (eds.). London.
1986 *Fenland landscape and settlement between Peterborough and March.* East Anglian Archaeology, Fenland Monograph 2. Norfolk.

Hansen, S.S.
1984 "Gudmund Hatt—The individualist against his time." *Journal of Danish Archaeology* 3. Odense.

Hayen, H.
1980 *Gedanken zum Schutz von Moor-Resten.* Oldenburg.
1981 *Moorarchäologie und Naturschutz.* Oldenburg.
1986 "Peat bog archaeology in Lower Saxony, Germany." In *European Wetlands in Prehistory.* J.M. Coles and A.J. Lawson (eds.). Oxford.

Hinchliffe, J. and R.T. Schadla-Hall (eds.)
1980 *The Past under the Plough.* Directorate of Ancient Monuments and Historic Buildings. *Occasional Paper* 3. London.

Hyenstrand, A.
1979 *Ancient monuments and prehistoric society.* Stockholm.

ICCROM
1985 *Preventive Measures During Excavation and Site Protection.* Rome.

Ilkjaer, J. and J. Lønstrup
1982 "Interpretation of the great votive deposits of Iron Age weapons." In *Journal of Danish Archaeology* 1. Odense.

Jackson, A.M.
1978 *Forestry and Archaeology. A study in survival of field monuments in south west Scotland.* Hertford.

Kelly, E.P.
1981 "A short study of the botanical zones on a ringfort at Simonstown, Co. Meath, used as an aid to the recovery of archaeological features." In *Irish Antiquity.* D.Ó Corráin (ed.). Cork.

Lambrick, G. (ed.)
1985 *Archaeology and nature conservation.* Oxford.

Lewis, N.A. and D. Miles
1985 "Archaeology and nature conservation: problems and potential at Cherbury Camp, Oxfordshire." In *Archaeology and nature conservation.* G. Lambrick (ed.). Oxford.

Louwe Kooijmans, L.P.
1985 *Sporen in het land. De Nederlandse delta in de prehistorie.* Amsterdam.

Lowe, P., G. Cox, M. MacEwen, T. O'Riordan, and M. Winter
1986 *Countryside conflicts. The politics of farming, forestry and conservation.* Aldershot.

Mallory, J.P.
n.d. *Navan fort. The ancient capital of Ulster.* Belfast.

Ministry of the Environment
1978 *Conservation of Nature Act.* National Agency for the protection of nature, monuments and sites. Copenhagen.

Muckelroy, K. (ed.).
1980 *Archaeology under water. An atlas of the world's submerged sites.* London.

Müller, S. and C. Neergaard
1908 "Danevirke." *Nordiske Fortidsminder* 1.

Poulik, J.
1960 "Découvertes dans le bourgwall de l'époque de l'Empire Grand-morave à Mikulcice en Moravie méridionale." *Nouvelles fouilles archéologiques en Tchécoslovaquie.*

Ravnsted-Larsen, L.
1983 "Fortidsminder og botaniske interesser. Gravhøje er andet end oldsager og skeletter!" *Antikvariske studier* 6.

Reinders, R.
1982 *Shipwrecks of the Zuiderzee.* Flevobericht 1. Rijksdienst voor de Ijsselmeerpolders 197. Lelystad.

Sankalia, H.D.
1974 *The prehistory and protohistory of India and Pakistan.* Poona.

Schwarz, K.
1977 "Denkmalschutz." *Jahresbericht der bayerischen Bodendenkmalpflege* 15/16, 1974/75.

Skaarup, J.
1983 "Submarine stenalderbopladser i Det sydfynske øhav." *Antikvariske studier* 6.

Solvang, G.
1983 "Fredningsplanlægning og kulturhistorie. Metoder og resultater fra et pilotprojekt i Vestsjællands amt." *Antikvariske studier* 6.

Thackray, D.W.R. and K.A. Hearn
1985 "Archaeology and nature conservation: the responsibility of the National Trust." In *Archaeology and nature conservation.* G. Lambrick (ed.). Oxford.

Thorsen, S.
1982 "Olhøj ved Asnæs. Lidt om en ganske almindelig gravhøj, og om dens tilstand gennem de sidste to hundrede år." *Antikvariske studier* 5.

van der Heide, G.D.
1959 "Dijkbouw door de eeuwen heen." In *Hondred eeuwen Nederland.* J.E. Bogaers, W. Glasbergen, P. Glazema, and H.T. Waterbolk (eds.). Gravenhage.

Vaud (Canton de Vaud)
1983 "Chronique des fouilles archéologiques 1983." *Revue historique vaudoise* 1983.

Vittery, A.
1985 "Wildlife and the Countryside." In *Archaeology and nature conservation.* G. Lambrick (ed.). Oxford.

Wainwright, G.J.
1985 "The preservation of ancient monuments." In *Archaeology and nature conservation.* G. Lambrick (ed.). Oxford.

Weber, L.
1984 *Konservierte Geschichte?* Kempten.

Weidmann, D.
1985 "Un manteau de terre pour protéger la chartreuse d'Oujon à Arzier." *Chantiers/Suisse* 16.

Wyss, R.
1979 *Das mittelsteinzeitliche Hirschjägerlager von Schötz 7 in Wauwilermoos.* Archäologische Forschungen. Zurich.

James A. Tuck and Judith A. Logan

ARCHAEOLOGY AND CONSERVATION: WORKING TOGETHER?

James A. Tuck is presently University Research Professor, Memorial University of Newfoundland, Canada and has been directing the excavation of a sixteenth-century Spanish Basque whaling port in southern Labrador. He has been one of the pioneers of on-site conservation in Canada.

In Canada, as well as in many other parts of the New World, the practice of employing conservators in archaeological field projects is less than two decades old. In fact, in many parts of the country the practice is still unknown. There are historical as well as financial reasons for this situation. Most archaeologists are trained as prehistorians and, owing to generally poor conditions for the preservation of objects—other than those made of stone, ceramic, and occasionally bone, antler, and ivory, little in the way of conservation was needed or so archaeologists believed. The literature is, however, sprinkled with statements about artifacts "turning to dust upon exposure," or being revealed "in a condition too friable to be recovered," and so on. Mishaps of this type were accepted by archaeologists as unavoidable. Yet it should be pointed out that this was not universally true. As early as 1929, Douglas Leechman noted that since "technical methods have reached their present perfection there is little excuse for the loss or deterioration of any specimen once it has reached the preparator's hands" (1931:127). Leechman might well have said, ". . . once artifacts are exposed by the archaeologist," for his article goes on to describe techniques for removing artifacts from their burial environments; temporary measures for stabilization in the field; handling and packing techniques, including the use of sphagnum moss for packing (which we have used extensively at Red Bay and whose antiseptic properties have recently been explained by Williams [1982:113–114]); and techniques for the treatment of most materials likely to be encountered in an archaeological excavation. Unfortunately, Leechman's work appears to have gone unnoticed, or at least unheeded, by Canadian archaeologists.

Only during the last two decades have archaeologists come to appreciate the services offered by conservators; unfortunately, on-site conservation specialists remain the exception rather than the rule. It was not until Parks Canada's Historic Sites and Parks Branch began large-scale excavations at the Fortress Louisbourg, Fort Beausejour, L'Anse aux Meadows, and elsewhere that conservation became an integral part of Canadian archaeology. Even then, its employment was limited to historical archaeology, largely because of the variety of materials found on sites of European origin, and because Canadian prehistoric archaeologists remained blissfully unaware of its existence.

Recently, archaeologists working on the northwest coast of North America were confronted with large numbers of organic objects preserved in waterlogged deposits (see Croes 1976). On some such projects, conservators apparently were present in the field, usually on the American side of the border. In Canada it appears as if the field problem and laboratory treatment of the material have been kept separate, although the material, fortunately, does not seem to have suffered from the absence of conservators in the field.

In 1972, the National Museums of Canada formed the Canadian Conservation Institute (CCI) with the mandate to provide advice, to carry out at least some limited conservation treatment on objects, and to research conservation techniques. The categories of materials included within this mandate covered a broad range and included fine-arts objects, paper, furniture, textiles, and ethnographic specimens. Archaeological conservation was not included in the original mandate of the CCI, partly because the objects recovered by archaeologists fall into one of the existing divisions, partly because most archaeologists in Canada worked at universities rather than museums, and, perhaps, partly because archaeological conservation is different in many ways from other branches of conservation.

An oil painting that has been hanging on various walls for four hundred years can be brought to a conservator for cleaning, restoration, and whatever other treatment is necessary. Barring some emergency resulting from a sudden change of environment, it does not matter whether the painting is shipped to the conservation treatment center today, tomorrow, or in a few weeks' time. By contrast, archaeological specimens snatched from a burial environment in which they have been preserved for hundreds or thousands of years require immediate attention. Weeks, days, hours, or even minutes of exposure to unfavorable ambient conditions can result in such severe deterioration that the object becomes useless for analysis and cannot even be restored to a state suitable for exhibition or other purposes.

The successful recovery and treatment of archaeological specimens requires specialized knowledge about a variety of materials: how they deteriorate in different burial environments, and how to prevent further deterioration during removal, storage, packing, and shipping until final stabilization can take place. Some archaeologists may possess such knowledge, but most do not. Those who do are archaeological conservators, and it is clearly best to have their involvement from the start of any excavation liable to produce materials that will require specialized treatment for survival. What follows is a description of one such project, carried out in eastern Canada, where field conservation has played a major role in an archaeological excavation.

Field conservation at the Red Bay site

The Red Bay Project's involvement with archaeological conservation parallels that of Canadian archaeology generally. Not until Memorial University began to investigate a large sixteenth-century Spanish and French Basque whaling station along the north shore of the Strait of Belle Isle, opposite the well-known L'Anse aux Meadows site, did the need for conservation become apparent (Tuck and Grenier 1981:180–190; Tuck 1985:50–57). Almost from the first, it was clear that the large amounts of iron, copper, lead, and other metals; the organic materials including wood, leather, baleen, bone; and the various other substances would require professional treatment if they were to survive even to the stage of analysis.

It is now clear that it was only through a coincidence that field conservators became involved in the Red Bay Project at the time they did. Archaeologists were aware of the newly formed CCI whose function—or so they understood—was to provide conservation assistance to archaeological projects in need of it. By the mid-1970s, the CCI had developed the capabilities for dealing with the conservation of archaeological materials. They were contacted in 1978 and sent a representative to the field within a few days. The following remarks are the result of the continuing association between archaeologists and the CCI.

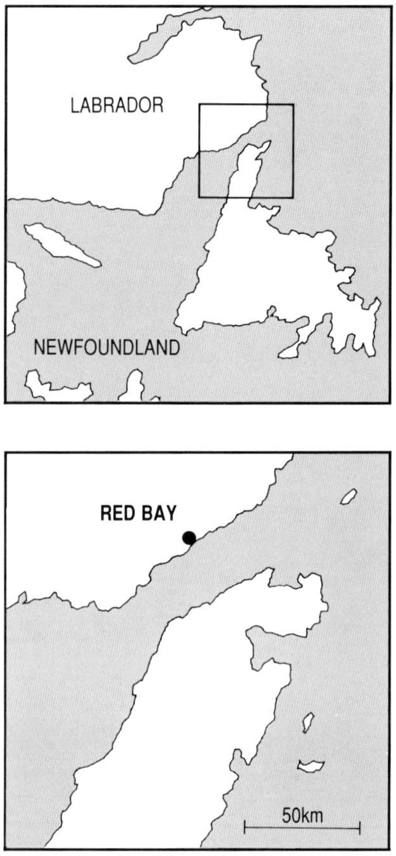

Figure 1. Maps of Labrador and Red Bay.

From an archaeological point of view, the experience has been a profitable one. Indeed, it seems certain, from the perspective of ten years of working with conservators, that the Red Bay Project could not have been carried out without the assistance of the CCI. From a conservation point of view, the project has also been productive. It stimulated research in a number of areas, particularly in the treatment of archaeological iron. This and other research, as well as new techniques for the removal, packing, and storage of objects, has provided information that will be of value to future field projects in archaeology.

The Red Bay Project has sometimes been touted as a model of cooperation between archaeology and conservation. In some ways this may be correct, but a more accurate statement might be that in the final analysis, the "Red Bay experience" provides some very good suggestions about how to coordinate an archaeological project involving field conservation. Remember that we have been at it for nearly a decade now; things did not happen overnight. While a great many of the initial problems have been solved, others—both specific and general—still remain.

Now let us turn to the field conservation procedures employed at the Red Bay site (fig. 1). Planning, it goes without saying, is the essence of any archaeological research project. Deciding *why* a particular site should be excavated, or *which* site should be excavated, is the province of the archaeologist, whose aim it is to gain understanding of the past. If the sites chosen for excavation are expected to produce features and artifacts that require conservation, some sort of initial testing of the site must be done by the archaeologist—preferably in consultation with conservation personnel—to determine what materials might be expected, in what quantities, from what environments. The organization of facilities, both in the field and the conservation laboratory, can then be designed to deal with an estimated quantity of known types of materials. At Red Bay, this phase more or less evolved; it was not formulated explicitly at the outset. We were fortunate to acquire the services of conservators early on in the project—just in time to make preparations for the following field season. Fortunately, the initial season was short and the field crew small, so the volume of excavated material was not as great as in the seasons that followed.

We have now established a more or less permanent field laboratory at Red Bay (in an abandoned oil storage shed) with facilities for the registration, cleaning, and storage of artifacts, as well as for the treatment of some materials and the restoration of ceramics and glass. A full-time conservator is present at Red Bay during the field season, courtesy of CCI. The conservator's duties include supervising trained assistants in many of the routine laboratory tasks, stabilizing the materials that can be treated in the field, restoring ceramics and glass, and assisting in a variety of field problems related to the removal of friable objects from their burial environments. In one instance, CCI staff members made a mold of a large portion of a trywork, where blubber was rendered into marketable oil. The resulting cast is now on display in St. John's, Newfoundland, where the public can take small pieces of epoxy resin for souvenirs.

The Red Bay sites have produced ample evidence of sixteenth-century activities in the form of artifacts, features, and burials. Conservators have not, for the most part, participated in the initial stages of excavation, partly because they have judged the archaeologists competent to do this up to a certain point, partly because their time is otherwise occupied with laboratory work, and partly because they abhor the idea of digging in pits in the cold and rain, assailed by blackflies. They do, however, monitor the excavation process in order to determine the best techniques for removal and to suggest how best to proceed with initial exposure to facilitate their application.

For example, one large "burial" feature, which turned out not to be a burial at all, was discovered during the 1982 field season. It consisted of the poorly preserved skeletons of about a dozen individuals whose bodies had been abandoned (apparently inside a small temporary structure) and never buried in the fashion typical of the sixteenth century. To complicate matters further, the skeletons were covered with woolen textiles bound up with roots that formed a solid mass measuring about 3 to 4 meters. Cleaning proved to be difficult. Although a number of techniques were tried, including flooding the area and gentle washing, the most effective method proved to be careful scraping with small bamboo and aluminum picks. Removing the skeletons and textiles proved even more difficult. Since they were so flattened as to be almost two-dimensional, traditional plaster or other reinforcing clearly was not possible. Their removal was eventually accomplished by isolating individual blocks of textile-wrapped bones, carefully excavating through the sterile gravel beneath the deposit, and transferring them to rigid supports cushioned with sphagnum moss.

Other burials were block lifted by more traditional means using car-body filler—a two-part polyester resin lighter than plaster, easier to handle, and considerably less messy (Logan 1986). Even though these skeletons were also nearly two-dimensional, the technique worked extremely well and could have been used on other similarly preserved skeletons. As it turned out, the amount of information obtained from such skeletons was negligible; those exposed subsequently were studied by physical anthropologists without disturbing what remained of the bones and then reburied in situ.

The lifting of other objects was also accomplished by conservators. Objects that were essentially three dimensional and could be pedestaled were removed using traditional block lifting techniques; plaster of Paris was normally employed. At times, however, paraffin wax was used in place of plaster because it set faster and more predictably and was extremely easy to form around the isolating layer of foil. In fact, its use was so simple that even archaeologists tried their hands at it with good results. Extremely degraded wood was sometimes lifted by applying melted PEG-4000 directly to the object, without the traditional isolating layer of foil. The material provides a certain amount of adhesion between fragments of the object; while there are problems with applying PEG-4000 directly onto a wet surface, the objects recovered in this way at Red Bay survived removal, were successfully treated, and now appear stable.

The removal of textile fragments (and, in at least one case, an entire garment) has been accomplished successfully by employing a method based on techniques used for the relining of paintings that cannot be impregnated with a resin (Logan and Segal 1985:17–19). Gauze impregnated with Rhoplex® AC-33 (an acrylic emulsion; approx. 30% solution in water) is dried and cut into small strips, which are applied to the textile by spraying them with acetone to soften the adhesive. The gauze adheres to the textile, forming a "nap" bond, which provides a somewhat flexible support. The specimen can then be lifted by loosening the soil beneath; it may then be transferred to a padded support for transportation to the field laboratory, where initial cleaning and packing for shipment take place.

Once at the field laboratory, all material is catalogued and given a permanent registration number, which follows the specimen through conservation, analysis, and into storage. Because of the large volume of material, this is done with the aid of a microcomputer from which the data is later transferred to the national specimen-record database.

Artifacts are first segregated according to material—metals, ceramic, organics, and so forth—and each class and subclass is treated individually. The following describes briefly what happens to each class of artifacts between the time they are excavated and their final treatment in either Ottawa or St. John's.

Ceramics, the least problematical artifacts, can safely be washed and dried. Numbers are then written on individual sherds, which are sorted into vessels, cross-mended, and, when possible, restored. Salts are leached from Red Bay ceramics by the abundant rainfall and thus pose no problem. Rather, the greatest problem might arise from "over-washing" the ceramics, which are often very soft, particularly the low-fired tin-glazed earthenwares called *majolica*. Restoration is something of a luxury because ceramics would normally be restored during the winter months, when excavation is impossible. However, since personnel are not available in St. John's for restoration, most of what has been restored has been done during the field season. This has proven valuable not only because the vessels come from the field with restoration complete and ready for analysis, but because the process continues to impress both visitors and the field crew.

Glass, for the most part, can be treated in a similar fashion. It should be noted, however, that a significant percentage of the sixteenth-century glass from Red Bay was manufactured with an imbalance of fluxes, which have leached out during the centuries in the ground. The result is a form of deterioration called "crizzling"—the most benign manifestation is appearance, and the most damaging is the deterioration of glass to a fine silica powder upon drying. It is possible to arrest damage in less serious cases by placing them in an environment with a constant relative humidity (in containers conditioned with silica gel to about 45% relative humidity, for example). The most extreme deterioration requires dewatering and consolidation. This has been done successfully with wet glass using acetone followed by toluene. The sherds are then brushed with a 2% w/v solution of the acrylic resin Acryloid® B-72 dissolved in toluene as they dry (Logan 1986). Glass in stable condition can be mended in the field, although final restoration with epoxy adhesives should be carried out under controlled conditions such as those at the CCI.

Some organics, including wood, leather, and textiles, are not usually treated at Red Bay. The treatments would be, for the most part, too complex or of too long a duration to attempt. Rather, such materials are stored on rigid and/or padded supports in fresh water, until they are shipped to St. John's or Ottawa for treatment. Bone is treated in the field by impregnating it with Rhoplex AC-33 under a slight vacuum and allowing it to dry slowly. Except in cases of extremely degraded bone, this procedure has been very successful. Repeated applications of the same substance, followed by slow drying, have been effective in treating baleen, the large keratin plates or strainers that hang from the upper jaws of baleen whales. Vacuum impregnation of baleen has not been attempted, primarily because the large plates are too big to fit in our dessicator.

Some metals are routinely treated in the field laboratory, while others are taken to St. John's or Ottawa. Since they essentially come from a wet environment, all metals are placed in fresh water immediately upon excavation and are kept wet until treatment is completed. Copper and lead, the most common nonferrous metals at Red Bay, are treated in the field if time permits. Copper is cleaned mechanically and impregnated with a 2% solution of benzatriazole in ethanol, which reacts with the metal to inhibit further corrosion. Objects are then coated with Acryloid B-72 in acetone to protect against fluctuations in humidity and handling (Logan 1986). Small lead objects are stripped of corrosion in a solution of diethylene triamine pentacetic acid and sodium hydroxide in water (see Logan 1986), rinsed in tap water, dried in acetone, and brushed and coated with microcrystalline wax. Large lead objects are treated by electrolytic reduction at the CCI.

Iron objects pose the greatest problem at Red Bay. Not only do they comprise the largest class of metal artifacts, but iron may be the most difficult metal to treat. This is especially true of Red Bay iron, which is highly mineralized and corroded, often containing no elemental iron at all. All such objects are radiographed, according to provenience, in the field laboratory with a portable X-ray unit. This allows archaeologists to look through the corrosion to see what the object might be, it gives at least a rough indication of the extent of disintegration, and, perhaps most important, it yields a permanent, full-sized record of the object (Logan 1984:14–17). Not all iron survives conservation. Prior to radiography, the iron is wrapped in gauze if necessary, and sandwiched between sheets of polyethylene screening. Teflon® strips with catalogue numbers are attached to each piece. Bags of iron can then be shipped to either St. John's or Ottawa in a wet state. Deterioration during shipment taking up to three weeks has not been noticeable.

Objects composed of more than one material, each requiring different and often incompatible treatment, remain a problem. Most composites are stored in deionized water until they are packed for shipment. From Red Bay, all were sent to the CCI for treatment, which required more skill than was available at Memorial University, and in some cases demanded additional research to determine appropriate methods of treatment.

This is a brief description of the system used at Red Bay since 1978. It should be emphasized that many of the procedures described above were developed specifically for handling and treating material recovered from the Red Bay site. Each burial environment is unique, and the procedures appropriate for one site might be wholly unsuitable for others. Conservation assistance should occur early on in the project so that materials and burial environments can be evaluated and suitable techniques for handling and treating objects can be determined.

Conservators and archaeologists

Since the inception of the Red Bay project, the collaboration of archeologists and conservators has been, almost without exception, a positive experience. In retrospect, it is probably fair to say that the excavations at Red Bay could not have been properly accomplished without the assistance of conservators. Excavations would have gone ahead anyway, but the outcome would have been far less successful. Yet in some ways the realization that conservation is essential to the excavation of European sites is disheartening. For example, archaeologists from Memorial University have recently discovered the site of one of the earliest English colonies in Newfoundland, and would like to pursue further research; however, it is questionable whether excavation is ethical, or even possible, without conservation assistance.

In Canada, most archaeology is done by faculty members and students in universities, whereas virtually all conservators are employed by museums or the federal government. In fact, to the best of our knowledge, there is not a single conservator employed by a Canadian university that also supports a program of archaeological research. There are reasons for this, of course, including the fact that at most universities, there simply are no positions for conservators. They do not fit into departments of anthropology, nor do they possess the "paper qualifications" necessary for tenure and promotion. Then, too, conservators are overqualified for most positions held by the technical staff of universities, such as machinists, welders, or carpenters.

This situation is not likely to change in the near future, especially given the financial constraints placed on Canadian universities.

Moreover, granting agencies are reluctant to divert archaeological research funds to the conservation components of the research grant proposals. We recently received an award from a well-known granting agency, the amount of which was reduced by precisely the sum requested for conservation services, materials, and supplies on the grounds that that function should be supported by the university or the government.

Some agreement in this respect must be reached among university administrators, granting agencies, the people actually carrying out archaeological research, museum curators, and the owners of the artifacts, or a good bit of archaeological research is going to be inadequate or simply not done at all.

Problems have arisen in the course of the Red Bay excavations, but they are mostly of a minor nature. There are often conflicts revolving around how specific objects ought to be treated or whether they should be treated at all. Excavation permits, for example, require *all* artifacts to be returned to the provincial museum "in conserved, that is, stable state" (Smart 1985:8). This could include thousands of iron nails or nail fragments, which are difficult and expensive to treat, but could be radiographed, photographed, otherwise recorded and described, and then discarded or reburied with some samples being retained for future analysis.

There are also occasional differences of opinion between archaeologists and conservators as to what sorts of treatments should be performed on particular artifacts. In Canada, for instance, the conservators' code of ethics discourages treatments that are not reversible or that will interfere with chemical analysis at some future date. This is all well and good, but it is distressing for archaeologists to see an iron object that might be saved by irreversibly impregnating it with epoxy, slowly crumble to rust that, while susceptible to chemical analysis, leaves nothing of the object for formal or stylistic analysis—of paramount importance to archaeologists.

Many of these problems can be solved if archaeologists have an understanding of one another's objectives. At Red Bay, mutual understanding often led to acceptable compromises on many issues. However, a recent statement by a conservator that she was pleased to see conservation required in order to obtain a research permit "because after all . . . it's the main concern" would be disputed by many archaeologists. For them, the reconstruction and understanding of past human behavior would be a more acceptable statement of the reasons we carry out archaeological research. Unfortunately, attitudes such as the one cited, whether real or perceived, have caused some archaeologists to reject entirely the services offered by conservation science. The "best" conservators we have had with us at Red Bay are those who have had some training and experience in archaeology and the reconstruction of prehistory or history. They have a better appreciation of field problems—from the difficulties involved in complex block lifts to excavating in the company of swarms of blackflies—as well as the archaeologists' reasons for excavating and the objectives of the research.

For their part, archaeologists should be aware of the importance of conservation science in the preservation of the archaeological record. All too often, archaeologists appear to have the attitude that, once they have studied the material, it is no longer of use. The fallacy in this sort of thinking is clear. Even attempting to obtain photographs for publication some years after a dig is done can be a sobering experience when you find that the specimens, in the worst case, have disintegrated beyond recognition. The problems faced by a student attempting to reanalyze such a collection—and this is happening at an increasing rate—are far worse. This situation could be overcome by the realization that the conservators' goals, even the extreme formulation of them quoted above, are not without merit. As those who remove objects from the places they have lain for hundreds or thousands of years, archaeologists have a responsibility to ensure that they remain available for study for as long as humanly possible. But, as we have already pointed out, no program of the study of archaeology in Canada includes any conservation component; virtually the same situation appears to exist in the United States. This is as unfortunate as a conservator's lack of appreciation for archaeology, for even if a "save the artifacts" ethic does develop, archaeologists will remain unaware of the importance of immediate post-excavation handling and how it can be accomplished efficiently, safely, and inexpensively. Thus, at least one message should be gotten across to archaeology students: Even if it is not possible to employ a full-time conservator in the field, a little planning, advice from conservators, and judicious use of time and locally available materials will result in a much more satisfying return and a collection that is far more valuable than it might have been otherwise.

In summary, the Red Bay Project has shown that conservators and archaeologists not only can but must work together to recover and preserve the material that constitutes the stuff of archaeological reconstructions. The experience has been of benefit not only to the particular interests of both parties but, more important, to the objects, features, and other materials and data collected during the course of the excavations.

References

Croes, Dale R. (ed.)
1976 "The excavation of water-saturated archaeological sites (wet sites) on the northwest coast of North America." *Archaeological Survey of Canada Mercury Series* 50. Ottawa.

Leechman, Douglas
1931 "Technical methods in the preservation of anthropological museum specimens." *National Museums of Canada Bulletin* 67. Ottawa.

Logan, Judith A.
1984 "An approach to handling large quantities of archaeological iron." In *Preprints, ICOM Committee for Conservation Seventh Triennial Meeting.* Copenhagen.

1986 "Conservation in the field: an example from Red Bay." In *Archeology in Newfoundland and Labrador 1985.* J.S. Thomson and C. Thomson (eds.) St. John's.

Logan, Judith A. and Martha Segal
1985 "A 16th-century costume." *Textile Conservation Newsletter.*

Smart, Rosemary
1985 "Conservation at the Newfoundland Museum." In *Archaeology in Newfoundland and Labrador 1984.* J.S. Thomson and C. Thomson (eds.) St. John's.

Tuck, James A.
1985 "Unearthing Red Bay's whaling history." *National Geographic Magazine* 168(1).

Tuck, James A. and Robert Grenier
1981 "A 16th-century Basque whaling station in Labrador." *Scientific American* 245(5).

Williams, Bryan
1982 "The healing powers of sphagnum moss." *New Scientist* 9.

Charles E.S. Hett

CONSERVATION MEASURES ON ARCTIC SITES

Charles E.S. Hett trained at the Royal Academy in London and at the Institute of Archaeology, University of London. He is presently Head, Conservation Artifacts, National Museum of Man, Canada and was previously Chief, Archaeology Division, Canadian Conservation Institute. He has wide experience at sites in the Arctic and in Mexico City, El Zapotal, Veracruz, Palenque, Bonampak, and Chiapas.

Archaeology on Canadian arctic sites is normally carried out by small crews working for a season that lasts a maximum of two months. The excavation of frozen sites is only practical when ambient temperatures are above freezing, allowing the excavation to follow the natural thawing of the upper layers of frozen ground, exposing deeper layers that must also thaw.

The size of crews and the length of the season is usually dictated by the remoteness of the site. Transport in these regions is expensive; access is usually by light aircraft, helicopter, or canoe. Enough equipment and supplies must be carried to the site to allow the crew to be self-sufficient: food, tents for working and habitation, and supplies for field work. Provisions must be carefully planned since additional equipment and supplies cannot be obtained in the field; at the same time, weight and bulk must be kept to a minimum. In this aspect, archaeology in the North is reminiscent of the pioneering days of field archaeology. These factors limit the extent of conservation work that is practical in the field and modify the traditional role of the conservator.

The cold environment

Permafrost is found on most northern land sites. It is defined as soil with a permanently frozen subsoil, covered by an active layer that thaws during the summer and freezes in the winter. The active layer supports biological growth and in some areas considerable vegetation, predominantly mosses and dwarf shrubs.

Precipitation is generally low, but the poor drainage of frozen ground typically results in sites that are wet when they are not frozen. While there is considerable variation in this pattern, it often produces soils that favor the preservation of organic materials by waterlogging or freezing.

Organic materials found in permafrost burial are often very well preserved. By contrast, similar artifacts found in the active (upper) layer may be in intermediate states of deterioration or have already disintegrated completely. Soil accumulation on sites is slow, and in the arctic archipelago, rare.

The origin of the burial environment will often determine the amount of deposition over a site and thus the quality of preservation of the artifacts. For example, houses that have a considerable superstructure and insulation form, on collapsing, a sufficiently deep deposit to create a permafrost environment for artifacts very rapidly. In ideal circumstances, this can happen within a few years of abandonment, providing a highly favorable environment for preservation. Middens also frequently form relatively deep deposits with poor drainage, favoring the growth of mosses, which provide good insulation and aid in the preservation of organic materials. Middens are frequently rich in deposits of sea mammal fats. The impregnation of organic artifacts with sea mammal fat promotes preservation by reducing the moisture content. Nonetheless, these deposits continue to deteriorate even when frozen, albeit very slowly. The smell of many middens attests to the viability of microorganisms.

The prehistoric material culture of northern Canada is distinct from other parts of the world in that organic artifacts not preserved in other environments form an important part of the archaeological record. These materials present special conservation problems. For example, their treatment may require the use of a freeze-dryer, which can only be found in a well-equipped conservation laboratory. Materials commonly found on prehistoric sites include baleen,[1] bark, wood, walrus ivory, untanned skins, polar bear canines, and "ceramics," which in use were consolidated with organic materials rather than fired. Wood (probably driftwood) was a more common resource than might be expected. The quality and ingenuity of woodworking is often fine, and, with the exception of house structures, artifacts are generally small. This is fortunate given the limited scale of field operations and the very long distances—often thousands of miles—artifacts must travel for research and treatment.

1. More commonly known as whalebone, baleen is similar to horn and forms the filter in the gorge of some species of whale.

Historic-period archaeology in northern Canada is pursued actively and provides a wide range of unusual conservation problems, again stemming from the preservation of materials that do not commonly survive in other environments. Many of the sites and artifacts investigated date from the period of British naval exploration in the mid-nineteenth century. Such artifacts are often found exposed on the surface. They generally have a high moisture content, and even standing wood remains (e.g., grave markers) have a moisture content above the fiber-saturation point, thus requiring long periods of seasoning. Almost all of the sites are close to the ocean. As a result, chlorides are present in abundance, and metals, though well preserved, are liable to suffer accelerated corrosion when exposed to higher temperatures. Some human burials made directly into permafrost have been so well preserved that, after recovery and thawing, autopsies could be successfully carried out on site. But many composite artifacts, such as barrels, present difficulties because individual components cannot be separated. Perhaps the most problematic artifacts are the many tin cans found with their contents still intact; they are usually painted and some even have paper labels. Apart from the treatment problems they pose, they can also constitute a health risk for the people who handle them.

The difficulty in accessing sites and the conservation problems posed by the environment have led conservators working in the North to focus on developing measures to ensure that artifacts arrive in sound physical condition at a conservation laboratory, rather than on attempting a great deal of treatment in the field. The work starts well ahead of the field season with the inclusion of a conservator on the interdepartmental committee that reviews applications for excavation permits. This early planning is intended to identify the amount of conservation assistance that will be needed and also to determine the provisions that will be necessary for the conservation of artifacts on their return. On a practical level, conservators can guide archaeologists in the preparation of appropriate field conservation kits and supplies. Archaeological and conservation units exist in widely diverse organizations so that cooperation across institutional, geographic, and bureaucratic boundaries is required. Fortunately, this cooperation does exist and works effectively.

Wet organic artifacts

The major efforts in field holding of wet organic artifacts are directed toward preventing the artifacts from drying out and discouraging biological growth. Since much organic material is not fully waterlogged, field holding must also maintain the exact moisture content of the artifacts at the time of excavation. Placing them in solutions in tanks can radically increase their moisture content. The most successful field holding has been achieved with low-temperature storage. Artifact registration should be integrated with conservation so that artifacts can be stored as soon as possible after excavation. Two principal methods of cold storage are employed, depending upon the physical environment of the excavation; both rely on the use of sturdy insulated containers. Camping coolers can usually be used to maintain a steady cold environment.

Subarctic environments, if well drained, allow pits to be dug to an adequate depth. If these pits are to be used for extended periods, perhaps several seasons, they can be shored up. Once a storage pit is well covered with insulating material (e.g., brush, peat, or rooty topsoil), it maintains a steady temperature around 5°-7°C, usually not exceeding 10°C, even when ambient temperatures outside range from 25°-30°C for several days, as can commonly occur. Cold pits of this sort have traditionally been used by indigenous peoples for storing meat.

In the cold climate of the high Arctic, however, the digging of storage pits is not practical, and other methods are needed to maintain a cold environment. In some areas, snowbanks remain year round in north-facing slopes and these can be used directly as a cache. More commonly, Freezer Paks® are employed. These are plastic packages containing a gel that can be frozen, maintaining a low temperature in an insulated container for an extended period of time. The Freezer Paks can be reused indefinitely and chilled in snow banks or meltwater that is close to freezing. Although it is difficult to protect coolers from any exposure to sun, since the summer sun in this region completes a 360° rotation during a twenty-four-hour period, the maximum possible protection should be sought. The setting-up of a holding system is the first task a conservator must attend to in the field.

The field processing of artifacts is the second concern of the conservator. The goal is to ensure that artifacts are registered, appropriately supported, wrapped, and housed as quickly as possible. It is reasonable to aim at having organic artifacts catalogued, wrapped, and housed in cold storage within twenty-four hours after excavation. The specific requirements for this processing must be discussed with the archaeologist well before the field season. The work usually requires a tent, workbenches, seating, and space for field conservation work supplies. Work space should be allocated for the controlled drying, cataloguing, and wrapping of artifacts. In practice, all work space is always fully utilized. For each project, these requirements must be addressed, and the additional weight in transport considered. Also important is the inclusion of waterproof labels and markers. Whatever field registration system is used, an individual label must be included in the wrapping of each artifact. The most common procedure is to use a waterproof marker on a label made of spun polyethylene paper. In Canada, Tyvek® labels are generally employed. Tests should be made to ensure that the markers chosen will withstand prolonged immersion in water and will write on tags that are damp or wet.

Proper field wrapping and packing of artifacts is crucial to their long-term well-being (Hett 1984). In much of the North, mosses are found on or near archaeological sites, and they make an ideal packing material for wet organic artifacts, providing both cushioning and a moist environment. Sphagnum moss is the most desirable material, since it is also an effective biostat at low temperatures. Where sphagnum is not present, other mosses seem to serve quite well. In some areas of the high Arctic where mosses do not grow, substitute materials must be brought to the site. Toweling, polyurethane foam, or polyester batting can be used to provide moisture retention and cushioning. In these cases, a biostat must be introduced. One of the best is a 75% solution of isopropanol, which can be sprayed on the artifacts before they are wrapped or, when further diluted, can be used as a wetting solution. The most common, commercially available wrapping material is polyvinylidene chloride film (Saran Wrap®). It has a low permeability to water vapor and liquid. Moreover, when this film is wrapped in a layer of aluminum foil, biological activity is further reduced, possibly owing to the exclusion of oxygen and light. Adhesive tape can be used to secure both the plastic and metal-foil wrappings, and field numbers can be recorded on the tape to facilitate the checking of packing lists.

Physical support, as well as cushioning, is required for a majority of wet organic artifacts. In most cases, adequate reinforcement can be provided by a flat support to which the artifact is held by the wrapping or by gauze bandages. This physical support may be provided by laminated polyethylene sheets (Coroplast®). Coroplast sheets are light, inert, and sufficiently rigid to support most artifacts. They can be easily cut with scissors or a sharp knife to shape a custom support for each artifact. Standard acrylic boxes, in a range of small sizes, and sealable polyethylene refrigerator containers are also useful, both for small artifacts and those in need of greater support.

Excavation

At times, the excavation of artifacts requires the attention of the field conservator. It is important for the conservator to spend a part of each day working as a member of the field excavation crew, rather than being occupied entirely with the processing and holding of artifacts. A great variety of environments may exist within a single site, and the conservator should become familiar with the microenvironments and the effects of excavation upon them.

Permafrost will melt only slowly and evenly in areas of poor drainage and this can make the excavation of individual artifacts difficult. To remove artifacts from permafrost, the most gentle method is to pour warm water around the object, melting the permafrost. Then once the water has chilled, a sponge is used to remove the water. This is a slow process, but it allows maximum control of a difficult situation; permafrost melts equally in all directions, and any attempt to excavate quickly will result in an inverted cone. This method of excavation requires sponges, gloves, and fuel to heat the water.

On sandy sites and in areas where topsoil and vegetation are cleared, the drainage and desiccation of artifacts in the soil will proceed faster than the excavation. As a result, an artifact that has had a constant high-moisture content for centuries will be recovered in a semidry condition. Desiccation can occur very rapidly, especially in conditions with twenty-four hours of daylight, fine weather, and strong winds.

In extremely dry areas, when excavation is not taking place, the site may need to be covered with damp moss, toweling, and polyethylene sheeting. In such conditions, when lifting fragile artifacts, it may be necessary to carry out excavation as one would on a wet site, arranging for water to be used as the principal excavation tool and also arranging for runoff and drainage. In one case of this sort, excavation of a baleen net was carried out using a plant sprayer with fine and coarse water jets.

Block lifting is not uncommon on northern sites, but is done with bandages and trays rather than the traditional plaster of Paris. The principal reason for this preference is perhaps the need to reduce shipping weight; also, the artifacts are usually small and densely concentrated, so that the use of a plaster-of-Paris block lift would risk damage to underlying artifacts.

The small size of most artifacts allows them to be lifted on a dustpan, which is part of the archaeologist's normal field equipment. Improvisation is required to make lifting supports for larger artifacts, and it is useful to have hacksaws, cold chisels, and a heavy hammer to modify any available sheet metal (of which old oil drums are a common source) into a suitable support.

Artifacts that can be treated and dried on site are numerous, and where the treatments are simple and drying conditions adequate, field treatment will facilitate the packing and transport of artifacts. Immersion in 15%-20% dilution of Rhoplex® AC-33 or Primal® AC-33 (an acrylic emulsion), followed by controlled drying, is used for artifacts made of bone and ceramics, which will not withstand controlled drying without consolidation.

A synthetic resin emulsion with greater adhesive properties and tack is required for baleen and antler, which require a different approach to treatment in that alternate drying under pressure and surface and subsurface applications of consolidants are necessary. This treatment demands a longer time and more constant attention than immersion and controlled drying. It becomes impractical toward the end of a field season, when the wet packing methods described earlier must be used.

Metals for the most part are best dried on site. In the case of iron, considerable judgment on the part of the conservator is required to determine the extent of corrosion. A knowledge of the appropriate laboratory treatments is necessary as well. For iron, if there is deep penetration of chlorides, the success of subsequent laboratory treatment will depend upon keeping the artifacts wet. Nonferrous metals can be safely dried.

The drying of artifacts facilitates packing considerably. Dried artifacts are wrapped in acid-free tissue paper secured with tape on which the field number is written. For most artifacts, the cushioning provided by the tissue will suffice for subsequent packing and shipping.

Health hazards

Several health hazards exist in handling artifacts from northern sites. The decaying contents of one-hundred-thirty-year-old tin cans have a smell that itself warns people handling them to take care. The principal hazard with these materials is ingestion, which can be avoided. These artifacts should be handled with gloves, sealed with at least two layers of polyethylene, refrigerated promptly on site, and frozen as soon as possible. In the laboratory, sterilization by radiation is the usual first stage of treatment. A common hazard is found in the handling of artifacts impregnated with sea mammal fat. People who handle these materials in the field and in laboratory situations have experienced a variety of ill effects. The symptoms noted include skin rashes, respiratory problems, unusual swellings, and general malaise. The origins of these complaints have not been precisely identified, but appear to be due to either allergies or the presence of microorganisms. The use of gloves, when surface cleaning and handling these materials shortly after excavation, appear to reduce hazards significantly. This cleaning has been done mostly with detergents, but upon reflection, the use of an ethanol or isopropanol spray followed by rinsing may be preferable. This is particularly true with artifacts made of skin, where the selection of detergents has to be made carefully in the laboratory following analysis of condition, which is done by methods not applicable in the field (Segal and MacDonald 1974:14–15).

References

Beattie, O.
1986 "Autopsy report on John Torrington." *Journal of the Canadian Medical Association*. In press.

Harrington, C.R.
1964 "The Cache on Dealy Island." *Canadian Geographic Journal* 64.

Hett, C.E.S.
1984 "The Role of Conservation in Northern Archaeology." *Musk-ox* 33.

Hett, C.E.S. and M.E. Weaver
1980 "The conservation of Captain Kellatt's Storehouse, Dealy Island, N.W.T." In *Conservation within Historic Buildings*. N. Bromelle, G. Thomson, and P. Smith (eds.). London.

Janes, R.R.
1981 "The preservation and ethnohistory of a frozen site in the Canadian Arctic." *Arctic* 35(3).

Logan, J.A.
1983 "Red Bay 1982 - Textile Discovery." *Textile Conservation Newsletter*.

Segal, M.
1983 "Treatment of the textiles from Red Bay, Labrador." *Textile Conservation Newsletter*.

Segal, M. and M. MacDonald
1984 "Current Research into the Conservation of Frozen Skins from the Arctic." *ICOM Committee for Conservation Preprints* 2:18.

Sutherland, P. (ed.)
1985 "The Franklin era in Canadian Arctic history 1845–59." In *Archaeological Survey of Canada*. National Museum of Man Mercury Series 131. Ottawa.

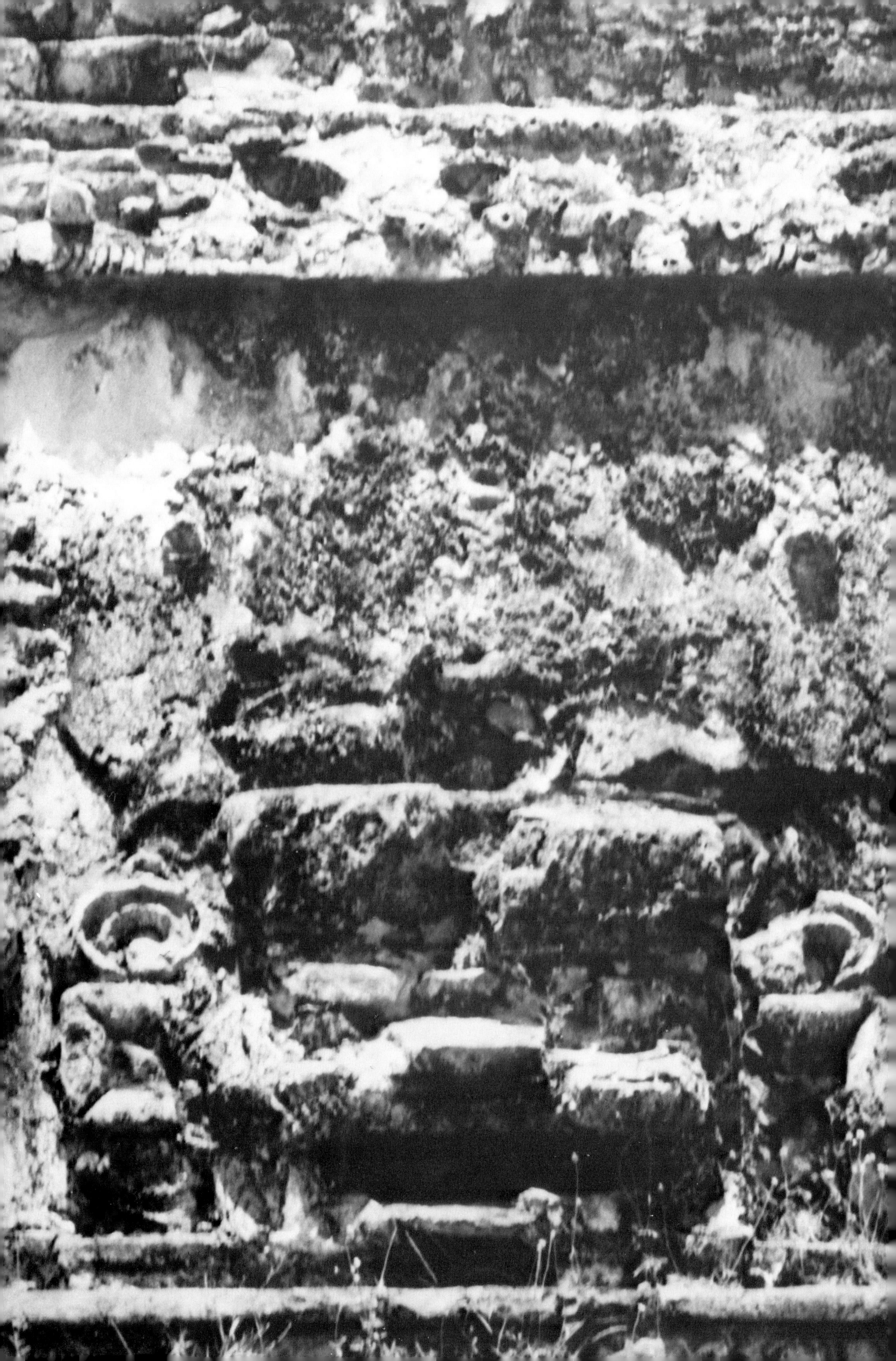

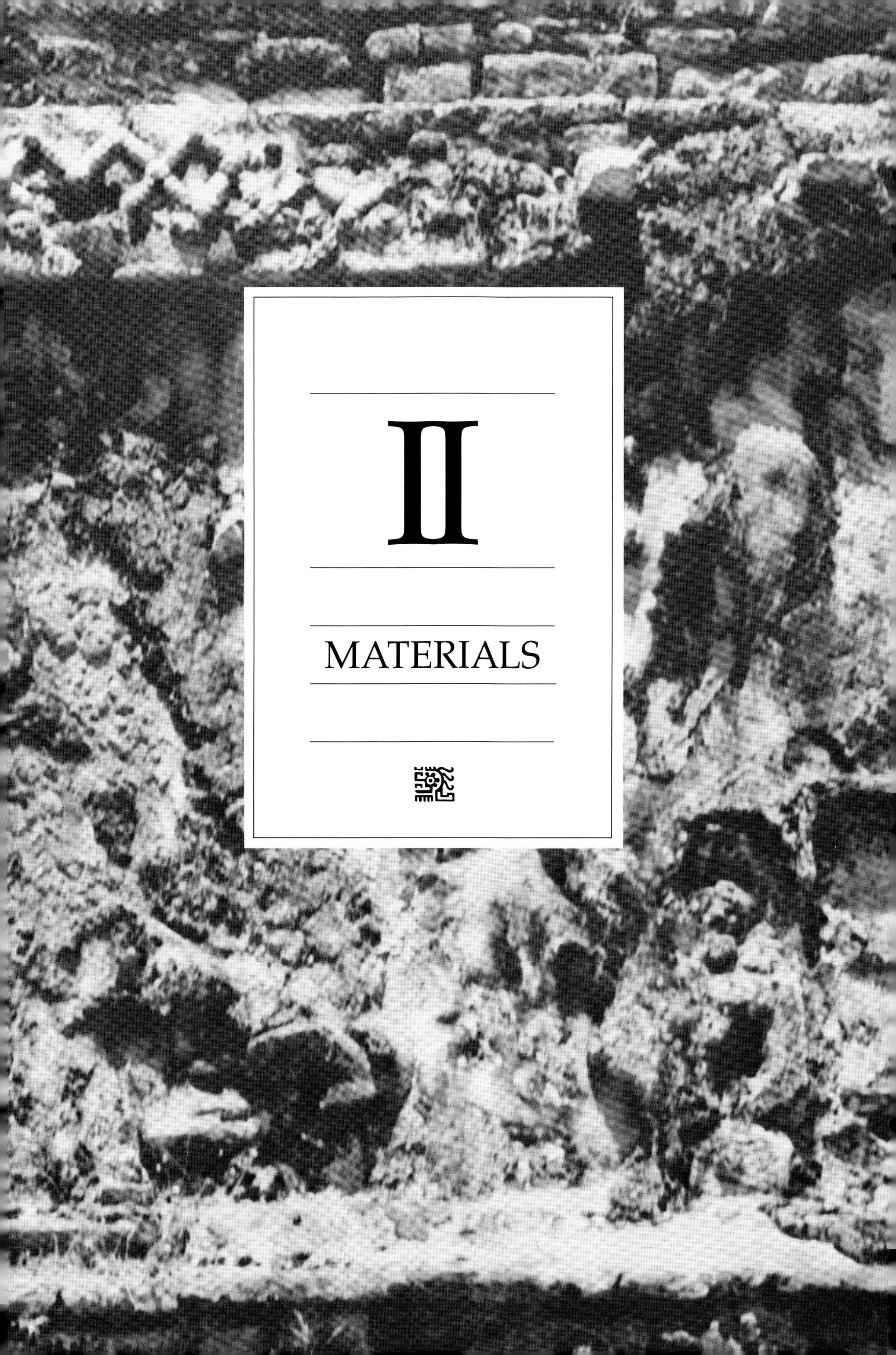

II

MATERIALS

Sharon Gordon Donnan

FIELD CONSERVATION OF ARCHAEOLOGICAL TEXTILES: A CASE STUDY FROM PACATNAMU, PERU

Sharon Gordon Donnan trained in textile conservation in the United States and Switzerland and is involved in textile conservation at the Museum of Cultural History, University of California at Los Angeles. She is also Textile Conservator to the Pacatnamu Project in Peru, which has yielded some two thousand textiles.

The Pacatnamu Project is a binational, multidisciplinary effort involving both Peruvian and North American archaeologists, physical anthropologists, ethnohistorians, faunal and floral experts, and a textile conservator. The Project is sponsored by the Museum of Cultural History at the University of California in Los Angeles and is funded by grants from the National Endowment for the Humanities and the National Geographic Society.

The north coast of Peru is crosscut by a series of rivers that descend rapidly from the Andes Mountains, flow through the arid desert, and empty into the Pacific Ocean. The rivers and their companion fertile valleys have provided excellent food resources for centuries.

Pacatnamu is located at the mouth of the Jequetepeque River, some 650 kilometers north of the capital city of Lima (fig. 1). The climate is mild and arid year round.

This extraordinary archaeological site is situated high on the bluffs, overlooking the Jequetepeque Valley to the south and the Pacific Ocean to the west (fig. 2). The large defensive walls and 30-meter cliffs and bluffs provide fortification for the sprawling architectural complex containing more than fifty truncated pyramids flanked by spacious courts, corridors, and elite quarters.

Beginning with Phase IV of the Moche style (ca. A.D. 500), the occupation of Pacatnamu spanned four epochs of foreign influence:

Figure 1 (below). Map of the north coast of Peru.

Figure 2 (below right). The site of Pacatnamu overlooking the Pacific Ocean.

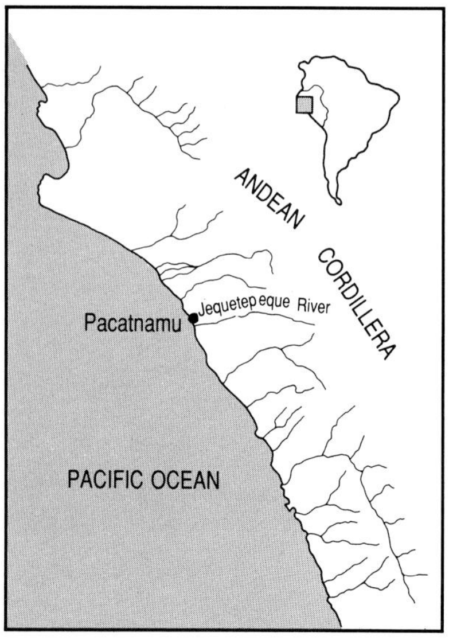

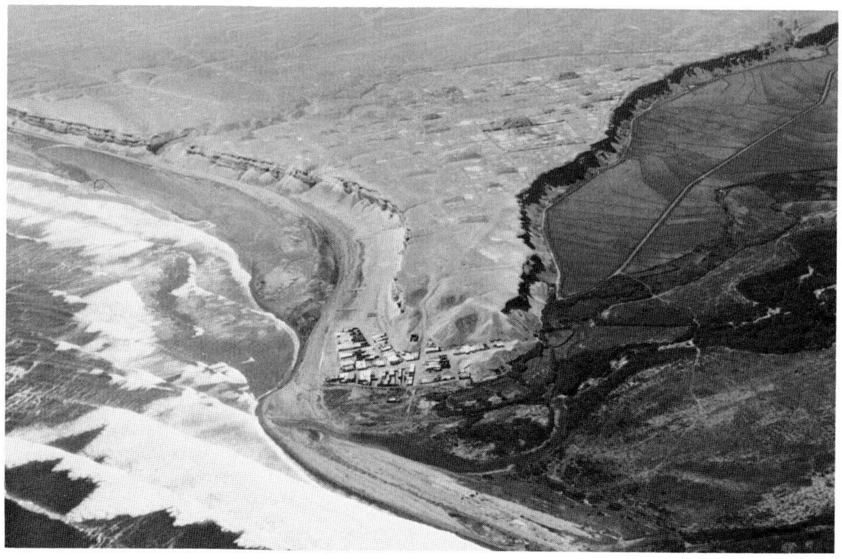

Huari (ca. 700), Chimu (ca. 1370), Inca (ca. 1470), and Spanish (ca. 1530). The impact of each of these foreign influences is discernible at Pacatnamu and, because of the remarkable preservation, it has been possible to compare their relative effects on architecture, religious practices, burial customs, population genetics, pathology, and diet. Pacatnamu offers "laboratory conditions" for the study of a major pre-Columbian center shaped by the dynamics of local development and foreign conquest.

Textiles have contributed immensely to the identification and understanding of both local and foreign influences. Thus far, excavations have yielded nearly two thousand textiles. These include tapestry-woven fragments with complex iconography, miniature textiles representing a variety of garment types, and the shrouds and garments associated with numerous burials. These important discoveries provide major insights into the role played by textiles in the ancient civilizations of Peru.

From the inception of the Pacatnamu Project, it was recognized that conservation must begin in situ. Not to do so would have endangered the textile as an artifact and risked the loss of important contextual and associated data.

The remote location of the archaeological site and the field laboratory creates unique problems requiring innovation and modification of standard museum procedures. The site can only be reached via an unmarked dirt road, which turns off the Pan-American Highway and cuts across rocky, desert terrain for 25 kilometers before reaching the outer wall of Pacatnamu. The road continues through the site, snakes along the edge of the cliff overlooking the intensely cultivated valley below, and terminates in the small village of La Barranca. The village is abandoned except for three months every year, when residents from the neighboring inland villages use it to escape the summer heat.

The proximity of the village to the site (a five-minute drive) makes it ideal as a base camp. A house in the village provides living quarters for a ten-person crew and also serves as a field laboratory. The second floor of the house is devoted to the conservation and analysis of excavated materials, which are brought from the site daily. Skeletal, ceramic, metal, and botanical collections are catalogued, analyzed, and drawn immediately after excavation.

The textile laboratory is located in the adjacent breezeway. Its openness provides abundant natural light, which is ideal for the close work of textile analysis and conservation. Wind can be a problem, so a temporary barrier of plastic sheeting and plywood serves as a shield. A piece of plywood covered with washable white contact paper and supported by sawhorses functions as a work table. Although the potable water trucked in weekly for household use has a high mineral content and is therefore unsuitable for wet cleaning textiles, we found that small bags of distilled water used for car batteries solved the problem. They can be purchased from gas stations along the Pan-American Highway at a cost of about nine cents per bag.

The first textiles processed in the laboratory were surface-collected from looted cemeteries outside the walls of Pacatnamu. A chronology for the site's occupation was based on these findings. The earliest occupation was approximately A.D. 400–500, and the site continued to be occupied until the Colonial period in the early sixteenth century.

Excavations inside the walls began with the largest pyramid (*huaca*), designated Pyramid 1. In attempting to map the quadrangle adjacent to this pyramid, we trenched along exposed walls to better define the architectural features. This revealed a far more complex architectural structure than we had expected (fig. 3).

Figure 3. Plan of the Major Quadrangle.

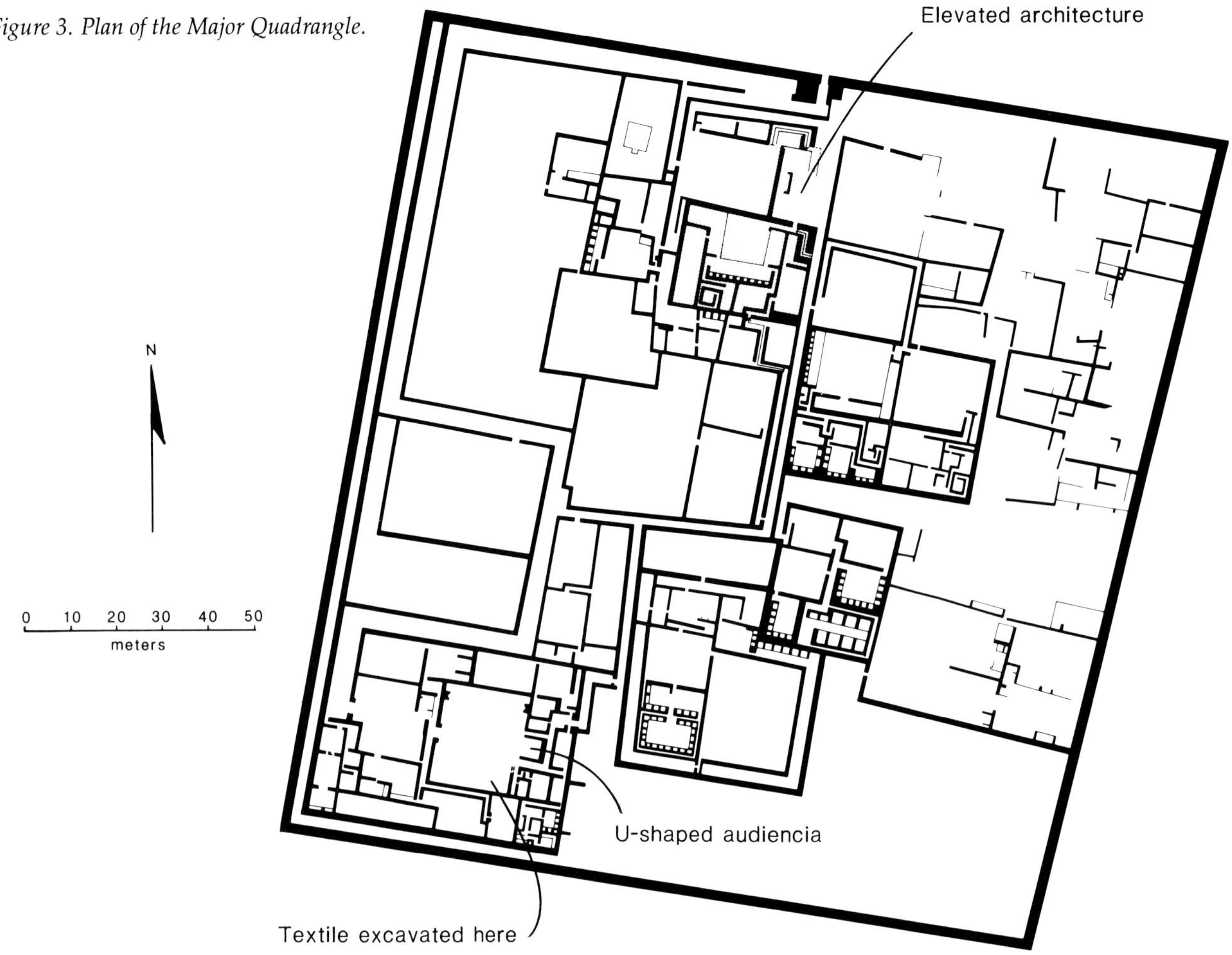

Figure 4. Miniature textiles were discovered in remarkably good condition.

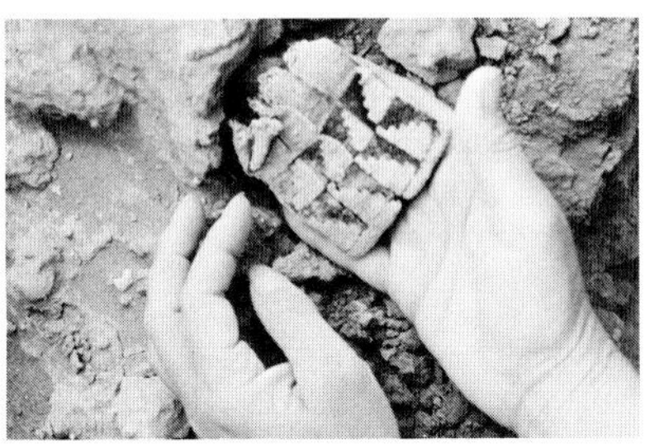

Figure 5. Textiles were cleaned and a condition report was filed on each one.

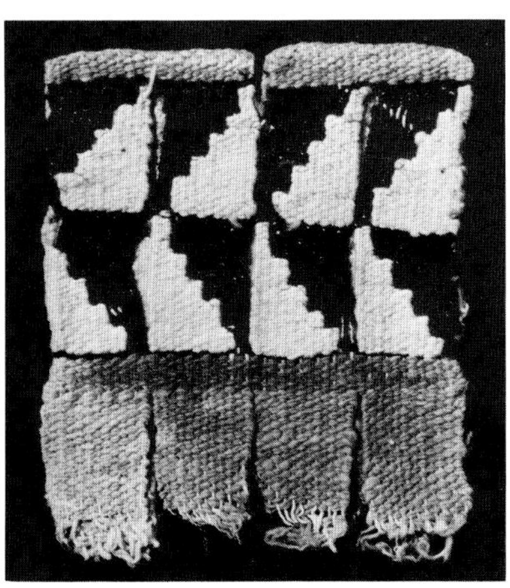

More than one thousand textiles were excavated in a small room in the southwest corner of the quadrangle (pls. 7-9). Many of these textiles were miniature shirts woven in a slit-tapestry technique with a tab fringe at the bottom and side seams sewn with an overcast stitch (fig. 4).

Immediately upon excavation, textiles are stored in muslin bags until the end of the work day. These bags protect the artifacts from the harmful effects of wind, direct sunlight, and rapid changes in temperature and humidity. They also serve as an excellent means for transporting the textiles to the field laboratory. The bags are numbered to correspond to the unit of excavation. This number is assigned to all textiles from that unit. In the laboratory, the textiles are carefully examined to determine if conservation treatment is necessary. After a light brushing on both sides to remove loose sand and soil, the textile is ready for analysis and storage. The following information is recorded on the textile data sheet: condition of collection; location; dimensions; description; construction; edge; finish; fiber spin and ply; thickness; ends per centimeter; and color of warp, weft, and sewing threads. Textiles requiring additional treatment remain in the lab and a conservation record sheet is prepared describing condition and treatment. For the most part, the miniature textiles were beautifully preserved and required minimal treatment (fig. 5).

One artifact that did require substantial conservation treatment was a miniature crown consisting of a cylindrical reed armature with a slit-tapestry, checkerboard textile covering (pl. 7, fig. 6). When excavated, the armature was broken and extremely brittle. We decided to remove the textile so that we could relax the brittle fibers of the armature in a vapor chamber. In the field, such a chamber can be created by placing a ceramic cup containing isopropyl alcohol inside a large Ziploc® bag. The bag, with artifact inside, must then be placed in indirect sunlight. Alcohol vapor relaxed the dry, brittle fibers so the armature could be manipulated into its original form (fig. 7). Tissue paper was temporarily placed inside the crown. Although somewhat revitalized in the vapor chamber, the armature remained weak and brittle. It was lined with Crepeline to provide support for its upright position.

The textile was wet cleaned in a solution of a detergent (Orvis®) and water. It was pin blocked, air dried, and sewn to an unsized fine silk fabric (Crepeline) for additional support.

We decided to reconstruct the piece and, because of the fragile condition of the armature, it seemed more prudent to use a facsimile. A basket weaver from a nearby village made an exact reproduction of the armature, and the textile covering was positioned around it (fig. 8). This substitution also allowed the original armature to remain unobscured for further analysis and research.

Deciding on an appropriate storage system for the entire textile collection was a complicated task. Several seemingly incompatible requirements had to be met. The textiles had to be visible from both sides to facilitate analysis and viewing. They had to be readily accessible to researchers during the three-month field season. And, of course, the textiles had to be well protected from dust, insect infestation, and general handling.

Four mil polyethylene Ziploc bags fulfilled all the requirements.[1] Seven standard-size bags were chosen with dimensions ranging from 5 centimeters to 45 centimeters. They were labeled alphabetically, "A" designating the smallest bag. The textiles in their individual polyethylene bags were then stored in numerical order (i.e., A-1, A-2, etc.). Small bags are sufficiently rigid to maintain an upright position when placed in a file cabinet; large bags had to be stored in boxes. The bag label and number and the excavation unit number were written on masking tape in indelible ink and placed on the upper-left-hand corner of the bag. The same information, typed on acid-free paper, was placed inside the bag in case the external number was lost or obscured. The Project maintains a storage catalogue, a textile catalogue, and an excavation-unit catalogue for cross-reference. This storage system, fashioned after the traditional filing system, permits rapid retrieval while meeting all the requirements of proper textile storage.

The excavation of burials and the unwrapping of mummy bundles is another important part of textile conservation on the Pacatnamu Project (fig. 9). Once exposed by the archaeologist, the burial is cleaned, and loose soil and sand are removed while it is still in situ. The removal of sand greatly alleviates stress on fragile fibers and reduces the risk of tearing. Soft bristle brushes are used for this purpose. Small picks and dental tools can remove hard-packed soil. The inventory of textiles in burials begins at this juncture, and in situ photographs are taken to record their exact position.

After it has been photographed, the burial is immediately transported to the field laboratory where further cleaning with brushes and dustpans proceeds. Removing soil and sand from the outer layer makes it possible to see the direction in which the textile is wrapped and how it should be removed. Unfolding each layer of cloth to expose the skeletal remains is a lengthy process. As the unwrapping proceeds, photographs are taken to record the position of each textile. Detailed notes and drawings also capture pertinent data. As each successive layer of textile is removed, more sand, soil, and other debris are brushed away. Accurate measurements can be obtained even from badly fragmented textiles during unwrapping and cleaning. One mummy bundle had outer shrouds, a shirt used as padding, and the skeleton of

Figure 6. Miniature crown in the condition in which it was found.

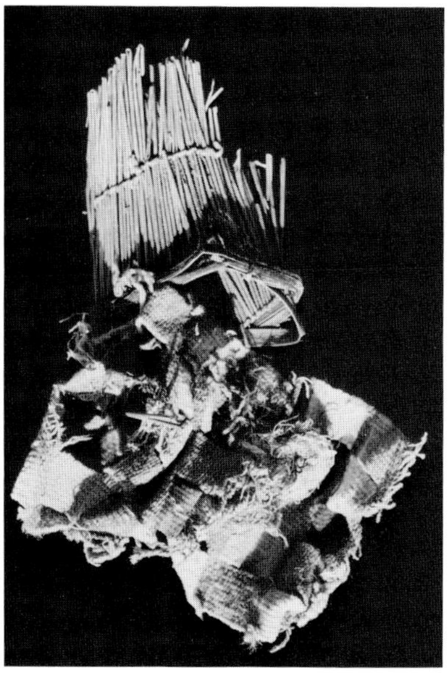

Figure 7. A reproduction of the armature (of the miniature crown) was made by a local basketmaker.

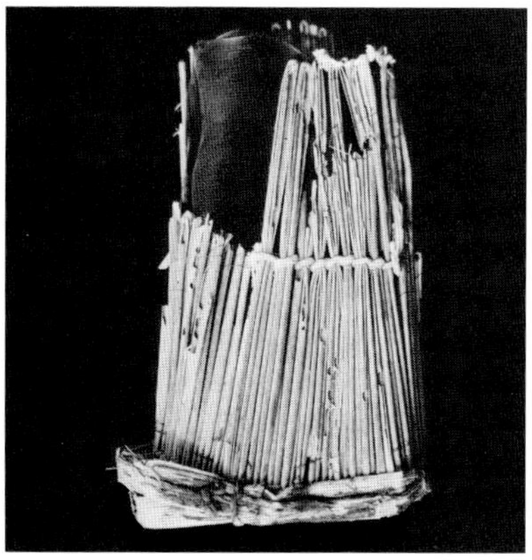

Figure 8. The original textile covers the new armature.

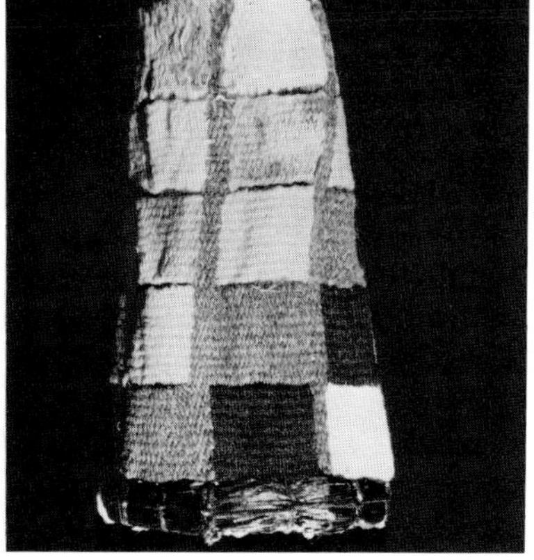

1. A primary concern with the polyethylene bags was the possibility of condensation. Careful monitoring revealed that condensation did not occur unless the bags were exposed to direct sunlight.

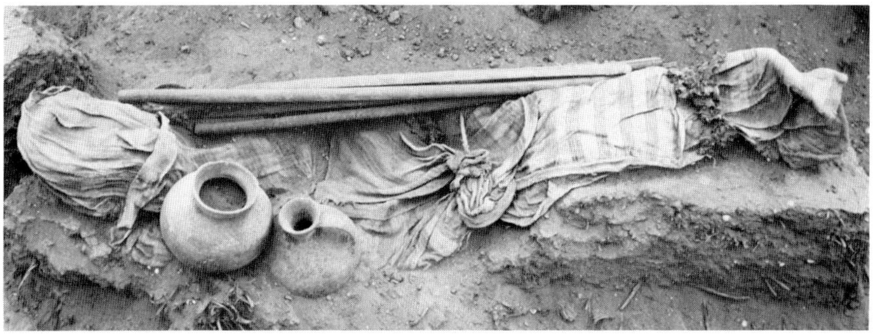

Figure 9 (right). A burial found at Pacatnamu.

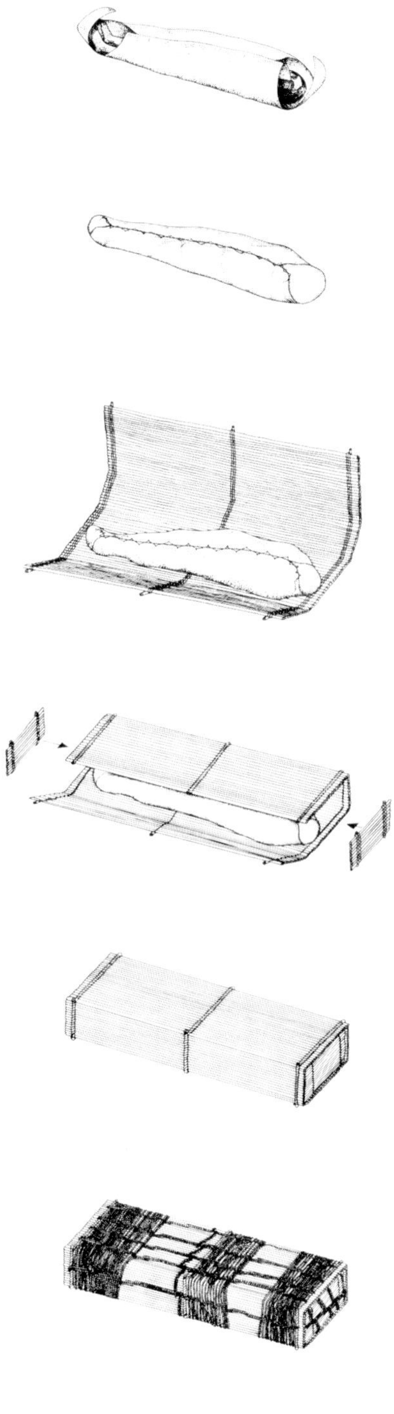

Figure 10 (below). Drawings indicating the construction of the coffin.

a very small female of advanced age. A head covering was wrapped around the skull. The exposed skull revealed a receding jaw with only one tooth. Her large shrouds and shirt, measuring more than 1.5 meters in length, were removed and rolled onto muslin covered tubes and wrapped in cotton sheeting.

During the 1985 field season, we unearthed an important, unlooted Moche cemetery, which led to the development of an alternative conservation technique. The cemetery yielded sixty-five graves, the largest number ever excavated from a Moche cemetery. They provided valuable data on funerary practices and allowed the archaeologists to reconstruct five distinct burial types reflecting different levels of social and economic status. The cane wrappings or coffins associated with the burials played an integral role in distinguishing the burial types. Types 1–4 were simply constructed and well preserved, but the single example of Type 5, a more complex cane construction, exhibited particularly poor preservation. Excavation proceeded slowly in an attempt to expose and yet preserve the position and form of the coffin. To understand better its construction and relationship to its content, we had a duplicate made (under the supervision of a representative from the Instituto Nacional de Cultura) of the original coffin in situ. The replica was constructed with painstaking precision. While the advanced state of deterioration left the ancient coffin almost useless as a research tool, the full-scale replica gave us a complete and accurate description of the coffin (fig. 10).

One textile in particular illustrates clearly the essential role of textile conservation on the Pacatnamu Project. This fragmentary piece of slit tapestry was perhaps the single most important artifact found during the first three years of excavation (pl. 10). Like the miniatures, it was found inside the Pyramid One quadrangle. After wet cleaning, pin blocking, and mending, the complex iconography was revealed.

Two high-status individuals seated on elevated structures under thatched roofs hold decorated goblets (fig. 11). The symmetrical notches on the tops of the thatched roofs are identical to those on some of the architecture depicted on Chimu ceramics.[2] The railed or elevated architecture where the two figures are seated resembles structures located near the entrance to the quadrangle. The depiction of two individuals sacrificing white llamas is particularly interesting because such llamas, carefully wrapped in elaborate textiles, were found ritually buried inside the quadrangle.

There are two weavers depicted on the textile, and many weaving materials were found inside the quadrangle. The multiple bobbins hanging from the loom on the textile imply the weaving of tapestry. Numerous balls of yarn, bobbins, and spindles were excavated, and the yarns were identical in fiber, spin, thickness, and color to those used in many of the unearthed textiles. This suggests that at least some of the textiles were actually woven in this area.

This remarkable textile, with its complex iconography and

2. Chimu was the predominant ceramic style on the north coast of Peru (A.D. 900–1500).

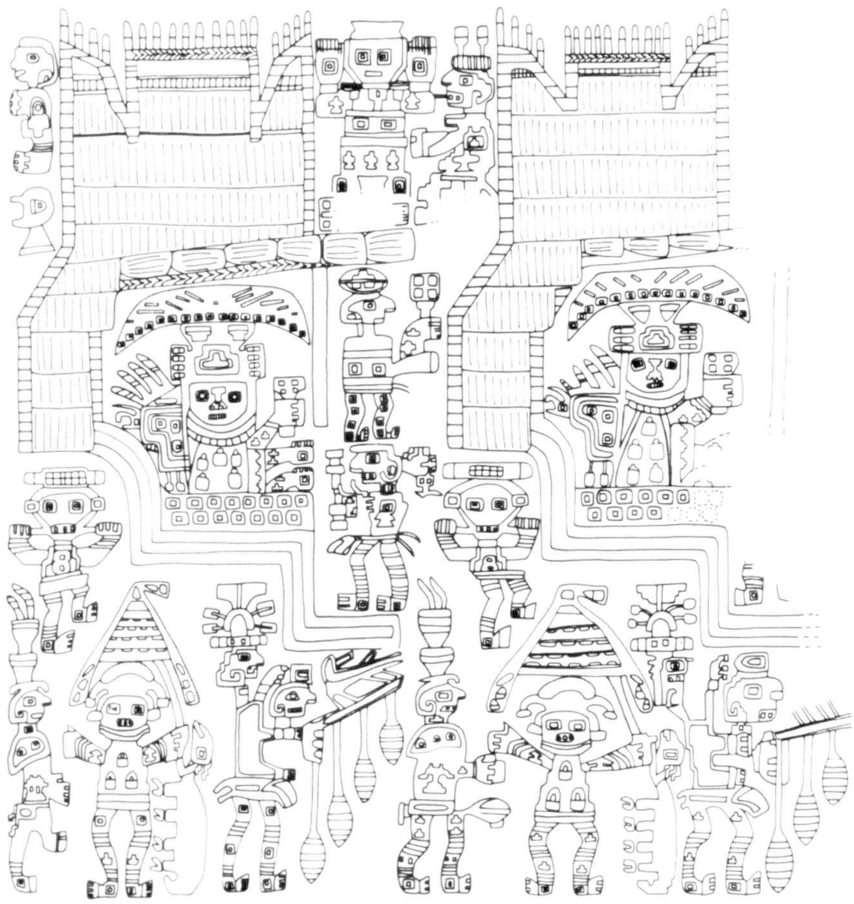

Figure 11. Drawing of the iconography on the elaborate textile found in the Major Quadrangle.

weaving techniques, depicts several individuals who appear to be engaged in ceremonial activities. The archaeological material excavated inside the quadrangle provides good evidence that these kinds of activities actually took place there. This small fragmentary textile provides a fascinating glimpse of how the quadrangle may have functioned, as seen through the eyes of the people who lived there in ancient times.

In the young and rapidly growing science of conservation in the Americas, textile conservation is a recent addition. This is not surprising since few archaeological zones in the world are as well preserved as Pacatnamu. Nonetheless, the growing awareness of the importance of textiles in reconstructing the past will lead to more and better in situ conservation in the future.

References

Donnan, C.B. and G.A. Cock (eds.) 1986 *Pacatnamu Papers*, Vol. I. Museum of Cultural History. Los Angeles.

Dowman, E.A. 1970 *Conservation in Field Archaeology*. London.

Finch, K. and G. Putnam 1977 *Caring for Textiles*. Oxford.

Keatinge, R.W. 1978 "The Pacatnamu Textiles." *Archaeology* 31.

Rowe, A.P. 1984 *Costumes and Featherwork of the Lords of Chimor Textiles from Peru's North Coast*. Washington, D.C.

Pamela French

THE PROBLEMS OF IN SITU CONSERVATION OF MUDBRICK AND MUD PLASTER

Pamela French studied archaeological conservation at the Institute of Archaeology, University of London and then studied the conservation of wall paintings on a Gulbenkian Foundation Scholarship to the Istituto Centrale del Restauro, Rome. She is presently at the British Institute of Archaeology, Ankara, Turkey and is also a part-time lecturer at the Middle East Technical University there.

Sun-dried mudbrick and mud plaster have been traditional building materials for thousands of years in many parts of the world—especially in areas that have long, hot, dry summers. In the past they were used to construct not only simple dwellings, but also large and elaborate monumental structures decorated with paintings, sculptural reliefs, and mosaics.

It has long been recognized that the conservation of mudbrick structures, whether complete monuments or excavated remains of buildings, presents problems not encountered in the preservation of ordinary masonry or burnt brick. However, it was not until the late 1960s that attempts were made to coordinate research into this problem. With the aid of the International Council on Monuments and Sites (ICOMOS), this effort led to the formation of the International Committee on the Conservation of Mudbrick. Since then, much interest in this field has been generated on national and international levels. Research programs are still in the early stages and, as yet, no definitive solutions have been found, though some results show promise. For the conservator and the archaeologist working in the field on archaeological excavation, the picture may seem unclear and confusing. It is hoped that this paper will serve to simplify some of the issues and outline the results of research to date.

Composition and deterioration of mudbrick

Mudbrick is essentially a mixture of clay, silt, and sand that is mixed with water and shaped, either by hand or in molds, to form bricks. The bricks are then allowed to dry in the sun until they have hardened. Sometimes filler material, such as chaff, straw, or hair, is added to the mix as a binder to reduce cracking as the bricks dry (Exodus 6:16).

The advantage of using a building material of this kind is its availability and low cost.[1] The precise composition of mudbrick varies from one region to another. Its durability, whether mudbrick or mud plaster, is largely dependent on the quality of the raw material used. A durable mudbrick, ideally, should contain a high ratio of sand to silt and clay and not more than 10%-15% clay to silt and sand (Brown and Clifton 1978:139–146). Clay and, to some extent, silt act as the binding media in the mix.

1. Villages in Central Anatolia that still use mudbrick have a designated area outside the village where the composition of the soil is thought to be the most suitable for brick making.

Clays, which are natural aluminosilicates, have a layered, platelike structure of linked aluminium silicate tetrahedra contained within layers of magnesium or aluminium hydroxide. Clay both absorbs and adsorbs water and becomes plastic. This action is accompanied by an expansion in volume that is reversed when water is lost from the clay. Some clay minerals expand more than others as they absorb moisture e.g., montmorillonite (an aluminosilicate with variable composition, which has a high cation-exchange capacity). Should there be an excess of clay, or should the clay contain a high percentage of such a mineral, the stability of the material would be severely impaired.

Groundwater in the region of a settlement will be heavily charged with soluble and sparingly soluble salts. Buried structures will have absorbed groundwater and thus an abundance of salts. When such structures are excavated and the wall surfaces exposed, the moisture evaporates. Evaporation takes place rapidly at first, as water from the exposed surface of the bricks and plaster is lost, and then more slowly from the interior of the structure as the clay particles gradually lose both adsorbed and absorbed water and the moisture moves to the evaporation zone. The clay component of plaster or brick shrinks, and soluble salts crystallize out of solution either at the surface or in pores below the surface of the structure. The effect of these two actions causes the surface to crack. If the newly exposed wall is then subjected to intermittent wetting from rainwater or condensation, followed by periods of drying, moisture will be absorbed and lost repeatedly.

The damage to vertical surfaces, especially at the base of walls where humidity is at its greatest, can be severe. The characteristic "undercut" appearance of mudbrick walls can be attributed to deterioration caused by salt movement from groundwater and by wind erosion (Torraca 1970:47–53). The tops of newly exposed walls are also vulnerable, as they are easily eroded by rain.

Conservation of mudbrick

Many excavated mudbrick structures are not architecturally important except as a part of wider research. After proper recording they are destroyed to expose earlier building levels, and the problem of conservation does not arise. If, however, the structures themselves are architecturally, historically, or archaeologically significant, or if they contain features such as reliefs or wall paintings, the situation changes dramatically.

In years past, the confusion surrounding the treatment of mudbrick and mud plaster, together with the large financial outlay required for such work, presented such a daunting prospect that many sites were left without the most basic precautions taken to ensure the protection of the excavated material. This desperate situation led to the drafting of various resolutions and recommendations by the International Committee on Mudbrick Preservation (*Third International Symposium on Mudbrick [Adobe]Preservation* 1980a:265,268; 1980b:281–284). The most important of these recommends that no archaeological excavation be undertaken on sites likely to produce mudbrick structures unless "a provisional conservation policy has been established and included in the excavation budget and agreed upon by archaeologists and the competent authorities." This is an admirable principle, but one that is not easy to follow because of the element of unpredictability that exists in most archaeological undertakings. In the event of a significant discovery, the excavator must have sufficient time, funds, and the inclination to take steps that will ensure at least the temporary safety of the material.

The essence of any conservation treatment, temporary or long-term, for material of this kind is to protect it from the effects of water once it is excavated. Backfilling excavated areas where mudbrick walls are exposed is one temporary solution that has been used extensively, especially on sites where there are wall paintings on mud plaster that need immediate protection until a suitable treatment can be planned (Chiari 1980:39–45).

Backfilling partially excavated *tepe* sites, however, can be impractical, especially areas in the upper part of the site or where the remaining walls are still high[2]. In these cases, adequate protection must be given to each individual structure to prevent erosion by rain, melting snow, and wind. Provisions must also be made to direct the movement of groundwater away from the fabric of the wall and to drain away surface water, which may form puddles at the foot of walls or in excavated trenches.

2. Also known as *tells* or *büyüks*, these mounds are the products of repeated building and rebuilding on the same site.

Any form of protection that may give rise to condensation, such as hanging polyethylene sheeting or a similar material over walls or excavated areas, must be avoided. Modern mudbrick dwellings are regularly maintained with new mud-plaster renderings, which protect the brickwork from weathering by acting as a sacrificial coating. This procedure has also been followed on archaeological sites, with varying degrees of success (Lewis 1980:109–118; Warren 1980:14). Capping the tops of walls to prevent moisture penetration leading to erosion and the formation of fissures and cracks is essential.

Providing open-sided, roofed shelters to protect an excavated area from the direct action of rain and snow can be effective if sufficient attention is paid to draining away the water and the possibility of erosion by the prevailing wind and wind-borne rain (Brown and Clifton 1978:139–146; Lewis 1980:109–118). These measures can be expensive and are usually undertaken as part of a program of permanent conservation.

An inflatable airhouse, though not suitable to protect a site permanently, can provide a stable environment while a conservation treatment is undertaken or while particularly sensitive materials, such as wall paintings, are uncovered (Weaver 1973:88–93).

The protection of vertical surfaces is often considered of secondary importance if the penetration of water at the top of the wall and at the foot has been dealt with successfully. The loss of material on these surfaces from erosion will proceed slowly and in the short term will be less dramatic. The surfaces of a structure at the time of excavation will be in a partially deteriorated state; the exposed surfaces may be rough and cracked, providing an enlarged surface area for the absorption of moisture from the atmosphere. The bond between rendering and brickwork (or the deteriorated face of brickwork and the structurally stronger substrate) may be weak, and the loss of decorative features, such as cornices, moldings, sculptural reliefs, and wall paintings is a very real possibility unless these surfaces are protected in some way.

Research on consolidation methods

The approach to the treatment of mudbrick, before the establishment of an organization to coordinate research, was rather haphazard and consisted of isolated attempts to restore mechanical strength and protect the material from weathering by consolidating it with synthetic resins. The results of these experiments were often poor and difficult to evaluate because of the lack of information concerning the composition of the material treated, the resins used, the method of application, and other variables, including climate and weather conditions at the time of treatment.

Research in the laboratory and, most important, on site has continued in the attempt to find a suitable material to restore mechanical strength and increase hydrophobicity. The penetration of resins depends not only on the chemical makeup of the material to be consolidated but also on the size of the resin molecules and the properties of the solvent solution (Lewin and Papadimitriou 1981). The solvent used directly influences the distribution of the resin within the porous structure of the mudbrick. Many of the early tests failed because the resin solutions did not have the right properties for the material to be consolidated. The resins were not deposited evenly within the pores of the mudbrick, but were drawn back to the surface by capillary action as the solvent evaporated. There they formed a thick, unsightly film that eventually became detached, taking with it an appreciable amount of the surface—a disaster when a wall painting was involved.

Many of the materials that have been tested and that are being tested currently are those that have been found useful in the treatment of deteriorated stone. They include both monomer and polymer systems of synthetic resins, silicone esters, and alkoxy silanes.

Acrylic resins have been favored in mudbrick and plaster consolidation over the last decade because of their stability, solubility in several organic solvents, consolidating ability, reversibility, and hydrophobicity, while allowing the passage of water vapor through the film. The monomer systems of these resins have been tested in the past—the low viscosity of the monomer making deep penetration of the consolidant more feasible. The process of polymerization in situ has proved problematic (Vassallo and Lewin 1981), and more attention has been focused on the polymers, particularly those of medium molecular weight (Butterbaugh and Pigott 1980:19–28). Results from several sites have proved the polymers satisfactory so far.

Using a low-percentage solution in a solvent that evaporates slowly and is made up of relatively small molecules, such as toluene, encourages deep penetration, and the resin film coats the surface of the pores. This protects the outer fabric of the wall against the penetration of moisture from the atmosphere, but still allows (in theory, at least) the passage of water vapor from the interior of the wall. Caution should be observed when selecting this resin for its hydrophobic properties, as recent tests of the material used to consolidate limestone showed a decrease in hydrophobicity and a possible change in the structure of the resin (Charola et al. 1985:739).

Encouraging results have been recorded using silicone esters and alkoxy silanes to impart water repellency to mud plaster and brick (Rossi-Manaresi and Chiari 1980). These materials, because of their low viscosity, penetrate easily and have the advantage that on curing, the polymers bond with siliceous materials in the mudbrick and improve cohesion (Lewin and Wheeler 1985:831–834; Nishuira 1981:699). Structural strength may not be dramatically altered; to increase this satisfactorily, an acrylic silane can be used or an acrylic polymer added to a silane solution (Bradley 1985:759–767). Good results have been recorded using ethyl silicate to treat volcanic tuff that is similar in composition to mud plaster and painted mud plaster (Rossi-Manaresi and Chiari 1980).

A cautionary note should be introduced here: There is a very wide range of materials available, and most experimental work has been carried out on stone, not mudbrick. While the results are promising, more research in the laboratory and in the field is needed, as the long-term stability and durability of mudbrick and mud plaster is, at best, uncertain.

One of the main points of disagreement among conservators concerns the depth a consolidant must penetrate into the brick or plaster to give adequate protection. Some maintain that the difference in the thermal contraction and expansion rate between the unconsolidated core of the material and the consolidated surface alone heightens deterioration and causes the surface to spall at the point where the two zones meet. The danger of splitting is increased with the deep penetration of a consolidant, when the boundary between zones is sharp rather than diffuse, and salts and water are allowed to back up behind the consolidated layer. Because of this, many conservators favor the application of a light, superficial, and sacrificial coating of water repellent or consolidant. This method requires regular inspection and maintenance and is not suitable for treating decorative reliefs or wall paintings.

For wall paintings on mud in open sites, there is no definitive solution, though some success, using mixed techniques on adobe friezes in Peru, has been recorded (Chiari 1980:39–45). Backfilling, or lifting and treating the paintings in the laboratory, while unsatisfactory, seem to be the safest answers at present.[3]

3. This does not apply to paintings within standing structures, or to buildings where it is possible to devise effective controls over the environment.

I realize that my comments may not be received with much enthusiasm by the delegates to this symposium on in situ conservation. But it must be recognized that, until quite recently, the conservation of wall paintings on mudbrick has too often been a matter of treating the symptoms rather than the disease. Now that we have begun to diagnose the many and varied causes of our problems, we are in a far stronger position to find effective cures. A great deal of research remains to be done, but at least we have good reason to be optimistic about the outcome.

References

Bradley, S.
1985 "Evaluation of organo silanes for use in consolidation of sculptures displayed indoors." *Fifth International Congress on Deterioration and Conservation of Stone.* Lausanne.

Brown, W.M.P. and J.R. Clifton
1978 "Adobe 1—The properties of adobe." *Studies in Conservation* 23.

Butterbaugh, D. and V. Pigott
1980 "Masca mudbrick/adobe conservation, interim report." *Third International Symposium on Mudbrick (Adobe) Preservation*. Ankara.

Charola, A.E., M. Laurenzi Tabasso, and U. Santamaria
1985 "The effect of water on the hydrophobic properties of an acrylic resin." *Fifth International Congress on Deterioration and Conservation of Stone*. Lausanne.

Chiari, G.
1980 "Treatment of adobe friezes in Peru." *Third International Symposium on Mudbrick (Adobe) Preservation*. Ankara.

Lewin, S.Z. and
A.D. Papadimitriou
1981 "Investigation of polymer impregnation of stone: Solvent transport of pre-polymerized methacrylate." *The Conservation of Stone II. Preprints of the Contributions to the International Symposium*. Bologna.

Lewin, S.Z. and
G.E. Wheeler
1985 "Alkoxy silane chemistry and stone conservation." *Fifth International Congress on Deterioration and Conservation of Stone*. Lausanne.

Lewis, H.
1980 "Experiments in mudbrick conservation at Tepe Nush-i-Jan." *Third International Symposium on Mudbrick (Adobe) Preservation*. Ankara.

Nishuira, T.
1981 "Conservation of old roof tiles for re-use." *The Conservation of Stone II. Preprints of the Contributions to the International Symposium*. Bologna.

Rossi-Manaresi, G. and
G. Chiari
1980 "Effectiveness of conservation treatment of volcanic tuff very similar to adobe." *Third International Symposium on Mudbrick (Adobe) Preservation*. Ankara.

Sengupta, R.
1980a "Restoration of proto-historic ruins of adobe, Lothar, India." *Third International Symposium on Mudbrick (Adobe) Preservation*. Ankara.

1980b "Resolutions from the 2nd International Symposium on the Conservation of Monuments in Mudbrick, Yazd (1976)." Ankara.

1980c "Recommendations approved in Ankara." Ankara.

Torraca, G.
1970 "An international project for the study of mudbrick preservation." *New York Conference on the Conservation of Stone and Wooden Objects*. IIC. London.

Vassallo, J.C. and
S.Z. Lewin
1981 "Investigation of polymer impregnation of stone II. In situ polymerization of methyl methacrylate." *The Conservation of Stone II. Preprints of the Contributions to the International Symposium*. Bologna.

Warren, J.
1980 "The form, life and conservation of mudbrick building." *Third International Symposium on Mudbrick (Adobe) Preservation*. Ankara.

Weaver, M.
1973 "The use of an inflatable airdome to produce controlled conditions for an archaeological site." *Studies in Conservation* 18.

Agustín Espinosa

CONSERVATION AND RESTORATION OF THE MURALS OF THE TEMPLE OF THE PAINTINGS IN BONAMPAK, CHIAPAS

Agustín Espinosa trained in the restoration of moveable objects. He is a professor of conservation theory and techniques at several institutions and has worked at the Interamerican Center for the Restoration of Cultural Heritage. He is currently President of the Mexican Association of Conservators of the Cultural Heritage and Director of Restoration of the Cultural Heritage, INAH, Mexico. He has been an adviser to UNESCO in various Latin American countries.

This paper presents some examples of archaeological conservation in situ, and also provides an opportunity to consider the effect of this sort of intervention policy on our cultural heritage.

In situ conservation is undoubtedly a victory of our time. We are struggling to leave behind the nineteenth-century vision that created museums for works of art (paintings, sculptures, archaeological artifacts) removed from their original locations. That practice is justifiable only as a last resort; that is, when no amount of safety, conservation, and maintenance measures will allow the works to survive in situ. In those situations, we must analyze all alternative means of treatment in the light of technical and historical knowledge, keeping in mind that the work should always return to its original location when suitable conditions exist. To this end, all relevant documentation must be kept, including architectural surveys, photographs, drawings, technical data on the treatment methods and materials employed, and location maps.

Conservation and restoration specialists are increasingly bound to participate in these decisions since monuments and their archaeological surroundings are generally taken to be part of an overall concept—not isolated, but seen together. Architectural elements, paintings, and sculptures must be viewed in their natural context if their historical and aesthetic relevance is to be fully comprehended. Studied in this way, they can reveal the major phases of cultural development of the civilizations that produced and used them.

The Temple of the Paintings

Bonampak represents an archaeological site where conditions favor the implementation of the principle of in situ conservation. It is located in the southern part of Mexico in the state of Chiapas, bordered by the state of Tabasco to the north and the Pacific Ocean to the south, with Guatemala to the east, and Veracruz and Oaxaca to the west. The archaeological zone is located in the municipality of Ocasingo, on the central *mesa.*

The Temple of the Paintings is reached by crossing the Great Plaza (*Gran Plaza*; fig. 1). The first thing that draws one's attention is the large stele, about 7 meters high, on which Montero worked in 1977. It stands out not only because of its great height, but also because of the sculptural richness of the figure incised in its center.

North of the Great Plaza are two groups of buildings on terraces at different levels, facing onto the plaza. The first terrace rises about 15 meters above the plaza. To the west is the Temple of the Paintings,

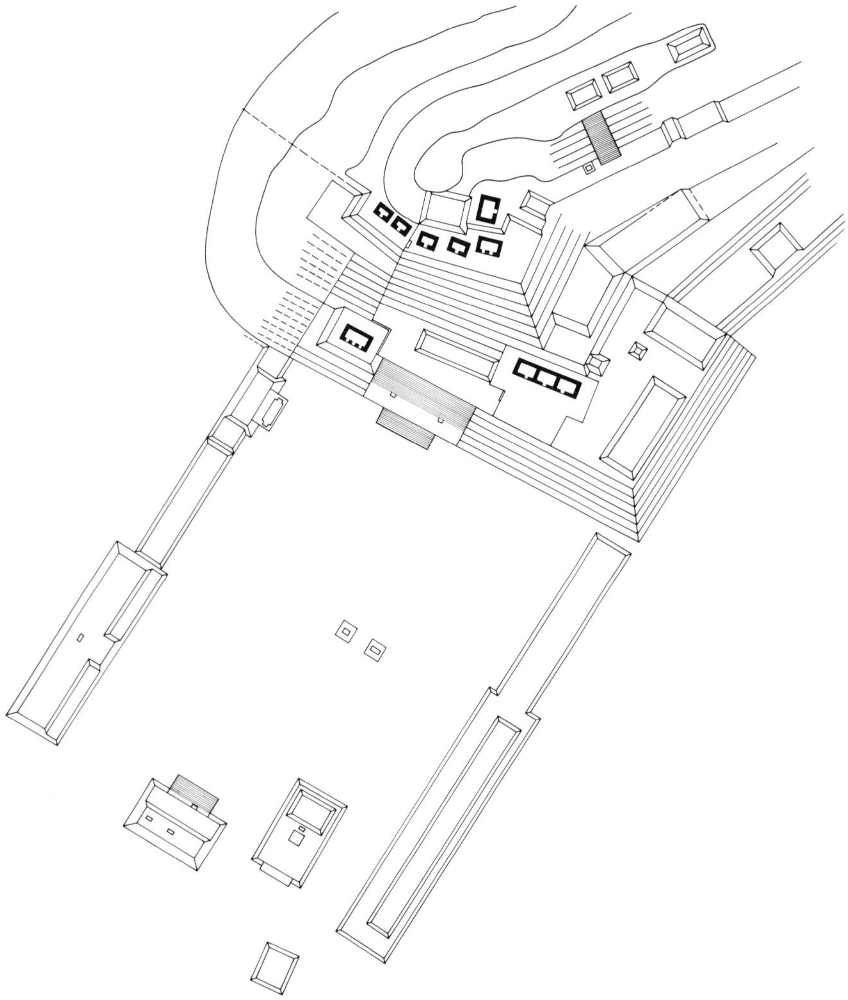

Figure 1. Site plan of Bonampak.

which can be reached by ascending a large stairway.

The building has three chambers, each with a separate access; inside these chambers are the important paintings. The elongated building is covered with a Mayan vault. An exterior frieze has the remains of molded stuccowork and three niches. The central niche contains the remains of a figure. Beneath the frieze, the wall plaster bears traces of polychrome decoration.

Above each chamber entry is a cut-stone lintel with a polychrome scene depicting two figures, one of which is a prisoner. Apart from the paintings, the interiors have rectangular benches along three sides.

Discovery of Bonampak

Although it is now generally known that Bonampak continued to be a ceremonial center for the Lacandon on into this century, it managed to remain hidden from civilization for a long time.[1] In February 1946, H. Carl Frey, an explorer who had been searching for ruins in the area for years, convinced one of the Lacandon to take him to Bonampak, thus exposing the archaeological site (pl. 13). In May of that year, Giles Greavile Haley, who was documenting the Maya culture for the United Fruit Company, discovered the paintings in the Temple.

The discovery, more than forty years ago, of a Maya temple with completely painted interior walls and vaults naturally caused great excitement. It was the first time anything more than mural fragments had been discovered, and the paintings were of a high artistic quality. More-

1. The Lacandon are the last remaining pure Mayans. They live in the forests of the state of Chiapas and number between 200 and 500. [Ed.]

over, the clothing, symbols, weapons, and musical instruments depicted provided extraordinary cultural documentation by shedding light on important aspects of the classic Maya civilization at its peak in the Usumacinta region in the late eighth century. The word *bonampak*, in fact, means "painted walls" and was the name given to the site by the English archaeologist Sylvanus G. Morley.

The Instituto Nacional de Antropología e Historia (INAH) and the Carnegie Institute were notified when the site was discovered, and the latter undertook the first exploration work (1946-1948), which resulted in a general survey of the zone and the Temple of the Paintings.

In 1949, the Instituto Nacional de Bellas Artes de México organized a research expedition with the participation of several specialists, including H. Carl Frey. Unfortunately, Frey drowned in the Lacanha River along with the Mexican engraver, Franco L. Gomez. This accident ended the expedition prematurely, its goal unmet.

It was not until 1960, when INAH took up the exploration work with archaeologist Raúl Pavón Abreu (he remained until 1964), that the majority of the monuments that can now be seen were discovered and consolidated.

Faced with the problem of conserving the paintings, INAH turned to UNESCO for experts who could assist in determining the best way to preserve the building in which they were housed. From 1961 to 1974, many national and foreign experts came to the site to study the problem of treating this important building.

The first camp was set up in 1961. An area near the zone was cleared to serve as a landing strip for light aircraft so that the conservation experts did not have to travel by mule. At the same time, work was begun to clear the bushes and trees that had overgrown the Temple of the Paintings. This resulted in the first climatic change the building underwent and may have caused some structural movement. The removal of the protective layer of vegetation also exposed the Temple to the effects of direct contact with rainwater.

The former Directorate of Restoration of INAH also became involved in the project that same year (1961). Their experts determined the extent of the deterioration of the paintings, caused by humidity and fungal growths that spotted the surface. Nonetheless, at the time of their discovery, the images probably showed an acceptable resolution since there had been a high level of humidity in the building. Drying over the subsequent years, however, caused salts to crystallize on the painted surface and obscure the images (pls. 11, 12).

In response to Mexico's request for UNESCO aid in 1961, Francesco Palessoni came to study the condition of the paintings and propose methods of conservation. He determined that the painting technique was "true fresco" and offered the opinion that the removal of the paintings was advisable. But he also suggested that before such a measure was taken, a team of international experts should be called upon to assess the problem. He specifically recommended the consultation of conservators from Italy because of their considerable experience with the restoration and preservation of mural paintings.

But while the UNESCO experts debated the best initial preventive measures, the paintings continued to deteriorate because of the action of rain and the growth of flora and microflora inside and outside the monument. In 1962, INAH (through the Department of Pre-Hispanic Monuments and the Directorate of Restoration) took the initiative to protect the monument with a provisional canopy of wood and *bajareque*, the traditional palm used by the Lacandon to build their homes. At the same time, to aid the evaporation of humidity, a drain was created along the interior perimeter of the building, air circulation was facilitated in the three chambers, and the long slow drying of its walls and structures began. The first attempts to consolidate the paintings were made using polyvinyl acetate emulsions and epoxy resins, although these failed because the materials were inappropriate, and the walls and plaster were too damp. UNESCO expert Franco Minissi, who came to the archaeological zone to advise on the conservation of the monument and its paintings, proposed waterproofing the building and controlling the relative humidity and temperature.

For Mexico, 1964 was a critical year not only for painting conservation but also for the training of specialists. Notable people in the restoration field, such as UNESCO advisors Leoneto Tintori and Paul Coremans, came to Mexico. After an exhaustive analysis of the condition of the paintings and the building, Tintori made the following recommendations and assessments:

1. An additional drain should be built at the rear of the building to speed the drying of the structures and also to stem humidity caused by capillarity.
2. It might be possible to detach the paintings once the building was dried, as Francesco Palessoni had suggested, but this would be difficult in such a small space.
3. The wall thickness could be reduced by removing the stuccoes and constructing fake walls around the exterior, leaving a 10-centimeter-wide air chamber between them. This would, however, break up the monument's architectural unity and necessitate the removal of the stucco reliefs decorating the exterior.
4. Air conditioning could be installed. This would keep the relative humidity at 60%–70% and the temperature at 25°–30° C.

Fortunately, most of these recommendations were not, or could not be, implemented. As a result, we are able to see the importance of archaeological conservation in situ and appreciate how it protects our cultural heritage. The first proposal was acted upon, and this intervention did greatly reduce the humidity. Had the paintings been detached and the building altered, it would have resulted in the loss of the integrity of the only archaeological building in the country with interior mural paintings of great historical and artistic import. The last proposal, the installation of air conditioning, simply was not feasible because of the high cost.

Also in 1964, UNESCO advisor Paul Coremans, then director of the Belgian Royal Institute for the Artistic Patrimony, examined the paintings. He took test samples and made proposals regarding cleaning, and issued the following recommendations:

1. The climatic conditions inside and outside the monument should be assessed.
2. The provisional palm-and-wood canopy should be replaced with a more stable one.
3. The foundation of the building and the terraces should be examined to determine their stability.
4. Fake walls, separated from the monument by an air space, should be constructed around the building, as Tintori had suggested.
5. The plaster that supports the pictorial layer should be secured to the wall.
6. Carbonate deposits should be removed from painted surfaces.
7. Color integration of the paintings should be undertaken.

Coremans, impressed by the cultural wealth of Bonampak and other sites in Mexico, promoted in UNESCO the creation of the Latin-

American Regional Center for Conservation Studies, which made possible the training of conservators in the field for a period of ten years. That influence has undoubtedly had a positive affect on the work now being done in Bonampak, and it also motivated INAH to create the National School of Restoration and Museum Science for conservation professionals and technicians.

After the highly significant events in 1964, many national and foreign experts visited the site. In 1971, Jaime Cama, Sergio Montero, Luis Torres, Charles Hett, Harold Barker, and Mihailo Vunjak met at Bonampak to discuss and plan an international meeting. They decided to invite foreign experts to propose a final project. INAH began lengthy negotiations with UNESCO to conduct the conference, which finally took place in 1974 in the Bonampak archaeological zone. The project was then interrupted for ten years, but in 1984, interest in the conservation and restoration of the paintings was renewed. It was decided that the problem should be addressed not only from the conservation point of view, but also from the perspectives of other fields of study, in order to broaden both the assessment of the problem and the possible solutions. Thus, restorers, archaeologists, chemists, biologists, epigraphers, architects, and photographers were invited to participate.

An analysis of the data and documentation material revealed that the main causes of deterioration were humidity and heat. These factors are inherent in a tropical setting with an average annual rainfall of 2,603.3 millimeters and a mean annual temperature of 24.6° C.

Seepage and capillary humidity, in the context of a tropical climate, resulted in flora and microflora growth inside and outside the building, and are the main causes of the destruction of the paintings. Capillary humidity moves up the walls, carrying soluble ground salts. This action weakens the cohesion of materials through expansion and hydration under changing temperatures. In addition, the constant and extreme variation in temperature between day and night causes repeated expansion and contraction, which contribute to the disintegration of construction materials. In fact, the effect of rainwater alone is enough to destroy the paintings. Fortunately, the protective measures taken in recent years have arrested the effects of humidity, enabling this important monument to be saved.

In 1984, experiments were conducted by various experts in an attempt to discover a way to strengthen and reveal the images and coloring that, because of the reduction in humidity, were obscured by salts that had crystallized into hard, thick layers of magnesium and calcium carbonates. A small area, treated before and after with water, was cleaned with a very dilute formic acid solution. This test produced negative results, probably because of the extreme hardness of the layers. The use of more powerful acids would have been risky, considering the nature of the material components of the paintings.

Traditional, that is, mechanical courses of action were tried next. Initially, sandblasting with different types of abrasives seemed favorable, but soon problems developed. If the equipment was used for more than an hour, it overheated and became inefficient. Also, the sand was truly effective only in soft areas and had virtually no effect on hard areas. For these reasons, sandblasting was abandoned.

Scalpels and X-Acto knives, which had been so effective in the removal of carbonates on mural paintings elsewhere, proved useless. These instruments would break when applied to the hard white layer that covered most of the pictorial surface, and could only be used in the cleaning process for finishing touches.

After much experimentation, it was determined that rotary tools equipped with diamond bits provided the fastest and best results. In removing the carbonates, there was an attempt to leave a thin layer that would allow the design and color of the work to show through. Finishing was accomplished with ultrasonic equipment, whose metal spatulas reduced the white layer in places where the diamond bits made no headway against the plaster texture. No single technique could be universally effective in this situation, because the layers varied in hardness from place to place, depending on the degree of crystallization.

The cleaning of the pictorial layer is obviously a most delicate process, since intervention occurs directly on the image. The removal of the carbonate covering requires not only manual dexterity, but also sensitivity to the artist's original intention. An erroneous interpretation can cause irreversible damage. Too much, or too little, cleaning can alter the intended aesthetic, affecting the spatial relationships and the designs involved in the depiction of figures, musical instruments, or clothing. To date, significant progress has been made in the cleaning of the paintings. As of 1985, 80 percent of the carbonates have been removed, a 150-square-meter area of pictorial surface has been cleaned, and Chamber 1 has been completed.

The reinforcement of the plaster (80 percent of which was detached from the walls) was accomplished at the same time as the cleaning. Hollow areas were located by knocking knuckles on the wall, and calcium caseinate was injected into the spaces between the plaster and the wall. Calcium caseinate is an appropriate reinforcing material since it consists of slaked lime (calcic caseinate) and a small percentage of polyvinyl acetate, and it integrates well with the limy material that constitutes the walls and plaster of the building.

Although color integration in archaeological painting is generally rejected in Mexico, in these paintings it was deemed appropriate because it promoted interpretation and understanding. Strict international norms of intervention were observed. Only those cracks that interrupted the design were colored, and the intervention was made evident by the juxtaposition of colored lines. Areas that were totally lost were dyed with a neutral color so that they remained in the background. Progress in this process is minimal compared with the cleaning and consolidation efforts.

The reaction of the synthetic materials used in conservation treatment and field restoration varied according to the climatic conditions in which they were used. After the removal of carbonates from some painted areas, a protective layer of Paraloid® B-72 solution was applied.[2] This acrylic copolymer resin is highly successful in the reinforcement of ceramic and archaeological objects under suitable climatic conditions. It did not prove effective on the Bonampak murals, however, because of the high humidity of the atmosphere. Its application altered the design and color of the surface; a whitish bloom appeared during the season following its use. Given the good cohesive condition of the pictorial and other layers, it was decided that the applied Paraloid B-72 solution should be removed; protection with another product could be applied if deemed necessary in the future.

2. See Appendix.

Upon completion of the work on the interior of the building, an attempt will be made to integrate the stucco elements covering the facade. The final phase of the project will be the construction of a roof to provide stability and obviate future damage to the monument.

Biochemical studies are being carried out to analyze the composition of the materials that make up the building and determine the steps that should be taken to control the flora and microflora covering the other buildings in the archaeological zone. Research is also being conducted, in collaboration with the Center for Maya Studies of the National Autonomous University of Mexico, on the meaning of the glyphs, which will undoubtedly reveal important data on the history of the Maya civilization.

Luciano Cedillo Alvarez

STUCCO: A REPORT ON THE METHODOLOGY DEVELOPED IN MEXICO

Luciano Cedillo-Alvarez is presently a restorer with the Directorate of Restoration of the Cultural Heritage, INAH, Mexico and has been a conservator since 1976. He has coordinated conservation projects in various archaeological areas, including Xel-Ha and Palenque, and has taught conservation of archaeological ceramics at the archaeological museum of Quibor, Venezuela.

This attempt to summarize stucco conservation activities in Mexico for the last twenty years must necessarily skip over many important aspects. Nonetheless, it is possible to offer an exposition and evaluation of the achievements and failures in this field. The work presented here is a compilation and analysis of the conservation and research activities of a number of investigators. In the Bonampak archaeological zone, there is the work of Jorge Angulo in 1963, Paul Coremans in 1964, Mihailo Vunjak in 1971, and the current efforts of the Directorate of Restoration, INAH. In Palenque, there is the work of Luis Torres in 1967, Jorge Acosta in 1968, Gloria Vera in 1969, Mihailo Vunjak in 1971, Charles Hett in 1972, Elsa Dubois in 1978, and, in 1981–1982, the work of a group of researchers and curators from the Directorate of Restoration: Alejandro Huerta, Beatriz Sandoval, Elsa Dubois, Daniel Camacho, Gloria Vera, Pablo Torres, Maria Eugenia Ritter, and Julio Chan. In the archaeological zone of Kohnunlich, we have the work of Luis Torres in 1969, Jaime Cama in 1969, Sergio Montero in 1977, and Gabriela Garcia and Luciano Cedillo Alvarez in 1978. Elsa Dubois and Rocio Jimenez, along with other valuable participants, worked in Xel-Ha in 1980–1981. Unfortunately, very little of this work has been published, although a great deal of information exists in various files in many places.

It is difficult to say when the term "stucco" first appeared in published studies of Mesoamerican archaeology. Nevertheless, it has been used for some time to describe mixtures of lime compounds and sand or other additives employed for various purposes. We find smooth surfaces, stucco reliefs, sculptured stucco, and floors. The use of lime as a construction material marks an important step in the development of Mesoamerican architecture. At present, it is not possible to determine where the first use of lime took place, or if it happened more or less simultaneously at different sites. Archaeologist Jorge Acosta places the first signs of lime use in the Late Preclassic period, about the year 600 B.C.:

> We do not know precisely the date when lime was first used in construction, but it must have been at the beginning of the Late Preclassic, because floors made of this material have been found in Monte Alban I as well as Monte Negro, and both date from about the year 600. We also have the example of the building called E-VII-sub of Uaxactun, which is decorated with large stucco masks. In Central Mexico, although it was used for floors and smooth surfaces, it was rarely used to join stones in walls, which were usually held together with clay.[1]

Stucco finishes were used throughout Mesoamerica in a variety of ways. They were used to make magnificent designs such as those at Palenque (fig. 1); to cover large surfaces as in Teotihuacan and Cacaxtla;

1. Paul Gendrop mentions other early sites with architectural monuments where stucco finishes were used; e.g., Building 5-D-sub 1-la at Tikal about the year 300 B.C.; and the Valley of Mexico, the place known as Cerro del Tepalcate (the Hill of Broken Pottery), between the years 600 and 400 B.C.

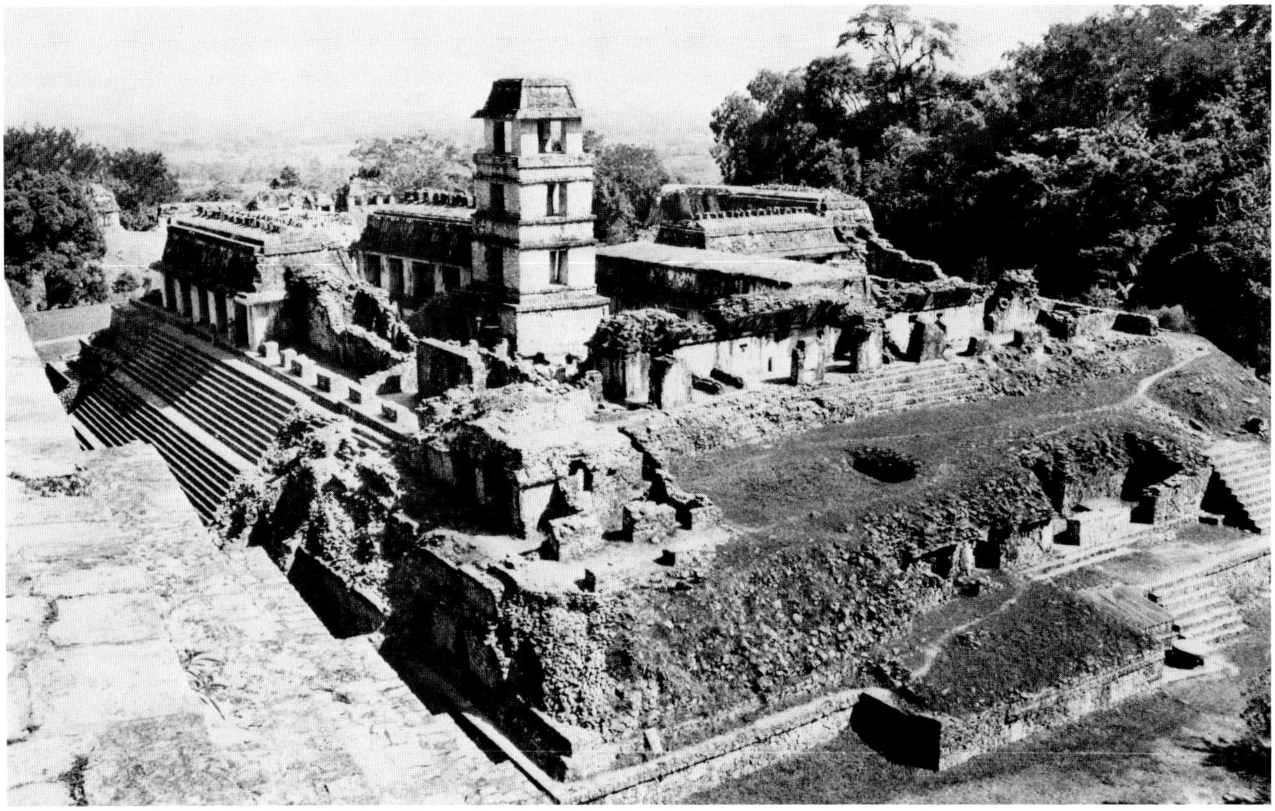

Figure 1. The Palace at Palenque showing the scope of conservation problems at that humid tropical site.

and to surface roads (called "white road"; *sacbe*), such as the one in Coba-Yaxuna in the Maya area, which is 100 kilometers long.

The workmanship of stuccoed surfaces is closely linked to the development of the architecture itself—the stucco forms what we could consider the architecture's skin. Only rarely are there stucco objects that do not exhibit this relationship. For this reason, to treat the problems of stucco conservation without taking into consideration those of the architectural structure to which they are applied, would lead to rather unsatisfactory results, as has happened in some cases.

Stucco conservation problems in Mexico

In order to fully understand the problems involved in the conservation of stucco in Mexico one needs to be aware of the realities. First of all is the sheer magnitude of the problem. Most archaeological structures have stucco finishes. Then there is the great number of archaeological sites in Mexico—several thousand, although there is as yet no exact count. In the southeastern part of the country, in the state of Yucatan alone, 1,117 sites were recorded for the Archaeological Atlas Project. Multiplying this figure by the average number of architectural structures on each site and its environs, we realize there are several thousand buildings and tens of thousands of square meters of stuccoed surfaces, including walls, floors, and reliefs.

Second, the diversity of geoclimatic conditions and the current state of preservation of stucco, especially in humid, tropical zones must be taken into account. Plenderleith, in speaking of the archaeological monuments of Mexico says:

> One must recognize that, unfortunately, there are large and important monuments, especially in the tropical zone, which are so disintegrated as to be apparently irreparable. In the rain forests of Mexico, frequently the only remains of a series of pyramids and temples are shapeless piles of earth.

Causes of change

Most reports that address the causes of the deterioration on sites focus on archaeological zones in southeast Mexico where the evidence of change is greatest. Most authors concur that humidity and temperature are the main elements that together give rise to complex interrelated changes. We can classify these processes roughly as physical (water, steam, rain, wind, temperature, salt crystallization), chemical (hydrolysis, oxidation, condensation, dissolution), and biological (the effects of insects, microorganisms, large plants, animals). Perhaps a "socioeconomic" category should be added (tourism, plundering, vandalism).

Stucco is very porous and allows water to circulate easily through the interior of its structure, accelerating the processes of change and deterioration. The relative volume percentage of porosity of the stucco in Palenque varies from 33% to 40%, and in Kohunlich, the relative porosity is approximately 14.8%.

The development of conservation processes

There is little information about the first attempts made in Mexico to conserve stucco, including the materials and techniques that were used. In referring to the archaeological interventions of the late nineteenth and early twentieth centuries, Ignacio Bernal writes:

> They only looked for "interesting" objects and monuments. Of course, we can't generalize, but there is an abyss between the immense care taken by Maudslay in copying or making casts of Mayan stelae, and the presumptuous carelessness of Batres, slicing up the Pyramid of the Sun in Teotihuacan, or of Seville, destroying the splendid facade of a Zapotec tomb in order to obtain some urns with which it was decorated.

Professor Eduardo Pareyon, when asked if he knew what materials had been used a long time ago for the preservation of stucco replied categorically: "Look, they only used cement to resurface and reinforce the stucco, which was the case in Teotihuacan and Tenayucan." Nevertheless, if it were not for this sort of resurfacing of defective spots in the stuccoes found during excavation, the losses would have been greater.

As new products appeared on the market over the years (mainly synthetic resins), they were used in an attempt to halt the deterioration of archaeological materials. With more good intentions than knowledge of the results of reinforcement, fixing, or patching, workers applied various products such as natural and synthetic varnishes, lacquers, and adhesives (lac, Duco® cement, UHU®, Multex®, Resistol®, Endurol®, Curasol®, Acritex®, etc.).[2]

2. For the chemical composition of these and other proprietary products, see the Appendix.

In 1961, with the creation of the Directorate of Restoration and the Catalogue of Artistic Heritage, headed by Don Manuel del Castillo Negrete, and the formation of the laboratory for preservation of archaeological materials in the Department of Prehistory under the direction of Professor Jose Luis Lorenzo, the road was opened in Mexico for the preservation of archaeological materials as an area of specialization. This was at that time a new idea to the Instituto Nacional de Antropología e Historia. From that point, however, a series of projects was developed and experience was accumulated, in some cases successfully and in others less so. At any rate, this provided the basis for what is now the Directorate of Restoration and the National School of Conservation at INAH, where several generations of curators have been trained.

The archaeological sites where restoration work on stucco was done include Bonampak, in 1963; Teotihuacan, in 1966; Palenque, in 1967; Lambitieco, in 1969; Kohnunlich, in 1969; Cacaxtla, in 1975; and Comalcalco, in 1978, among others. The following summarizes the procedures developed at those sites and the materials used.

It is important to note that the geoclimatic situations in each of these zones vary from hot and humid conditions in Palenque, Bonampak, Kohnunlich, and Comalcalco in the southeast, to hot and dry conditions in Teotihuacan and Cacaxtla in the Central Plateau and Lambitieco in the Valley of Oaxaca (fig. 2). These characteristics influenced the preservation of the stucco, its response to excavation, and the results of attempts at conservation.

It is also necessary to point out that when this work was started, there was no previous experience to draw upon and thus no basis on which to predict the results. As happens in many developing countries, technology and materials are imported from abroad to meet needs as they arise; only with the passage of time are techniques modified or alternative methods developed. In the area of conservation, Mexico was no exception.

On some sites conservation was done after archaeologists had completed excavation, discovery, and initial treatment. In other cases, there was joint participation, albeit with some problems in defining the limits of activities and working out personal relationships, which fortunately have been smoothed over time.

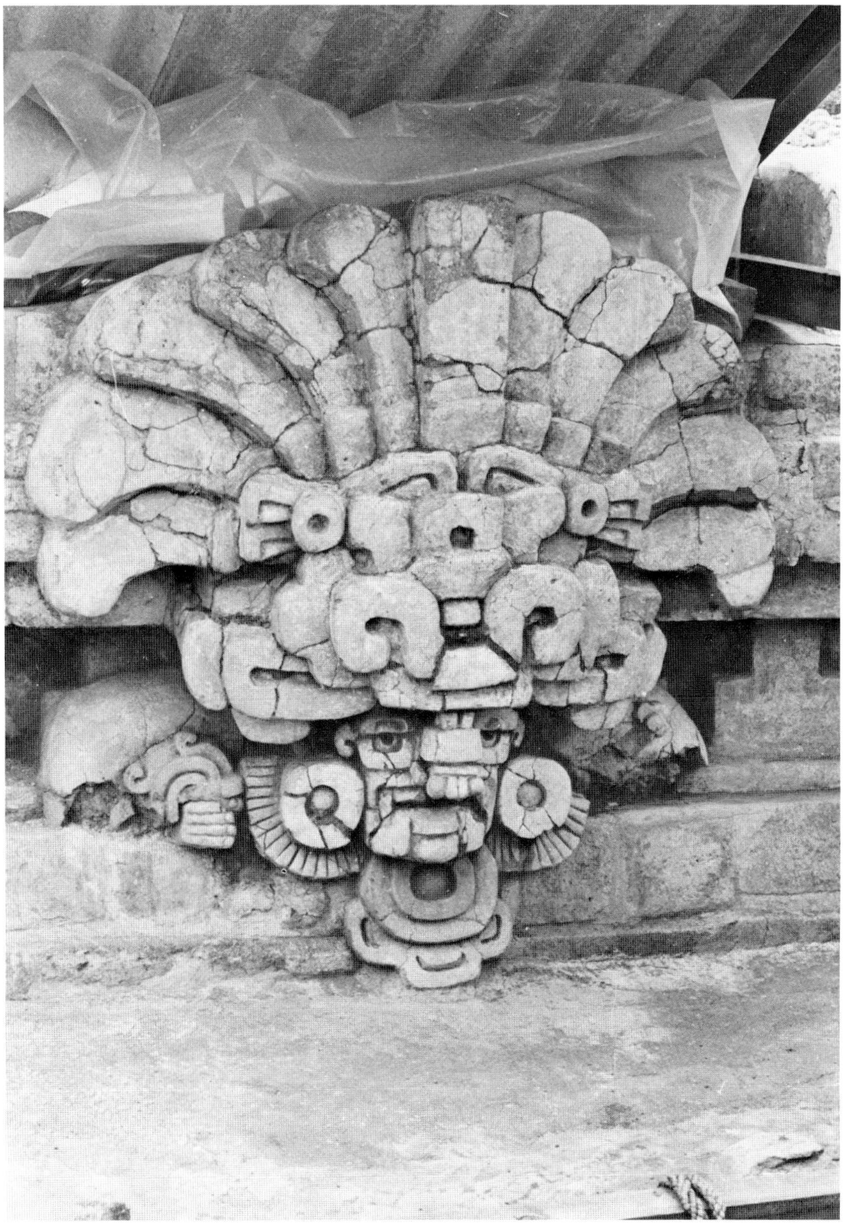

Figure 2. A stucco "mask" from Lambitieco shows clearly the attempts at consolidation.

Methods of protection

One of the first measures taken for the preservation of stucco was to build roofs or temporary covers to protect against sun and direct contact with rain. The materials used were sheets of cardboard, asbestos, acrylic, palm leaves, or whatever was at hand at the moment, and depending on the budget allocated for the project. At the same time, work was done to channel water, waterproof, and reinforce architectural elements.

Unfortunately, protective coverings and roofs were limited to the areas considered most important, that is, where there were designs in stucco. Places with smooth surfaces and stucco floors—although they were investigated—remained without protection because of budgetary constraints. In time, the temporary shelters were changed and improved in places such as Bonampak, Cacaxtla, and Kohnunlich. Although their aesthetic qualities are very debatable, nobody denies their effectiveness as an aid to conservation.

Preparing work areas

To facilitate research and conservation activities in the field, areas were prepared which, despite lack of water, electricity, and furniture, did make the work easier. On some projects, there was collaboration with other specialists, including chemists, biologists, archaeologists, and architects. The minimal amount of laboratory equipment available for on-site research had a beneficial effect: It forced the participants to communicate with each other about conservation procedures. This joint project, as a side effect of its original goal, broke down many of the barriers that separated specialists from different fields who work on the same project.

Conservation processes and materials

The following steps were usually taken in the conservation of stucco:

1. Cleaning and removing dirt and roots
2. Collecting and placing detached fragments in their original positions
3. Resurfacing and binding cracks, fissures, gaps, and the edges of the stucco
4. Injecting mixtures to fill cavities
5. Cleaning and removing salts and microorganisms
6. Consolidating
7. Waterproofing
8. Fixing colors
9. Applying biocides

These same methods are used in the conservation of wall paintings, with similar results. What follows is a discussion of only the most important processes, which are emphasized because of the problems they present.

Resurfacing and binding

In order to secure stucco fragments in danger of becoming detached from the wall, various pastes and mixtures were used to repair cracks and fissures and to reinforce the edges of the stuccowork. The pastes and mixtures included materials ranging from the traditional lime with sand, volcanic rock, and white rock (*sascab*) to synthetic resins and filler materials.

The materials employed for this process were: (1) cementing agents (white cement, hydrated lime [commercial], or slaked quicklime); (2) filler materials (sand, volcanic rock, *sascab*, soil, ground stucco, white rock, or powdered rock); or (3) synthetic adhesives (Mowilith® DM4 and 15% DRL, Endural®, or Primal® AC-33).

A plastic surface film resulted from these pastes, particularly those containing the highest amount of synthetic resins. This film contracted and separated, leaving the filler material without strength. Several authors make reference to this problem.

The migration of an adhesive toward the exterior is caused by evaporation. The polymerization of resins usually takes place through a loss of solvent, which occurs on the surface layer of the mixture or paste. This was confirmed by numerous tests on mending pastes that were carried out during the Xel-Ha project in 1980 and 1981.

> According to Charles Hett in his 1972 Palenque report: Preliminary tests of strengthening agents were carried out in order to determine their setting ability in conditions of high relative humidity . . . upon examining the results of the setting of the previously chosen material, it was deemed fit to substitute a lime mortar for the acrylic emulsion-based paste in the areas where the humidity made the setting of the emulsions impossible.

The best results were obtained with traditional mixtures of slaked lime with sand or some other material such as powdered rock. As a rule, a ratio of 3:1 is recommended for the lime mixtures; however, in order to ensure good results, the filler material should be tested in each case.

Injecting mixtures to fill cavities

Various mixtures in liquid form were injected into existing cavities between the wall and stucco to anchor stucco that could become detached. Existing cracks and fissures were used to penetrate the stucco layer, holes were drilled into the affected areas with very fine bits. The mixtures usually contained calcium caseinate alone or with synthetic resin additives. The materials mentioned are white cement, slaked quicklime, calcium carbonate, 20% and 50% Primal AC-33, Mowilith, Curasol®, and Paraloid® B-72.

The results obtained using calcium caseinate in areas with high humidity, such as southeastern Mexico, were difficult to evaluate, since a proliferation of microorganisms was observed in the residues, despite the addition of a preservative (fig. 3). Isolated fragments of stucco covered with an intermediate coat of calcium caseinate were tested to evaluate their resistance to the effects of weather. There was a physical

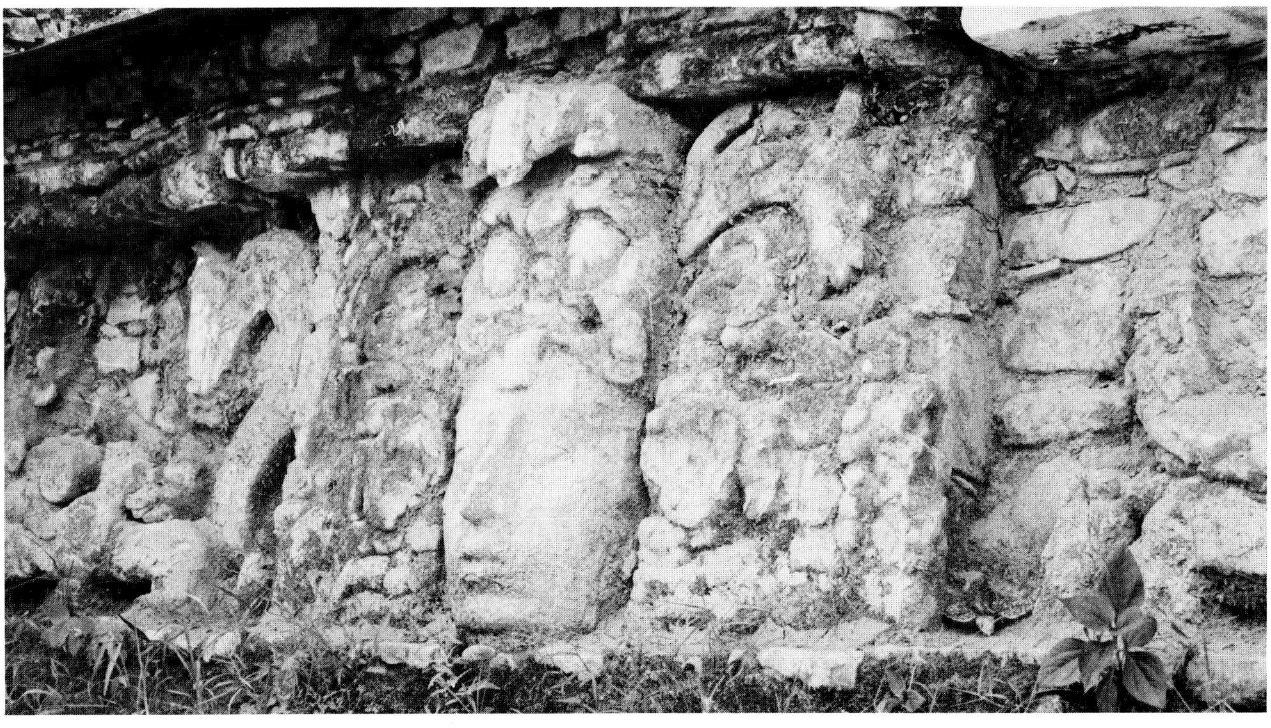

Figure 3. The destructive effects of flora and microflora on a stucco "mask" from Palenque.

change in the caseinate, which resulted in the separation of the fragments. The mixtures containing synthetic resins produced the same results as the mixtures with synthetic resins used in resurfacing. Finally, in the laboratory, a sedimentation from a mixture of calcium caseinate and synthetic resins resulted. It is possible that this separation of heavy and light materials occurs at the moment of injection.

The experiments done in 1981 saw positive results with mixtures containing slaked lime, two types of dolomitic sand, a minimal amount of Portland-type white cement, and an initial setting retardant. A mixture of these materials produced a product with the same porosity and hardness of the original stucco. The long-term effect of treatment with this material remains to be determined.

Consolidation of stuccowork

A number of products have been used to consolidate and strengthen stucco that has started to crumble. Most common are synthetic resins, such as polyvinyl acetates and acrylic copolymers. They may be applied using a dropper or a brush; injected with venoclysis equipment, syringes, or spray pumps; or the fragment can be immersed. The materials cited are: Mowilith DMIH and DRL (50%), DM4 (4%, 10%), lime water, Primal AC-33 (2%, 5% 15%, ethyl silicate OH and ethyl silicate H (added as waterproofing), Endurol, Paraloid B-72 (3%, 4%, 5%), and Resistol® 850 (5%, 10%, 20%).

After some time, the reinforcement with synthetic resins resulted in some detachments of the surface layer of stucco in which the resin was deposited. The greatest damage was noted at archaeological sites with high relative humidity. At the dry sites, damage was found in those areas where the stuccowork was exposed to humidity during the rainy season.

Vunjak, in his report concerning the conservation problems at Palenque, writes:

> The synthetic resin emulsions (polyvinyl acetate), especially in Palenque's humid conditions, do not offer any guarantee of endurance. . . .The treatments with various chemical products—specifically those of Acosta in 1968—applied in order to strengthen and waterproof the stuccowork and stone slabs of the temples, are, without a doubt, more harmful than they are beneficial to the conservation of these works, especially when considering them over a long period of time.

Consolidations with ethyl silicate have been successful in some cases; but a longer evaluation period is necessary before it can be assessed. At present, there is no proven solution to the problem of consolidating friable stuccowork, especially in humid tropical zones.

Fixing the polychromy

When the polychromy was turning to dust and in danger of falling off, it was adhered with a synthetic resin applied to the surface with soft hair brushes. The materials used for this process were: Endurol (10%), Primal AC-33 (3% and 5%), Paraloid B-72 (3%), and Curasol (5% and 10%). The observations were the same as those made for the synthetic resins used in the consolidation process.

Application of biocides

For the control of flora and microflora such as algae, moss, lichen, liverworts, ferns, and vascular plants, several biocide products were applied superficially with a brush or sprayer. The materials were: ammonia (5% and 10%) in water, Cetavlon® (4%) in water, pentachlorophenol (4% and 5%) in ethyl alcohol, sodium pentachlorophenate (2%, 4% and 9%) in water, sodium pentachlorophenate (20%) in ethyl alcohol, cupitretramine sulfate, and Diuron® (1.5%) and Bromacin® (2:1) in water.

In general, these biocides have been only partially successful. For example, the high concentration of sodium pentachlorophenate diluted in ethyl alcohol did not prevent a subsequent invasion of microflora.

Recently, biologist Pablo Torres has experimented with a mixture of Diuron and Bromacin at 1.5%. He apparently had good results in Cacaxtla and the Templo Mayor, but variable success in Palenque.

General observations on the procedures

Up to now, not only in Mexico, but throughout the world, no satisfactory procedure has been found for the conservation of stuccowork in situ. Nevertheless, protective coverings and other measures taken against humidity (e.g., drains, channels, and waterproofing systems) are important measures in the conservation of stuccowork.

Field experience with synthetic resins has demonstrated that they cannot solve all our problems, as was believed a few years ago. On the contrary, the indiscriminate use of these materials has had serious consequences. It is, however, useless to generalize, since each situation is unique.

Conclusions

In Mexico, there are some research projects (although they are inconsistent) designed to study the alteration processes of archaeological materials in situ and to seek out methods and techniques that will guarantee their conservation. Despite the pronouncements that new explorations should not proceed until the conservation of those materials already unearthed is accomplished, excavations continue and new finds are revealed. For this reason, the work of conservation must continue unabated. Direct intervention is necessary in view of the danger of losing material of great value, but there is insufficient time to look back and compile the results of past experience. Clearly there is a need to establish, within the areas of conservation and archaeology, a permanent research program dealing with these critical problems.

References

Bernal, Ignacio
1963 *Teotihuacán: Descubrimientos, Reconstrucciones.* Mexico.

Gendrop, Paul
1984 "La Problemática de la Conservación." Introducción editorial in *Cuadermos de Arquitecura Mesoamericana* 3. UNAM. México.

Sergio Arturo Montero

THE CONSERVATION OF ARCHAEOLOGICAL PAINTING

Sergio A. Montero has been a professor of restoration at INAH's School of Restoration for several years. He has been involved in the restoration of mural paintings at Teotihuacan, La Barca, Bonampak, and Ethiopia's Historic Route, among other sites.

Archaeological investigation stops neither at the collection of data obtained in excavations nor even at their interpretation, since new discoveries can enhance or so radically modify the initial interpretation that a site may become the subject of permanent investigation. Thus a single piece of evidence, whether a simple household utensil or an elaborate artistic creation, holds the potential to reveal the degree of cultural, economic, political, and social development of an ethnic group over the ages. Such a study can enlarge the understanding of our cultural roots in the context of the development of our present civilization, clarify the bases of our national identity, and provide clues to the future. This compels us to pass on to future generations these pieces of evidence of our history, with all the richness of their authenticity, and complete with the changes with which time has marked them.

Since humans and their thought processes are products of nature, it is imperative that, as well as carrying out archaeological research, we conserve all aspects of human cultural expression. They must be preserved and appreciated not only as artifacts, but as witnesses to the development of human thought processes and man's attempt to understand and dominate nature.

Accepting this idea as a departure point—and bearing in mind that objects are composed of matter, subject to universal laws of movement and change—we must then accept that their conservation is not a simple task, but a thoroughly complex one. In order to achieve it, the combined knowledge from various scientific disciplines is essential. This knowledge may explain the particular laws that govern the changes over time in many types of material and the influence of the environment on the mechanisms of these changes—changes that might be considered defects when they affect objects held to be cultural treasures.

Today, the conservation of cultural property is no longer as empirical as it once was. It has become a technical profession solidly supported by scientific developments. The multidisciplinary character of the task gives rise to the need to establish common criteria that unite and coordinate the input from all the various fields, beginning with archaeology, that must take part in the conservation of cultural property. At the same time, there is a need to define the boundaries of activity of each specialist, limiting each one to his or her area of expertise.

The absence of such criteria in the past is understandable, because conservation as a professional discipline was not yet in existence. As a result, other professionals became conservation workers, with practitioners of various disciplines being pressed into service according to the apparent aptness of their skills (painting for the painters; buildings for the architects; archaeological objects, including monuments, for the archaeologists), a situation that led to the loss of valuable historical evidence.

As a specific and exact discipline, conservation must cope with precise and specialized problems that defy broad generalized approaches. Though all objects are composed of matter, there are different types and states of matter. Hence, organic materials will not behave the same way as inorganic materials, and various climatic conditions will affect even the same type of material in different ways.

If we further consider that the composition of archaeological objects may include materials of diverse characteristics and origins, we realize that their interactions with the environment will cause changes that are both diverse and complex, making each one a specific case that requires an individual solution. Conservation should follow the same principle as medicine: One must treat the patient not the symptoms. For example, different precautions are taken when a tooth is extracted from a diabetic than from a nondiabetic, and antibiotics would never be administered to a person who is allergic to them.

Furthermore, we cannot consider objects, simple as they may be, in isolation and removed from a context, be it chronological, historical, or physical. The object, in and of itself, cannot tell us anything beyond its appearance. This certainly is the case with mural paintings, in which aesthetic, historical, and material components are united. In addition, their physical aspects must be considered: the many materials that may figure in their composition, as well as their physical relationship to the building or place in which they are found. This relationship, moreover, is not simply physical: It also has aesthetic and historical dimensions.

A mural painting, therefore, must be regarded as an integral part, not only of the building in which it is contained, but also of the townscape—and even of the landscape—that surrounds it. All interventions in mural painting conservation, then, should come from an understanding of this concept and focus on the need to conserve the unity of the complex and the function for which it was created. Should we fail to adhere to this principle, wall paintings become transfigured into movable objects more akin to museum pieces. Nonetheless, this is only feasible on a theoretical basis, especially regarding archaeological mural painting, because insurmountable obstacles still prevent us from honoring it in practice.

The problems of wall painting conservation in Mexico

Archaeological mural paintings, and more specifically the pre-Hispanic mural paintings that concern us, are found in buildings that had ceased to be functional when the conquering culture was imposed. The majority of these cases were far from the new centers of settlement, at times in regions that were remote, inaccessible, and desolate, making the protection and maintenance of these murals extremely difficult.

Moreover, it is not a question of trying to preserve five, twenty, or a hundred archaeological sites; the problem is far greater than that. In Mexico alone, thousands of sites exist, located in diverse, extreme climatic zones, where—because one is not able to intervene on their behalf—they are being completely destroyed by natural phenomena. The lack of specialized techniques in conservation and a shortage of financial resources limit the capacity for action, not only to deal with conservation problems, but to train technical personnel. It is not my purpose here to criticize governmental action in this area, since the situation is a natural consequence of a need that was recognized only a short time ago. It would be of no use if tomorrow millions of pesos were granted to us for conservation, as it takes years to properly train technical personnel.

On the positive side, in the last decade the authorities have supported and encouraged the development of this discipline. As evidence of this initiative, Mexico can now rely on a center charged with the conservation of cultural property, in the form of the Directorate of Restoration of the Instituto Nacional de Antropolgía e Historia (INAH). It comprises a group of trained personnel that compares favorably with, and in certain areas surpasses, those in European centers of longstanding tradition and prestige. Although we still have a long way to go when it comes to resolving our problems, given the country's economic situation, there is no point in complaining; rather we should accept the current state of affairs, and concentrate on proposing solutions that will allow us to establish a policy for the conservation of our cultural heritage.

To expound in further detail on this topic, however, would take me far from the theme of this discussion, so let us return to the conservation of archaeological mural paintings. At the outset it must be said clearly that the physical conservation of archaeological mural painting, as of any other object of cultural heritage, is the responsibility of specialized professionals. We have seen the destruction of valuable objects through oversight, the ignorance of the basic principles of preservation, and occasionally the investigator's overconfidence when tackling matters that really lie outside his field.

In the excitement following the discovery of a find, an archaeologist may be moved to proceed rapidly with the excavation without properly analyzing the risks. This is, from a human point of view, very understandable, but from the perspective of conservation, very dangerous. For example, when uncovering a tomb that for centuries has had stable environmental conditions, the abrupt change in atmospheric pressure or relative humidity can cause objects to degrade, even those of such theoretical stability as ceramics and metals. Accidents of this kind have in fact occurred many times. Wood objects, for example, that have been long-submerged in water crack, shrink, warp, or even turn to powder when they are dried quickly.

Mural paintings, if preserved in situ, survive precisely because they have been buried. The arbitrary removal of plants that grow on ruins and have over the ages become part of the structure can cause irreparable damage, as can the alteration of the microclimate through deforestation. An example of this can be seen at Bonampak, where the deforestation of the ruins and the plaza seriously altered the moisture conditions inside the chambers. As a result of this and later attempts to dry out the chambers, the salts on the surface crystallized, forming a thick layer that made the appreciation of the paintings impossible. The removal of this layer would be a very complicated, delicate, and risky undertaking.

Even more serious has been the empirical treatment of objects, since any intervention requires a deep understanding not only of their nature and use, but also of the materials and methods to be applied, expertise that only a specialist can provide. Although a willingness on the part of archaeologists to undertake conservation measures may be praiseworthy, their inexpert knowledge has often caused them to apply treatments that have caused much destruction, which they now lament. We have seen mural paintings, Teotihuacan vases decorated after firing, and many other objects whose state of conservation was magnificent at the moment of their discovery, but which have been destroyed by ill-informed treatment. To one of the most outstanding paintings of Teotihuacan, that of the *Templo de Agricultura* (Temple of Agriculture), for which data exists showing that it was almost complete, archaeologists applied a synthetic varnish with the intention of protecting it. As a result, the painting was almost completely destroyed because changes in temperature caused the varnish to gradually peel off, taking fragments of paint with it. When we were entrusted with restoring this painting, less than 30 percent of the paint surface remained, and it was impossible to eliminate the cause of the deterioration because varnish is an irreversible synthetic resin. Many other paintings and many post-firing-decorated Teotihuacan vases underwent the same procedure, with the same fatal results.

For these reasons—given the enormous quantity of cultural property that needs protection, and considering the obstacles that prevent us from providing adequate protection—it becomes clear that archaeologists should refrain from carrying out new excavations and instead focus their efforts, combined with others from related disciplines, on conserving and protecting that which has already been discovered. Failing this, even more will be lost.

Every mural painting that has been excavated needs specialized attention immediately, and all efforts should be focused on its conservation in situ. This means employing all means that guarantee its protection, including security and permanent maintenance, thereby preventing situations that could threaten the work of art. The measures to be taken should be reversible, so that they do not inhibit future work. Conservation techniques applied using these criteria offer us the best results that can be expected at present; but at some future time these may be improved, since science may offer us better weapons. The future possibility of applying improved treatments should therefore not be inhibited.

Protective structures

In the case of mural paintings, methods of a remedial nature should be applied as a last resort. Preventive methods that help stop the agents of deterioration are always preferable. It is useless, for example, to strengthen mural paintings that are coming loose because of water filtration; instead, we should eliminate the source of the moisture. Doing so, however, may cause other conflicts, often of an aesthetic nature. For example, one measure that can stop rainwater from damaging an archaeological mural painting (generally outdoors) is to place a protective roof over it, but there is no question that the site will be altered aesthetically. Thus, in order to keep the paintings in situ, we are forced to weigh the alternatives: allow the rain to destroy the paintings, or sacrifice the appearance of the site. No roof should be considered unless the design takes into account both function and appearance so that the site is altered as little as possible. Moreover, the solution should be regarded as temporary, one that in the future, if a better protection system is found, can be dismantled without endangering the site. In the meantime, a roof permits conservation not only of the mural, but also of the site.

In any case, a roof will not guarantee successful conservation without constant attention and constant maintenance. The roof itself is susceptible to deterioration, and conceivably could even aggravate the situation. Here, more than in any other situation, the coordinated action of archaeologists, restorers, and all those who have some responsibility in the work is of the utmost importance. The site managers should set down the structure's material-conservation needs, which should also be approved by the specialists. In this way, one could avoid problems that arouse professional sensibilities. With such an arrangement, the restorers would not take over the archaeologist's work or invade other fields, nor would they take entire credit for the work. In these situations, all of us should behave professionally in order to reach our common objective, which is the fullest understanding and the best possible conservation of cultural property from the past.

Sites that cannot be protected

I have emphasized that, given the variety of materials that compose an artifact, as well as the diversity of conditions to which it is subjected, it is impossible to generalize about treatment; rather one must treat each object in a specific fashion according to its needs. The Bonampak case, then, is only an example and cannot be taken as a universal model. Needless to say, the conservation of archaeological mural painting does not begin and end with the construction of a roof. The case of the mural painting in the Totanac archaeological site of the Higueras, Veracruz, comes to mind. Here the threats of destruction came not only from rain, but also other sources, among them the instability of the structure itself (the monument was constructed of limestone rock and sand, and was subject to constant settling once excavated); the nearness of the sea and the resulting presence of strong winds laden with salts, causing hydroscopicity; and the difficulty of providing security and protection in an area vulnerable to vandalism. These hazards could not be avoided by simply providing a roof.

Furthermore, several layers of pictorial superimpositions were discovered at the site—in some places as many as twenty-four—which promised a source of archaeological information of extraordinary importance. Aside from any technical considerations, this single fact plainly justified the procedure that was adopted, one that in other cases we would have judged to be improper: namely, the peeling off of the paint layers, one by one, and then exhibition in the museum after they had been given new supports (fig. 1). This was the only practical solution, dictated basically by the needs of investigation, which guaranteed at the same time the conservation of the painting. In this particular case, even though the technical problems of conservation in situ could have been solved, had that approach been followed strictly, the archaeologist would have had to forego the highly valuable information obtained by separating the layers.

The course followed was to adapt this approach, seeking to keep intact the concept of mural painting. As it turned out, the markings of time were not obscured; moreover, the assembly was constructed in such a fashion that if at any time the protection of the site became feasible, the layers could be replaced and conserved in situ in approximately their original positions.

Figure 1. The procedure for separating layers of superimposed paintings and providing them with new supports.

a. Three remaining layers of paintings, after several have been detached.

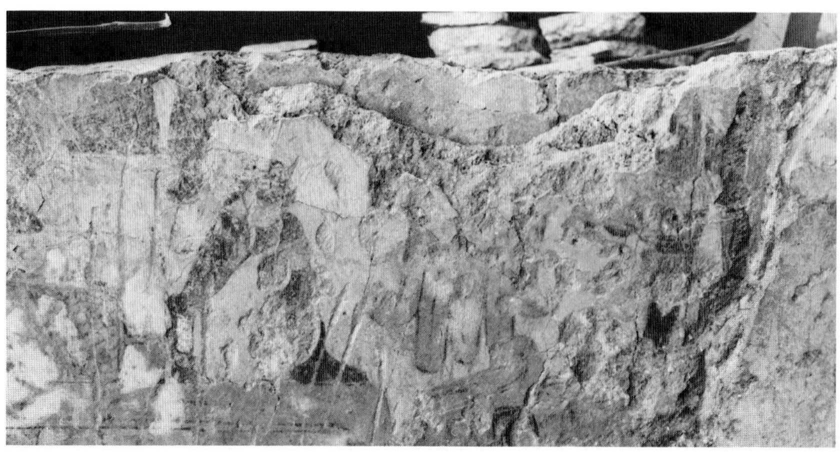

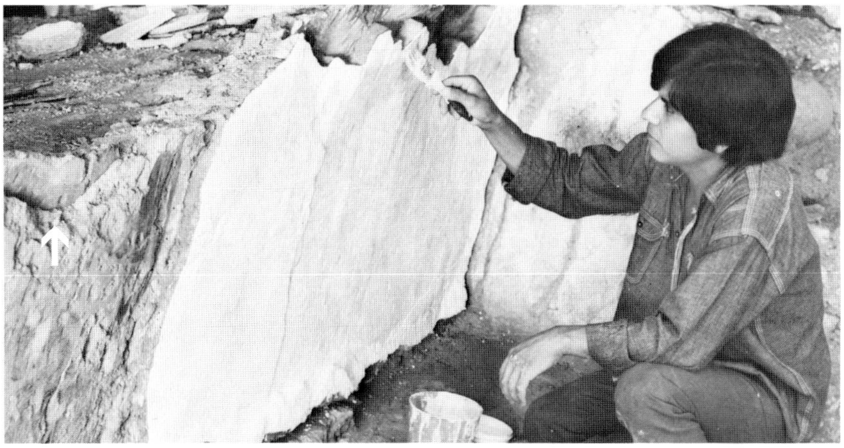

b. The application of a facing prior to the removal of a mural.

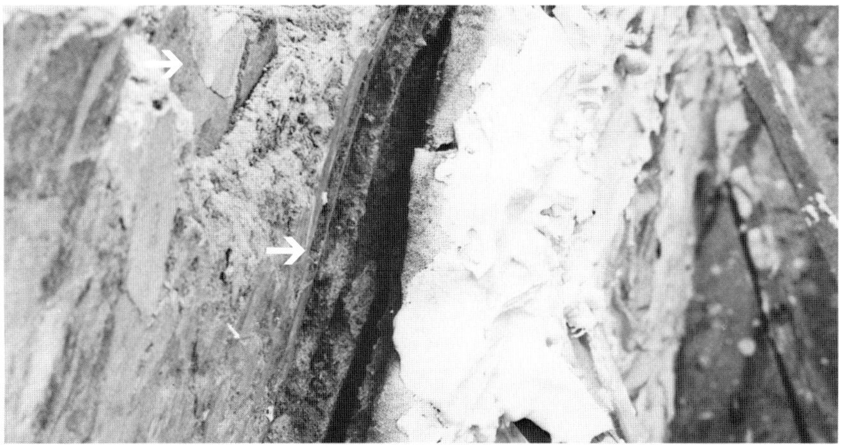

c. A temporary plaster support makes possible the removal of a mural layer.

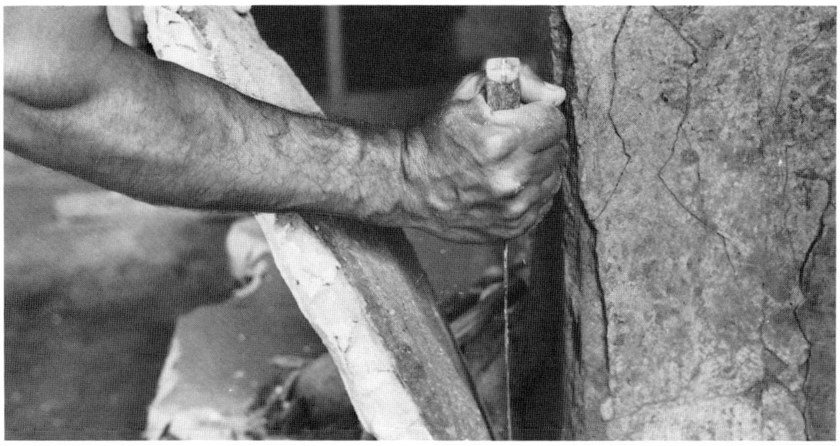

d. The removal of one layer exposes another painting.

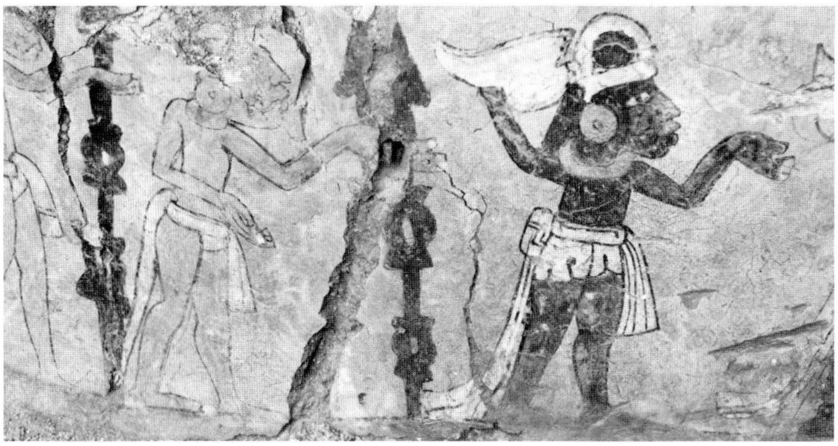

e. A painting immediately after exposure.

f. The same painting seen in Figure 1e several minutes after exposure. The build-up of carbonates due to rapid drying is already visible.

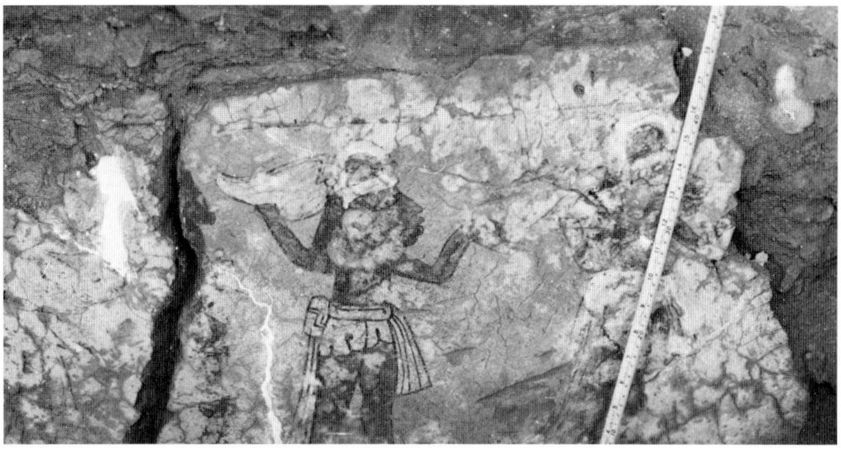

g. The application of a facing.

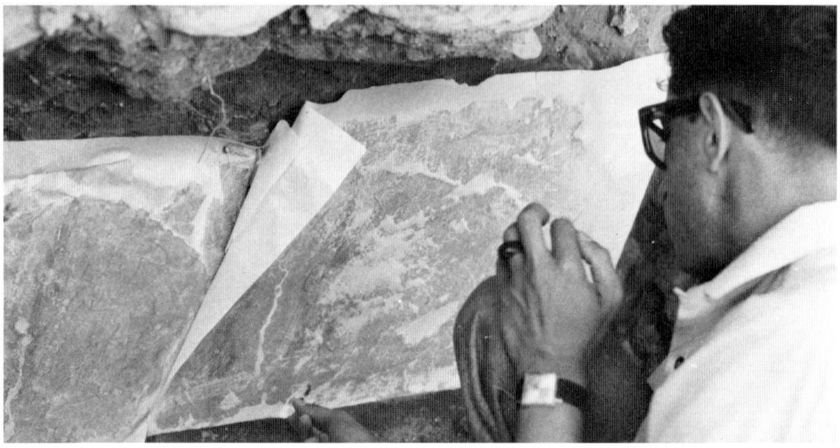

h. In the treatment facility, the painting is prepared to receive a new support.

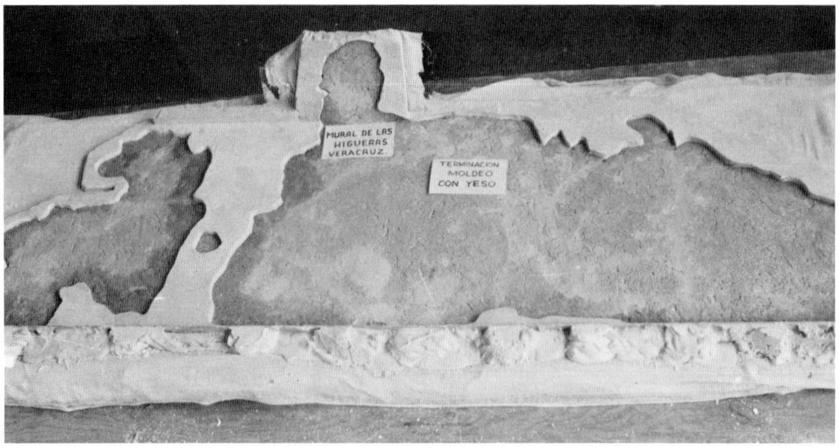

i. The final stage of support construction.

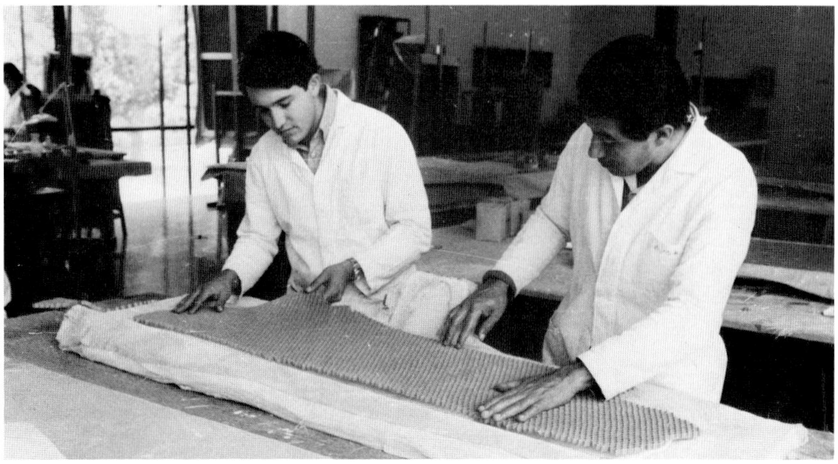

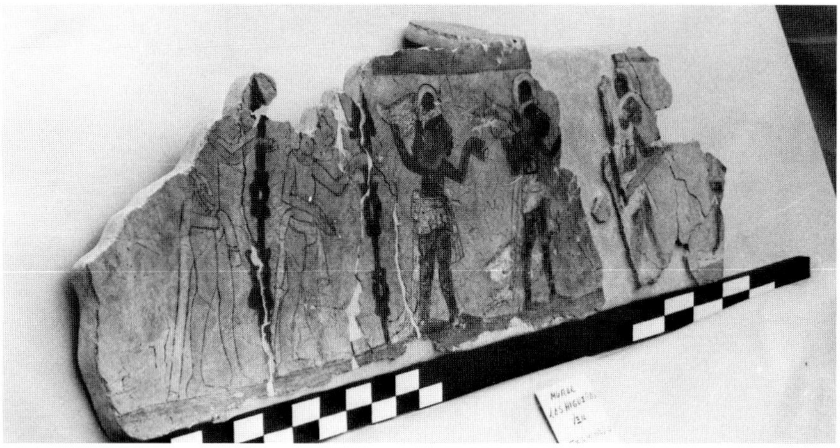

j. The mural painting with its new support. Such a mounting would allow the painting to be reinstated in its original location if conditions permitted.

Outline for action

We have selected as examples two extreme cases. Others could have been cited reflecting diverse conditions that merited varied solutions—ones that could not be adopted *a priori* but required a specific and detailed analysis, resulting in a diagnosis and an individually tailored treatment. We should not generalize, much less should we establish dogmatic criteria. The conservation of mural painting demands the participation of technical specialists whose work is inextricably tied to those disciplines and techniques that serve the wider objectives of archaeological study.

To conclude, I believe that some general outlines could be established that would serve as a basis for developing a workable practice for conservation; not to safeguard the particular interests of any discipline nor to establish its supremacy above others, but to guarantee the effective protection of our cultural heritage. As a first step I would suggest the following measures:

1. Periodic meetings whereby knowledge, ideas, and experience could be exchanged and criteria for action unified. Archaeologists, restorers, and other specialists would participate. Thus the restorer might gain the fundamental knowledge of the importance of the discovery from the archaeologist's point of view. The archaeologist, at the same time, could become aware of the preservation problems and the changes that might take place in the materials, which might curb his eagerness to press on with excavation before specialized personnel have had the opportunity to carry out rescue work.
2. The participation, at the beginning of any project, of all specialists in order to be prepared against any eventuality and hence to be able to solve problems expeditiously.
3. To analyze and be self-critical of the results of each project so that errors may be avoided in succeeding projects, and techniques can continue to improve.
4. To respect the site management staff and the technical maintenance directors as experts on the conservation of the site and to report on a timely basis any anomaly to the specialists, not with the desire to shift the blame, but to ensure collective cooperation aimed at the protection of the site.
5. To allocate duties among the specialists so that each one can perform to the best of his or her ability. The effort must be a joint one, its sole aim being to benefit the conservation of our cultural heritage.

James A. Spriggs

WOOD, WITH AN EMPHASIS ON THE PRESERVATION OF IN SITU STRUCTURES

James A. Spriggs holds a diploma in archaeological conservation from the Institute of Archaeology, University of London and is presently Head of Conservation for the York Archaeological Trust, England. He has been deeply involved with the planning and development of the Jorvik Viking Center, as well as being Conservator and Laboratory Manager on excavations of Mayan sites in Belize, Central America.

Wood, because of its many physical and chemical properties, lends itself admirably to a great variety of uses. It is reasonable to estimate that in many pre- or non-industrial societies, more than 80 percent of all buildings, means of transport, equipment, tools, and domestic utensils were made of wood. Any of these items surviving into the "industrial era" might fairly be regarded as being to some extent archaeological, even though they exist above ground. We are therefore increasingly faced with a conservation problem of immense size and considerable diversity, even if only a fraction of our heritage of timber monuments are to survive in perpetuity.

As is the case for all organic materials, the forces of nature are against us. Those entrusted with the care of wood structures in situ are literally battling against the elements, for when exposed to air and water, wood easily falls victim to wood-destroying insects and the various forms of fungi upon which they feed. However, if the wood surface is protected from exposure to air and water, it can survive for hundreds, if not thousands, of years.

Figure 1. Stave church at Norgund, Norway.

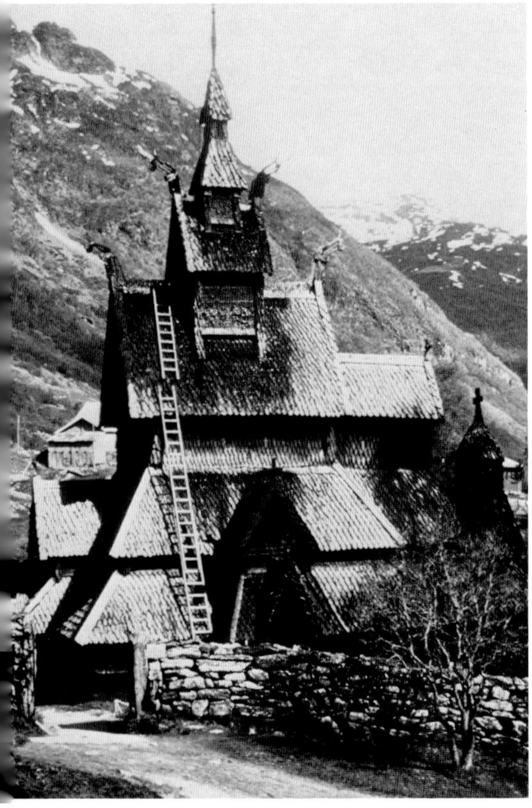

Apart from timber that has been preserved accidentally over the ages by freak burial conditions, long-term survival can be achieved by several traditional techniques. One of these is the rapid removal of water from the vicinity of the wood. This method has been applied to many old Norwegian stave churches, where unusually wide eaves keep water away from the vertical timbers that form the walls of the building. A second traditional technique is to provide surface protection against incursion of water, as can be seen in the same stave churches where wood roof shingles are periodically treated with oil tar (fig. 1). Sometimes protection is provided inadvertently, as in the case of wood granaries supported on staddlestones. The choice of timber species is also important, as some are more naturally resistant to decay than others. To these traditional techniques of timber conservation can be added these realities: Nothing is really permanent, one cannot win over the forces of nature, and continual repairs and, sometimes, total replacement are inevitable. As far as the curator or custodian of an ancient timber structure is concerned, the only reassuring aspect of this fatalistic attitude is that the traditional knowledge and skills associated with building in wood are kept alive—something I hope we would all agree is almost as important as the preservation of the products of that knowledge and skill.

With these thoughts in mind, I have chosen four case studies from England in order to illustrate a few current problems of in situ timber conservation and the ways in which they are being solved.

The church door at Stillingfleet, Yorkshire

(Addyman and Goodall 1979)

This twelfth-century door is of oak wood with applied decorative wrought iron depicting boats, human figures, and other motifs (fig. 2). It is considered *important* partly because of its age, and partly because of its art historical and technological significance.

Its present *function* is as an occasionally used door, and it serves to keep the weather out of the church.

The current *problem* is that the surviving original wood is rotting between the joints and the ironwork is corroding. Earlier repairs that sealed cracks in the door are themselves deteriorating and no longer keep out the wind and the rain. Much of the lower part of the door was replaced with new wood in the last century.

The *solution* is to advise the parish council to bring the door into the church for safekeeping and to hang a modern door, more efficient at keeping out drafts, in its place. The original door will be comparatively safe and stable inside the church until a more acceptable alternative can be found.

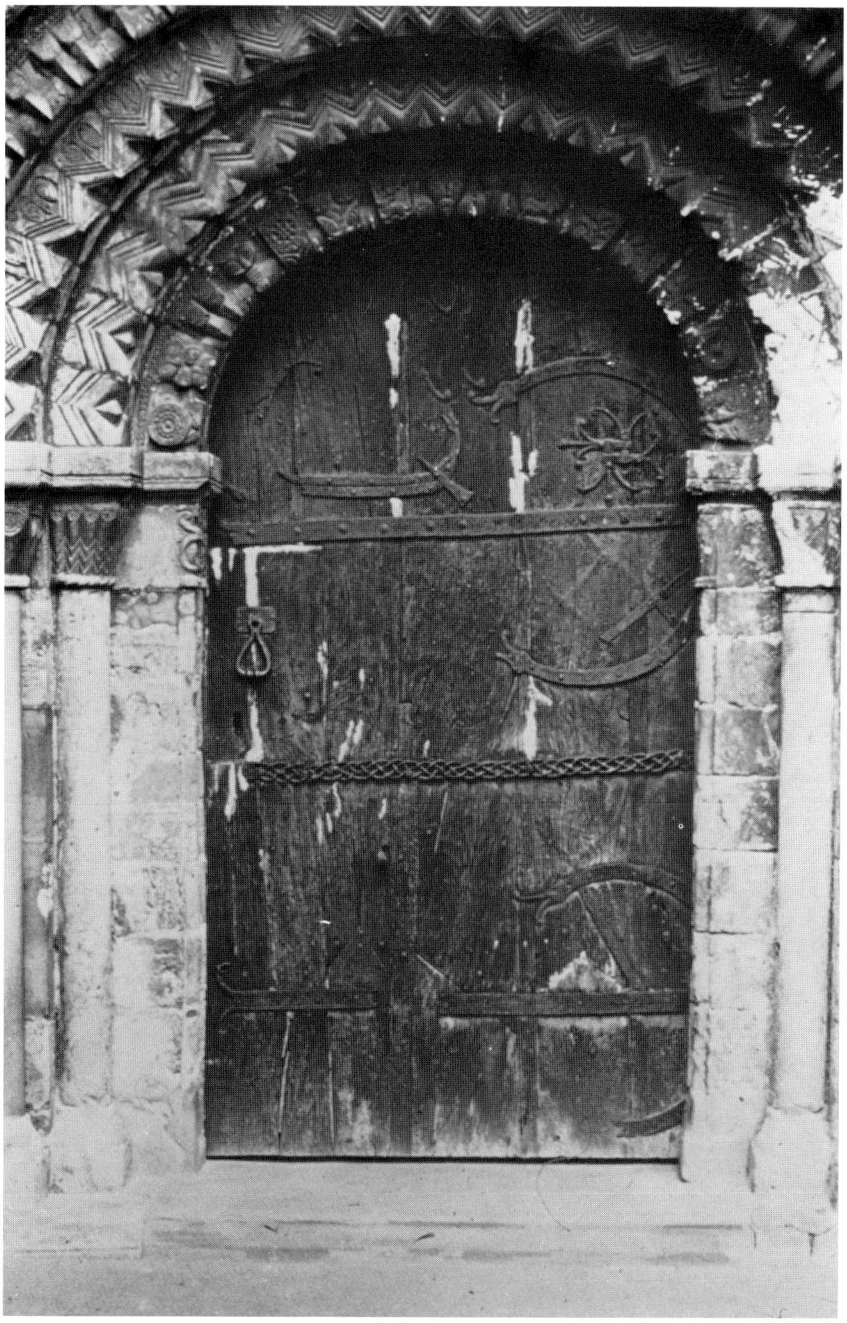

Figure 2. Twelfth-century door at Stillingfleet, Yorkshire, England.

The maypole at Barwick-in-Elmet, Yorkshire

(Kightley 1986)

This is a 26 meter-high pole (85 imperial feet) that stands in the center of the village. It is traditionally made of several sections of well-seasoned English larch, spliced and bound with iron bands. Since World War II, seasoned Norwegian spruce treated with biocides has been used, good lengths of local larch having become unavailable.

The *function* of the pole is obscure, but most probably it was associated with fertility rites and was almost certainly pagan in origin.

It is *important* because it is one of the only survivors of a tradition once common throughout England. It is the center of a triennial festival at which the pole is lowered for repainting, and the rotten butt end is repaired. The re-erection of the pole after maintenance is a major local event (fig. 3).

There is no real *problem* with the conservation of this pole since traditional methods of preservation are being employed, including occasional total replacement. In this instance, the traditions surrounding the maintenance of the pole are probably of more interest and significance than the pole itself.

Figure 3. Raising the maypole at Barwick-in-Elmet.

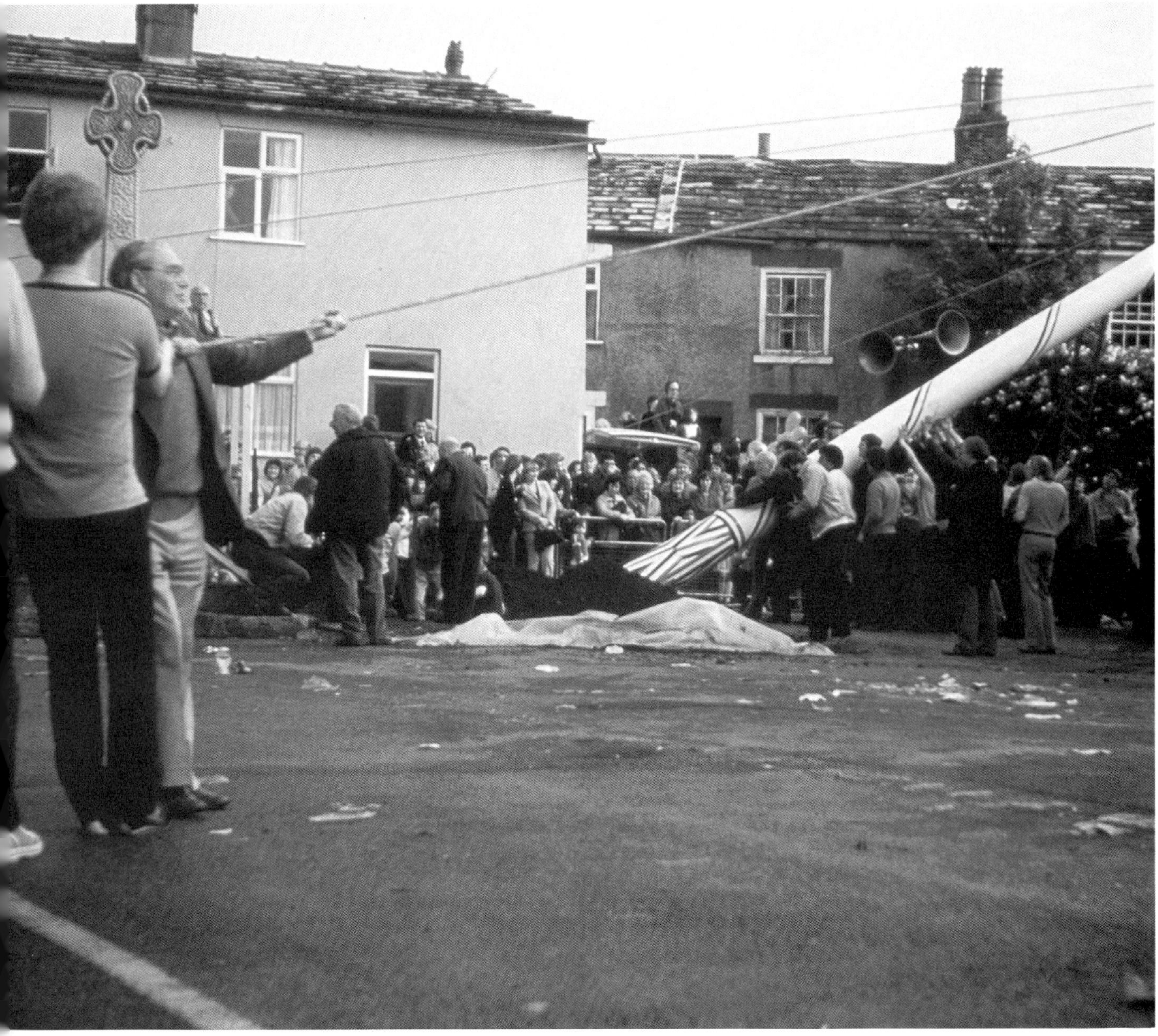

The warships HMS Victory and Mary Rose

The *importance* of the HMS *Victory* is that she was the flagship of Admiral Lord Nelson at the Battle of Trafalgar (fig. 4). She is also the last surviving all-timber warship of her period.

Her present *function* is to serve as the official flagship of the British Navy. She is still used for naval ceremonies and is saluted by all ships entering and leaving Portsmouth Harbor. HMS *Victory* is also a very popular tourist attraction.

The *problem* is that the oak wood of which she was principally built is suffering from dry rot, wet rot, and death-watch beetle infestation, to the extent that most of her original timbers have already been replaced at least once. No recent chemical treatments against either insect or fungus attack have been successful.

The *solution* is to replace all rotten wood with an iroko wood laminate incorporating thermosetting resins for the "knees" and other structural members, and with teak wood for the decking and the decorative elements of the bow and stern.[1] Old timber is also being used where appropriate, and the ship is gradually being restored to her as-built appearance of 1803.

1. Iroko is a hardwood exported from the tropical rainforests of West Africa.

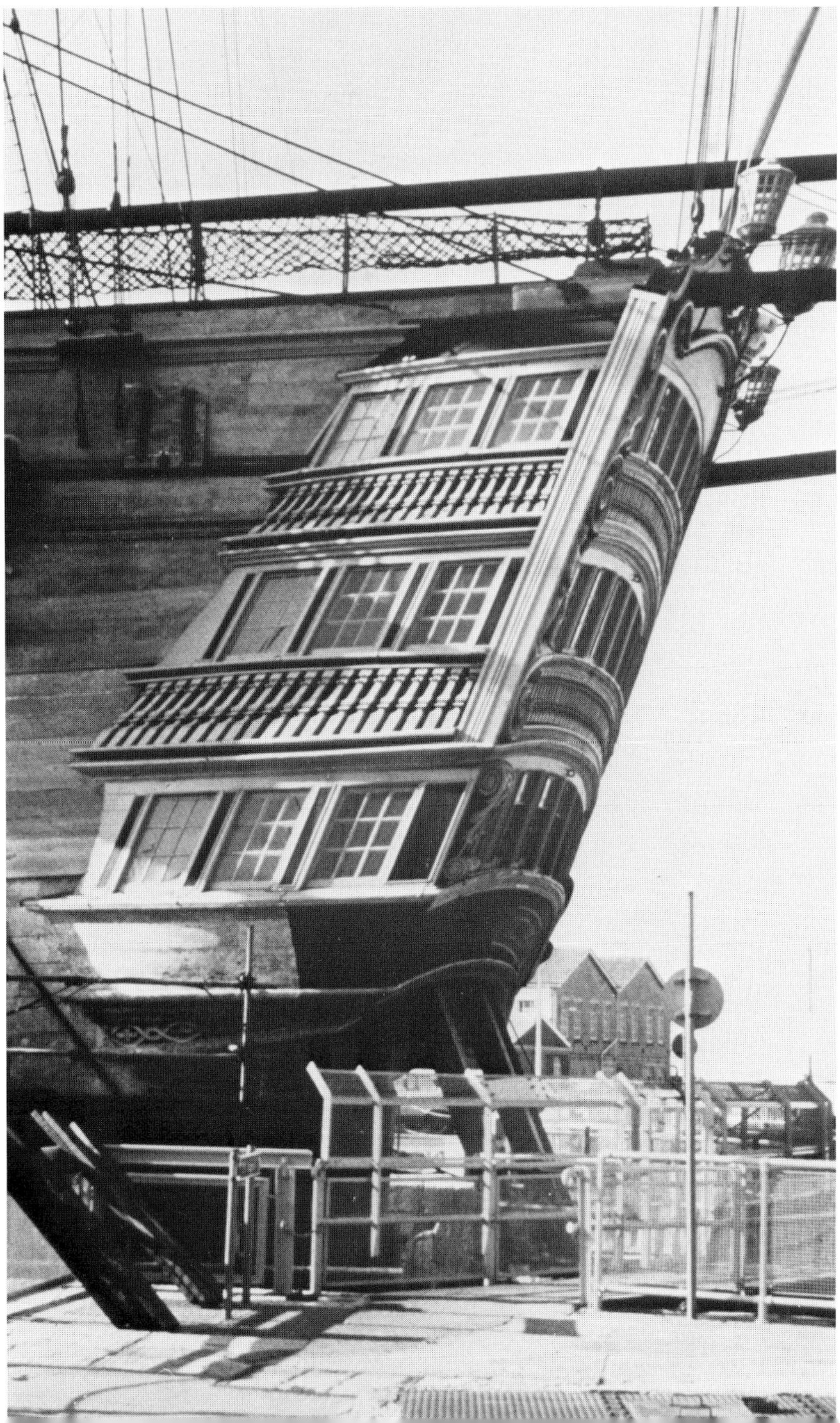

Figure 4. Stern section of HMS Victory.

Immediately beside HMS *Victory*, in the next dry dock, are the remains of the *Mary Rose*, another naval flagship that was recently raised from the seabed off Portsmouth (Rule 1983). Two-hundred-fifty years older than the *Victory*, she went down while going into action against the French navy in 1545. Although this sixteenth-century warship is not being conserved in situ exactly, her present home in the naval dockyard is highly appropriate from historical and practical points of view. The *Mary Rose*, in contrast to the *Victory*, is not being restored to her as-new condition, but is quite deliberately being kept in the physical condition in which she was found. Timbers found loose on the seabed are being replaced in their original positions (when this is known), but no other additions or alterations are being made, apart from stabilizing the timbers of the ship to permit display indoors (fig. 5). It would be possible to rebuild the missing half of the ship with modern materials and to replace the masts and rigging, but in this case (as with most other objects regarded as being strictly archaeological), this is considered unnecessary and unethical. However, the *Mary Rose* will remain a museum exhibit and as such will be displayed indoors. The whole approach to her conservation and restoration has been determined by this fact.

Figure 5. The Mary Rose under water sprays.

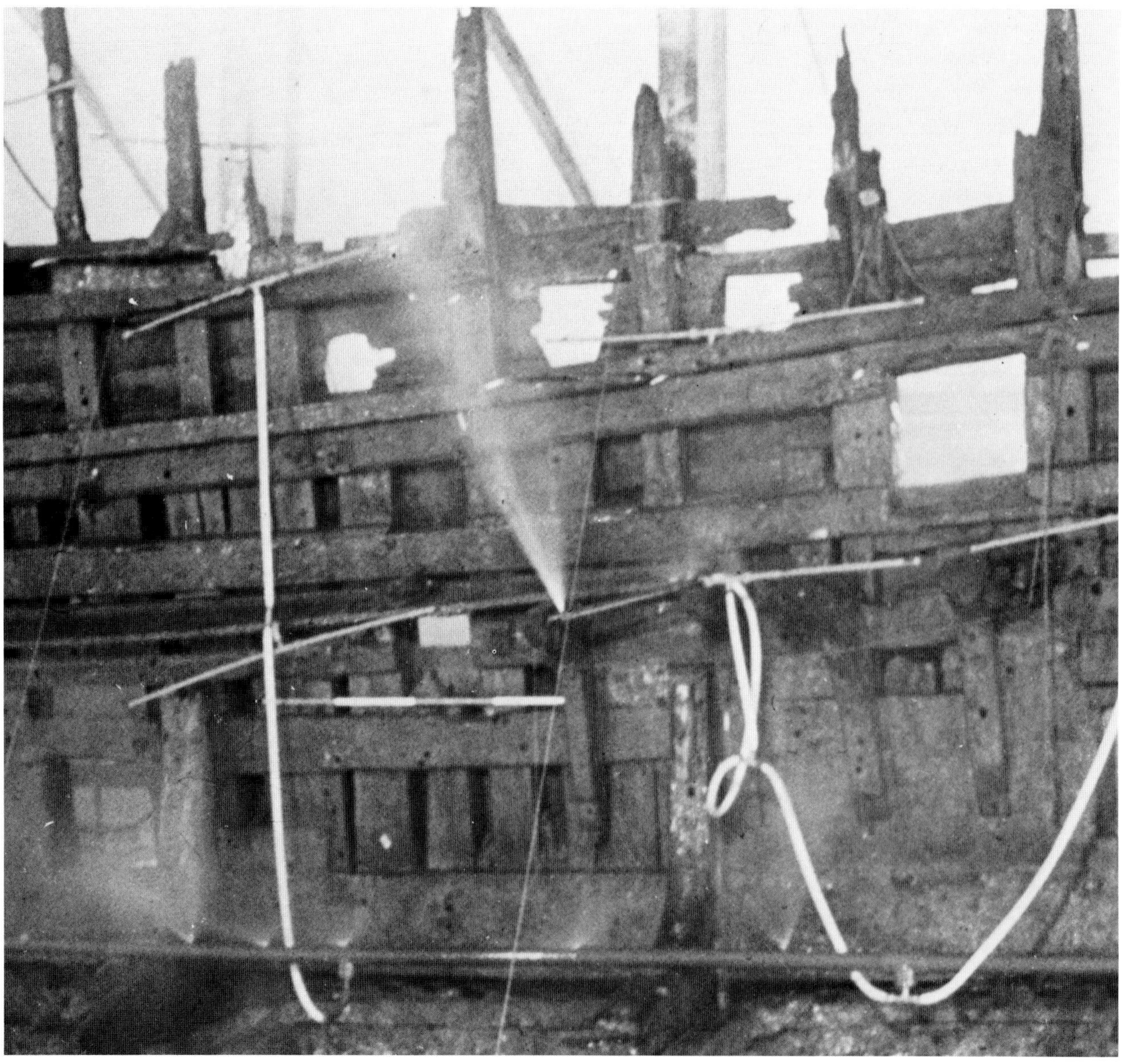

The Viking Buildings, York

In contrast to the other case studies, the following is purely archaeological, in the normally accepted sense of the word. These tenth-century houses and workshops are of oak wood post-and-plank construction. They were preserved primarily because oxygen was eliminated from the burial deposits by water saturation.

The *importance* of these structures is that they are the only known survivors of an architectural style from a period of British history underrepresented by artifacts, and they are valuable for the technological information they contain.

Their *function*, in contrast to the other case histories, is largely as a source for research, but they are also potentially valuable as museum display material.

The *problem* was that the wood was fragile and liable to cracking because of air drying after excavation. Conservation efforts were directed to preserving the wood so that it could be displayed in situ in the Jorvik Viking Center, a museum that was to be built on the site of the excavations. Since traditional techniques could not provide a *solution*, modern conservation practices were necessary. After two years of storage under water, the timbers were stabilized by replacing the water with polyethylene glycol wax, now a fairly standard procedure. The preserved timbers have now been re-erected as they were excavated, within a few feet of where they were originally found. Their stability is maintained in an air-conditioned environment and they are checked periodically with photogrammetry (fig. 6). This situation possibly illustrates the most ideal case of in situ conservation and display of timber from an archaeological excavation.

Conclusion

These case histories demonstrate the wide diversity in approach and attitude to the conservation of timber monuments on the part of the authorities responsible for their care. On the whole, it seems that traditional techniques are still used in favor of those that are more scientific. Despite continual advances in the development of fungicides, insecticides, and resins for consolidation and surface protection, very few new techniques or materials have been found to be effective in protecting against, or as a cure for, the various types of degradation found in timber monuments exposed to the elements. The only sure way of protecting timber is to isolate it from the elements with some form of cover. In the majority of instances, this is neither practical nor desirable.

There have been, of course, considerable developments over recent years in treatments for the protection of new timber—various forms of artificial seasonings and copper-chrome arsenate compounds to name two. For the treatment of rotted and fragile wood, mention should be made of the French Nucleart program (Ginier-Gillet et al. 1985:125–137). Resin monomers are introduced into the wood object and catalyzed by exposure to gamma radiation, producing a strong but irreversible result.

There have also been great advances in the technology of surface coatings, some of which are undoubtedly of interest in the context of timber protection. But caution in the wholesale application of these techniques and materials is needed, for it is possible that the use of some of them may well compromise the ethics of conservation. The same could be said of many of the traditional techniques of conservation mentioned earlier, and the curator, conservator, and restorer must tread warily among the choices of traditional and modern techniques for the in situ preservation of wood. There are no easy answers. Attitudes about the reasons for preserving archaeological or ethnographic materials change with time. As the traditional knowledge of the purpose and use of an artifact or monument dwindles, it is often replaced (or possibly displaced) by a sense of its purely artistic value or quaintness, in a society with little understanding of the historical context in which the object was conceived. The problem of conserving timber artifacts, especially standing monuments, has possibly as much to do with our present attitudes toward conservation and restoration, as it has to do with the real technical difficulties that undoubtedly exist.

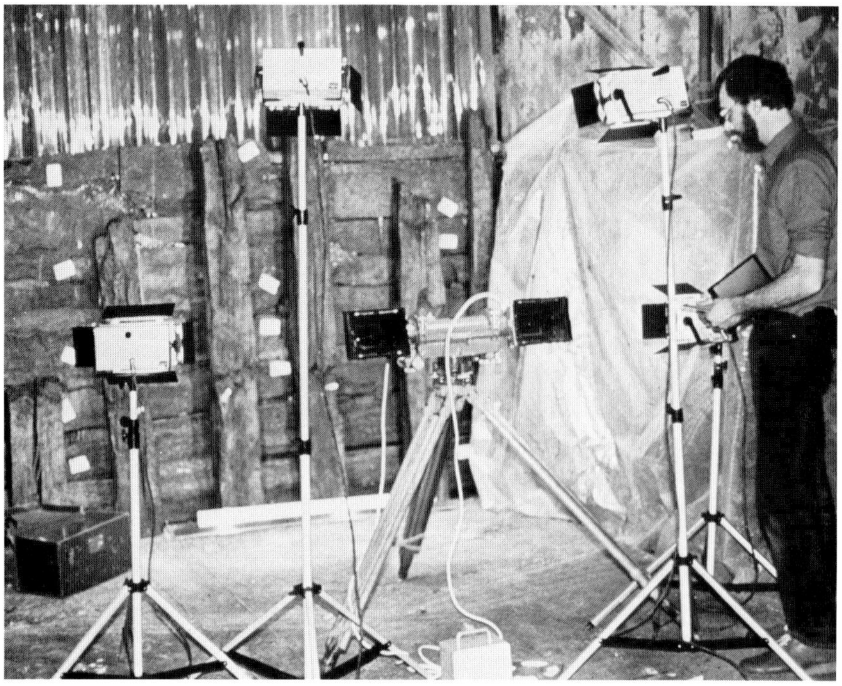

Figure 6. Photogrammetry applied to wall of Viking building, Jorvik Viking Center, York, England.

Since the uses to which wood can be put are so very diverse, there are many different types of craftsmen, conservators, and restorers, not to mention archaeologists and curators, responsible for the care of objects and monuments made of wood.

Architects renovating old timber-framed buildings may sometimes use techniques that would make museum restorers wince; archaeological conservators use methods and materials unknown to joiners and carpenters. Because all these people have such very different backgrounds and work in such different spheres, the lack of contact between them can only perpetuate the division between traditional and scientific approaches to the conservation of wood. Part of the answer to the problem of wood conservation in general could be to create opportunities for those working in these various spheres to meet and share their knowledge and to discuss their differences in approach and method. International bodies, such as the International Institute for Conservation of Historic and Artistic Works (IIC), the International Council of Museums (ICOM), and the International Centre for the Study of the Preservation and Restoration of Cultural Property (ICCROM), have of course held meetings in the past on wood-related topics. Reference could be made, for instance, to a paper by the former director of ICCROM, Sir Bernard Feilden (1984), on the ethics involved in the conservation of timber structures. This is very relevant to our topic and specifically highlights the need for better communication and understanding between craftsman and conservator. This is one possible way forward and could provide a most useful and interesting subject area for a future conference. Timber conservation is a wide-ranging problem, not restricted to any one country or part of the world. The issues, as illustrated above, are not merely technical, but have social, cultural, historical, political, and even religious implications, which all combine to make this a fascinating area for further discussion.

Acknowledgments

I am grateful to Commander P. Whitlock, Naval Historian to HMS *Victory*, and to Mr. G.H. Shinn, Polemaster of Barwick-in-Elmet, for information contained in this paper.

References

Addyman, P.V. and I.H. Goodall
1979 "The Norseman Church Door at Stillingfleet, North Yorkshire." *Archaeologia* 106:75–105.

Feilden, B.
1984 "A Possible Ethic for the Conservation of Timber Structures." Appendix to *Conservation of Timber Buildings*. F.W.B. Charles and M. Charles. London.

Ginier-Gillet, A., M.D. Parchas, R. Ramière, and Q.K. Tran
1985 "Méthodes de conservation developpées au Centre d'étude et de traitement des bois gorgés d'eau, Grenoble." *Proceedings of the Second ICOM Waterlogged Wood Working Party Conference*. Grenoble.

Kightley, C.
1986 *The Customs and Ceremonies of Britain*. London.

Rule, M.
1983 *The Mary Rose*. London.

Spriggs, J.A.
1984 "Treating the Coppergate Structures." In *Archaeological Papers from York presented to M.W. Barley*. P.V. Addyman and V.E. Black (eds.). York.

Peter van Geersdaele

MOLDING IN THE FIELD USING PLASTER OF PARIS

Peter van Geersdaele trained at the Victoria and Albert Museum and the Institute of Archaeology, University of London. He is presently Deputy Head, Conservation Department, at the National Maritime Museum in London, and was previously Assistant Chief, Conservation Division, National Historic Sites, Canada. He is a Fellow of the International Institute for Conservation.

The largest plaster mold ever made in the field may be the Sutton Hoo ship impression. Sutton Hoo is an ancient Saxon burial ground situated near the Deben River at Woodbridge in Suffolk, England. The site consists of numerous ancient burial mounds on an expanse of sandy heathland. In 1939, the largest mound was excavated by Basil Brown, an employee of E.M. Pretty, the owner of the land. He soon discovered a system of rusted iron rivets in situ. He determined they were part of a boat of considerable size.

Recognizing that he had found something extraordinary, Mr. Brown called a halt to the excavation, and the British Museum and the Ministry of Public Building and Works were called upon for advice as to how to proceed. The task ahead was one of unprecedented difficulty. No one in England had ever excavated a burial deposit remotely approaching this in importance and complexity. The only remains of the ship were the impression left by the decayed timbers in the sand.

The length of the ship was estimated to be 27 meters, its greatest beam was 2.25 meters, and its depth midships was 1.37 meters. This scale led the excavators to the initial conclusion that the burial was a Viking ship, contemporaneous with the well-known ship burials of the ninth and tenth centuries at Oseberg and Gokstad in Norway, and Ladby in Denmark. It soon became apparent, however, that this burial belonged to an earlier age and was unparalleled in British archaeology.

In July 1939, gold jewelry, coins, silver vessels, weapons, bronze bowls, parts of cauldrons, and other miscellaneous objects were found in a burial chamber amidships. This was the richest treasure ever discovered on British soil and the most important archaeological record yet found in Europe dating to the period of the migration of the Germanic peoples (fifth through seventh centuries A.D.).

Unfortunately, war intervened, and the near-perfect impression of the ship with its iron rivets still in place was filled with bracken for the duration of the hostilities. The heathland, with its ancient mounds, was used by the War Department as a training ground for tank drivers. The continuous use of the site as a tank-run did irreparable damage, which was only discovered when the ship remains were reexcavated in 1966.

In May 1967, the conservation laboratory working for the Department of Medieval and Later Antiquities of the British Museum was informed that a mold of the ship impression was to be made in order to produce a fiberglass replica. Although the remains were in an extremely damaged condition, a cast would still reveal the size, proportions, and profile of the ship; this had never before been achieved. In addition, as it was necessary to excavate below the level of the ship impression, a cast would provide a permanent record of the extent of the remains.

Preparing the mold

Various methods of making the mold were considered. Because cost and time were factors, plaster of Paris was chosen as the molding medium (fig. 1). Experiments in the laboratory and the field revealed that a barrier of paper towels between the sandy surface of the boat and the plaster was most effective in protecting the impression during molding. The plaster would be reinforced with metal electric-conduit tubing and plasterers' hessian cloth. Two lifting handles of conduit tubing would be attached to each mold section for lifting. The iron rivets still in situ would be protected by polyethylene sheeting and modeling clay to prevent them from being pulled away when the mold was lifted.

Materials were estimated and ordered, and these included 6 tons of plaster and 250 meters of conduit tubing. As there was no water supply on the site, four storage tanks were installed and topped-up when required.

In July 1967, molding began. We agreed that the burial chamber was the best place to start because it had the least detail, and it would give us an opportunity to get used to the molding method. First, the most experienced molder in the team of six worked on the ribs of the ship, and then the rest of the team molded the center of the burial chamber. Retainers for the side of each mold-piece were made of polyethylene tubes filled with sand, in varying lengths up to 1 meter. These contained the wet plaster while each section was made.

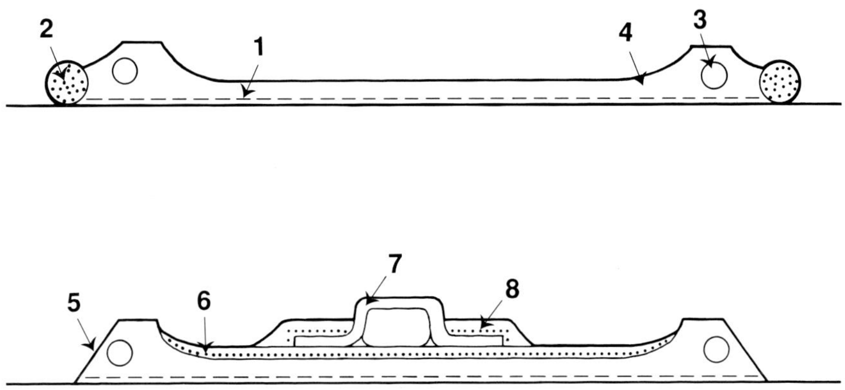

Figure 1. The method of making a mold piece using plaster of Paris as the medium.

Top: First stage in making a mold piece.

1. *Wet paper toweling*
2. *Sand-filled polyethylene retainers*
3. *Steel electrical conduit tubing*
4. *First coat of plaster*

Bottom: Final stage in making a mold piece.

5. *Edges trimmed to 60°*
6. *Hessian and plaster coating*
7. *Lifting handle made of bent tubing*
8. *Plaster and hessian securing handle*

The first mold-piece, approximately 0.75 meters square, was made at the center of the burial chamber. The pieces got larger as work proceeded. The sides of the burial chamber were more difficult to mold and took much longer to prepare, as there still remained many rivets in situ.

Sausage-shaped retainers were positioned where required. Polyethylene sheeting and clay were placed over each rivet, and wet paper towels were pressed down with a paintbrush. The paper was kept wet by constant spraying with water from a high-pressure garden sprayer. Then plaster mixed to a creamy consistency was applied to the surface.

As the plaster thickened, the sides of each section were built up to a thickness of 6 centimeters. Conduit tubing cut to size was then placed around the sides of the section for reinforcement. Four pieces of tubing were applied, two bent into right angles at each end so there would be no break in the metal around the edge of the mold.

Then, more plaster was applied to the section before the first layer had set. This was most important, because if the plaster had been allowed to set, there would have been little bonding between the two layers of plaster. To strengthen the center of the mold section, hessian cloth dipped into plaster, was laid over it in 30-centimeter squares. The two lifting handles of conduit tubing were held in position with hessian cloth and plaster.

The edges of the section were cut to an angle of 50° to 60° before the plaster had set, avoiding an otherwise difficult and time-consuming operation. It was essential that each edge was smooth, without undercuts that could trap the next section. Throughout the molding process we followed a plan wherever possible whereby the last mold-piece made would be the first to be removed, so that each section trapped the one molded previously (fig. 2).

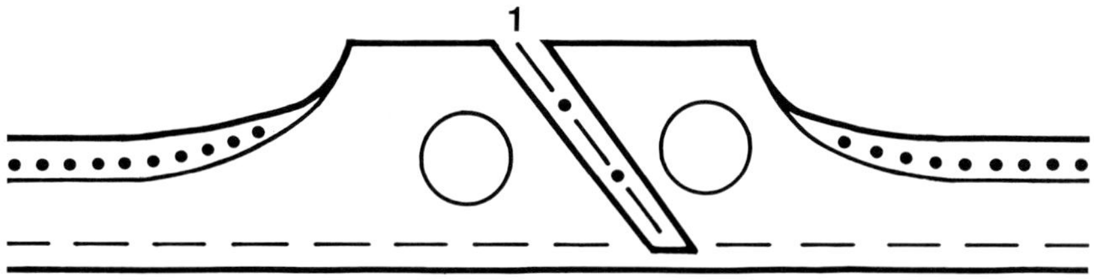

Figure 2. Junction between adjacent mold pieces. Shellac and release agent act as a separator (1).

When set, the plaster edges of each section were treated with a shellac solution and a release agent was applied. The next section was then prepared and molded. It was found that the rivets could not be covered with clay too far in advance, as the clay tended to dry and contract, thus defeating the purpose of the covering; each section had to be molded as we went along. After several sections were complete, we lifted one to see how things were going; the result gave us every reason to continue.

Molding the ribs of the boat, where there was a greater likelihood of damage when the mold was removed, was a more tedious process (fig. 3). The ribs were very fragile, and each one had to be molded in seven to nine sections. Each piece was designed so that when it was removed it would not damage the rib. As previously mentioned, this operation preceded the work of the main party of molders who were making the larger sections. The larger pieces were made to fit over the smaller rib-pieces. Consequently, when the mold was later reassembled for casting, the larger sections held the smaller ones in position. After the burial chamber was completed (twenty-two sections), each mold-piece was numbered and plotted on a plan drawn to measurements taken from a central datum line.

The stern of the boat was then molded in exactly the same way as the burial chamber. Because archaeologists were anxious to excavate the rivets and boat-fill, the mold of the burial chamber and the stern had to be removed before the work on the bow began. Unfortunately, therefore, no photograph exists of the boat completely covered with its plaster mold. Finally, the bow section of the ship was molded in the same manner as the stern and burial chamber, the whole operation taking three weeks and three days to complete. The mold was then removed and transported to storage to await reassembly for casting.

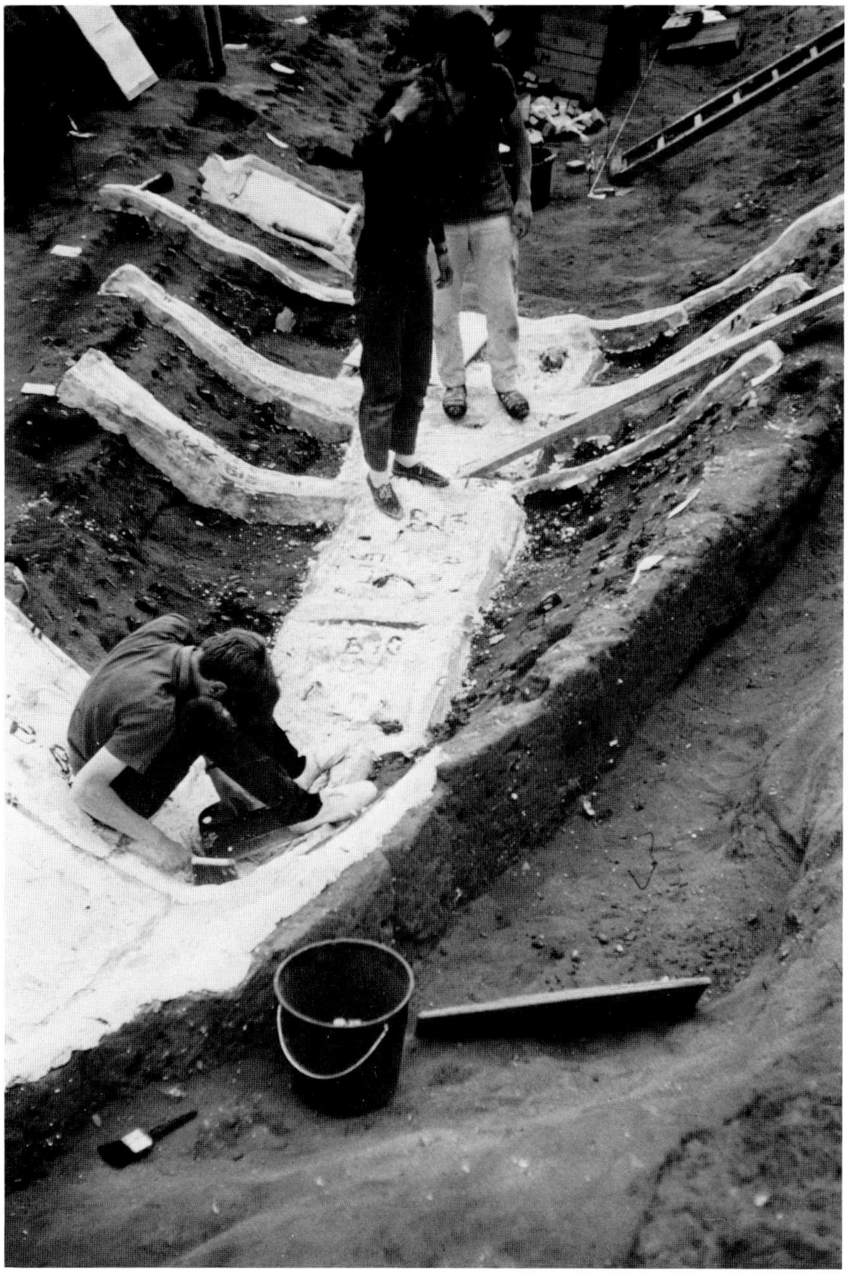

Figure 3. Molding the fragile ribs of the boat.

Reassembling the mold

It was estimated that a space at least 30 meters long and 6 meters wide was required to make the replica of the ship impression. Fortunately, such a room was available on the ground floor of the building housing the mold. This room also had two sets of double-doors opening onto the roadway. These were vital, as they would provide the ventilation required when casting with polyester resin in large quantities.

Support for the upturned mold was accomplished with twenty-five oil drums (1 m x 0.7 m) and wood beer crates borrowed from a local brewery (fig. 4). A datum line was painted on the floor and, using the plan and measurements taken in 1967, the reassembly of the mold began with the stern section. Progress was slow, and it was also very hard physical work.

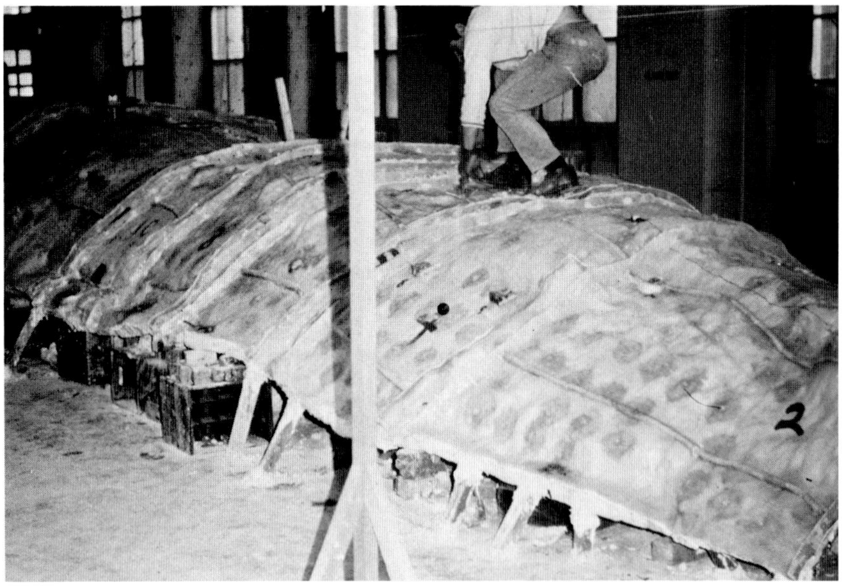

Figure 4. Oil drums and wood beer crates support the upturned mold.

The mold was assembled section by section, with great attention to each join. Each section had to be supported and connected, on the underside, to its neighbor with plaster and hessian cloth. Then 6 x 6-centimeter timber legs cut to size were secured both to the underside of the mold and to the floor with more plaster and scrim. As the mold grew higher in the burial chamber area, the oil drums were used as supports.

All the joins between sections were then filled with plaster to make the mold very smooth. Because of its size, the mold had to be cast in sections. As in the molding process, each cast section was butted up to its neighbor in order to achieve a perfect join. The ribs of the ship looked as though they might be difficult to remove from the mold because they had slight undercuts. Therefore, it was decided to cast these first, with their sides angled, so that the pieces butting up against them could be removed, leaving the rib pieces in position. The ribs could then be removed as the mold was taken down.

In order to cast the ribs and other sections, plaster walls against which to cast the resin were prepared. These were made by placing 4-centimeter-high clay bands across the mold, along each side of a rib, at an angle of approximately 60°. The mold surface to which plaster was to be applied was treated with a shellac solution and a release agent so that the plaster wall could be removed easily after it had served its purpose. Plaster was then applied to the clay bands to a thickness of about 2 centimeters. When it had set, the clay was removed, leaving walls of plaster against which the resin could be cast. This process was repeated along the entire length of the mold wherever required.

Making the casting

When the plaster walls were in position, preparations for casting began. Two coats of Crystic® release agent No. 1 were applied to a section.[1] When dry, they were followed by an application of Simoniz® wax, which was then polished. Crystic release agent No. 2 was then applied and left to dry for 24 hours. The gel-coat, a very thin layer of polyester resin, was brushed onto the section (60 gm per square meter). The purpose of the gel-coat was to prevent surplus resin from running down the sides of the mold. When the gel-coat was in a "green" state—that is, when it was still soft, but could be touched without its sticking to one's finger (roughly half an hour)—Crystic resin 405 was applied.

1. For the chemical composition of this and other proprietary materials, see the Appendix.

The resin was brushed generously over the gel-coat. Woven rovings (shredded fiberglass matting) were then pressed into the holes representing the rivets. Fiberglass matting pre-cut into 30-centimeter squares was then laid down and stippled until it had absorbed the resin. Each square overlapped the next by 6 centimeters, until the whole section was covered. The process was repeated to form a double layer of matting. To strengthen the cast, 2-centimeter "D-shaped" paper rope was laid across the section. More resin was then brushed onto the rope, and to hold it in position, a layer of fiberglass matting was stippled over it. Finally, metal leg attachments were secured in position with more resin and fiberglass matting. These leg attachments (commonly used by do-it-yourself hobbyists) were used to secure the removable wood legs that would support the finished cast.

This procedure was used throughout the casting operation, except in the areas where the plaster walls prepared for casting the ribs were removed, leaving fiberglass walls against which to cast. Here the fiberglass wall had to be trimmed, leaving a sharp edge on top. Crystic release agent No. 1 was not required on the fiberglass edge, only the Simoniz wax and Crystic release agent No. 2. Two sections were cast each day, and two prepared for casting the following day. Casting took three weeks to complete.

Before the removal of the cast from the mold, the edges of the walls and cast were trimmed. Holes 1.25 centimeters in diameter were drilled through the walls adjoining sections at intervals of 23 centimeters to accommodate the bolts that would hold the sections together (fig. 5). All the leg attachments were numbered and recorded onto a plan of the cast. Legs to support the finished cast were made. Because these would be of different lengths, a method of measuring was devised, since measurements for the legs had to be taken while the cast was still on the mold.

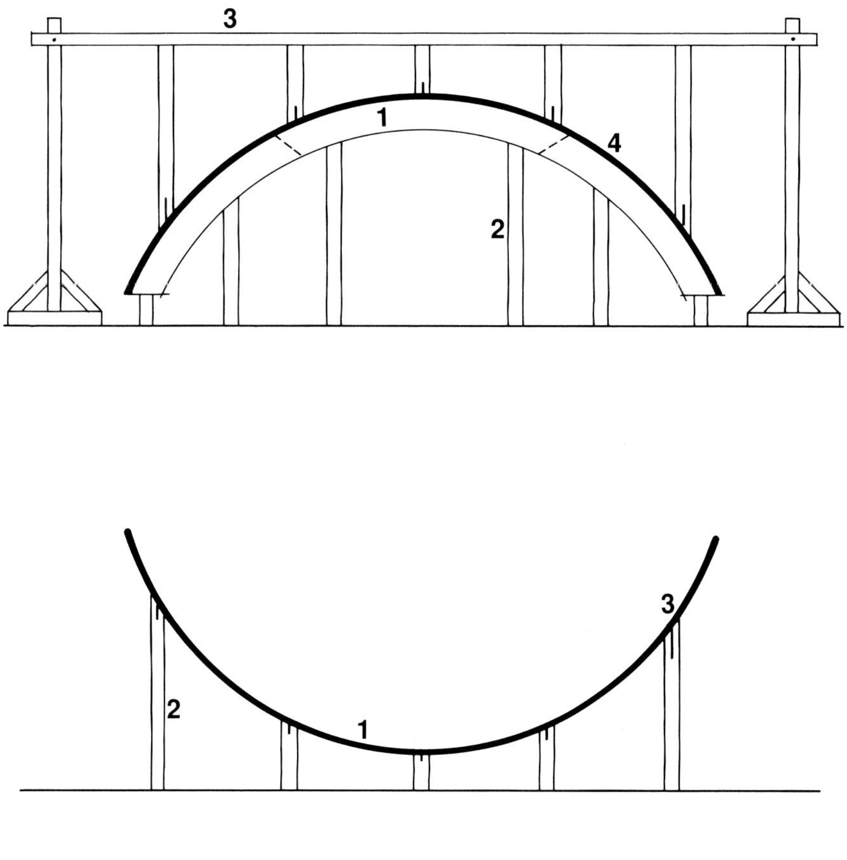

Figure 5. Preparing the cast.

Top: Assembly during casting.

1. *Reassembled mold pieces*
2. *Wood supports*
3. *Guide rail for trimming legs*
4. *Polyester and fiberglass casting*

Bottom: Casting as displayed.

1. *Polyester and fiberglass casting*
2. *Supporting legs*
3. *Table leg bolt integrated into casting*

A 6 x 6-centimeter movable timber frame 2 meters high and 6 meters wide was constructed over the mold. This represented the floor upon which the legs would eventually stand. One 2-meter-long leg was made of 6 x 6-centimeter timber with studding on one end. This was then screwed into a leg attachment by means of the studding. The wood frame was then moved up to it, and a measurement taken on the leg to the cross-piece of the frame. The crossbar was checked with a spirit level as each measurement was taken, to ensure accuracy. Each measurement was noted along with its appropriate leg-attachment number, and a leg to size.

Everything was now prepared for the removal of the cast from the mold. A knife blade was run along between the fiberglass walls to release the vacuum between them. The edges of the cast section were then pried up with chisels and long wood levers. Once the air had broken the vacuum between the cast and the mold, the sections were released fairly easily.

The reassembly of the cast, which began with the burial chamber section, proved to be the easiest part of the whole operation.

More recent work on molding vessels

The method used to mold the waterlogged timber remains of a late tenth-century boat found on the Kent marshes near Graveney in 1970 was similar to that used to make the Sutton Hoo ship impression. The boat—10 meters long and 3 meters at its maximum beam—was to be molded in such a way that after the waterlogged wood had been treated, the mold could be used as a form for reshaping the strakes after preservation. However, this did not happen; instead, the mold was reassembled and a fiberglass cast was made from it.

Owing to the fact that the boat lay in the excavations of a drainage canal to be flooded almost immediately, the molding operation had to be carried out in four days. To save time, the ribs of the boat were removed prior to molding the hull, and they were molded at a later date.

A canvas canopy was erected over the site to protect it from the adverse October weather. The working conditions were worsened by the necessity of continually spraying the wood with water in order to prevent it from drying. The water rapidly turned the clay soil into a heavy mud, which surrounded the whole area.

The first section of the mold was made at the forward end of the bow. As there had been no time for experiments beforehand, it was decided to mold this section, lift it, and, if successful, place it back into position. The mold was a success and the method was repeated for each of the nineteen sections until the waterlogged remains were entirely covered by the plaster piece-mold.

Only the area of wood to be molded immediately was exposed from its protective polyethylene sheeting. This area was continually sprayed with water until the application of the plaster of Paris. The wood surface was examined for the presence of undercuts, and any found were eliminated by pushing modeling clay into them. Paper toweling was used as a separator between the wood surface and the plaster. Owing to the wet conditions, it was possible to lay the dry toweling on the wood where it rapidly absorbed moisture and could then be stippled into position with a paintbrush.

Clay "walls" 2.5 centimeters wide and 5 millimeters thick were laid, outlining the area to be molded. They were shaped with one or two hemispherical indentations to provide keys for the eventual reassembly of the mold. Plaster of Paris mixed to a creamy consistency was applied over the exposed section, the sides being built up to a height of 6 centimeters. Each mold section was reinforced exactly as at Sutton Hoo. A release agent of soft soap and petroleum jelly was applied to the angled edges of the mold before the next piece was made.

Again, the mold was constructed so that the last piece to be made was the first to be lifted. Three sections were completed on the first day, six on the next day, and ten on the third day. The mold was cleansed of excess plaster, the sections numbered, and a plan of the mold drawn.

During this molding operation, it was discovered that the paper toweling separator was not necessary. The wet muddy surface of the wood acted as a separating agent for the plaster. This technique, unfortunately discovered too late to help in molding the Graveney boat, was later used on molding a waterlogged wood platform discovered on an industrial site at Trois Rivières, Quebec, Canada. Before molding, the wood was given a light clay wash, and plaster was applied directly onto the object's surface. The sharpness of the mold and the finished cast were far superior to both the Sutton Hoo and Graveney casts.

The most outstanding molding achievement in recent years must be that undertaken in 1983 by the Wet Site Section of the Conservation Division Parks, Canada. A section of the Basque whaling ship *San Juan* was molded under 10 meters of water, using a polysulfide rubber molding compound on FMC301 and a mother-mold of plaster of Paris. The rubber mold and resin cast will be used for interpretive purposes as well as being a permanent record of one portion of the *San Juan*.

Conclusion

Plaster of Paris has the advantage of being easy to obtain, it does not require a mother-mold to retain its shape when removed, and it is not expensive. On the negative side, if a mold made of plaster is large, it will be very heavy and if one does not take care to work tidily, the site will be covered with fragments of plaster. Other molding materials that have been used in the field include rubber latex and polysulfide rubber. Because both these materials are flexible, they require a mother-mold of fiberglass or plaster, which has to be applied before removal, to retain the shape of the mold. These flexible materials certainly simplify the process of making molds and also provide a lighter matrix and, although excellent for smaller tasks, their high cost may be prohibitive when preparing molds of the size I have just described.

Acknowledgments

I would like to thank Dr. R.L.S. Bruce-Mitford, Mr. J. Langhorn, Mr. A. Prescott, Miss G. Joysmith, Mr. N. Williams, Mr. G. Adamson, Miss A. Evans, Miss Y. Crossman, Ms. E. Pye, Ms. S. Davison, Mr. P. Lawson, Mr. L. Murdoch, Mr. R. Marion, Mr. S. Holm, Mr. J. Lauder, and Mr. T. Daley for their contributions to this paper.

Elizabeth E. Peacock

ARCHAEOLOGICAL SKIN MATERIALS

Elizabeth E. Peacock trained in conservation at museums in the United States and Europe, and received a Masters in Art Conservation from Queen's University, Canada. She is currently Conservator, University of Trondheim Museum, Norway, and Conservation Consultant to the Conservancy for Historical Remains and Ancient Monuments of Svalbard and Jan Mayen.

Archaeological skin materials are complex, heterogenous systems. They are generally fragile, contaminated, physically weakened, and chemically unstable; and they are never robust. In common with other organic materials, preserved skin, when excavated, may begin to deteriorate from the moment it is uncovered. This process will continue, and perhaps accelerate, unless knowledgeable and skilled action is taken during excavation, as well as post-excavation, to prevent it.

Archaeological skin conservation is a subdiscipline of archaeological conservation that has been neglected until now. It has relied upon the traditions of a craft and has not attained the status of a formal discipline based on a scientific approach to preservation (NIC 1984a; Fogel and Raphael 1984:6–15). There exists no reliable body of knowledge within this specialty. The information in the available literature, including some recent publications, is out of date. That which is available addresses the treatment of archaeological leather; there is little or no information concerning the conservation of other archaeological skin materials.

Skin products—including leather, parchment, and gut, among others—are in themselves complicated structures. They are the physically and chemically diverse products of innumerable empirical methods of treating animal skin; animal skin is itself a complex organ of the body based upon the protein collagen. When it is subjected to the physicochemical factors of a preserving archaeological environment, the complexities are compounded.

Preserving environments

The preservation of buried objects made of skin materials occurs in a narrow range of microenvironments. Desiccated, frozen, and terrestrial wet archaeological sites are noted for frequently providing a fortunate combination of physicochemical factors (Reed 1972). The present discussion is limited to the terrestrial preservation of skin materials in their organic form; neither pseudomorphic forms, such as the replacement or mold formation of skin by metal compounds, nor marine environments are considered here.

Desiccating conditions are characterized by scant precipitation, high evaporation rates of surface moisture, and good drainage. These conditions are found both in high-temperature regions of the subtropical climatic zone and low-temperature regions of the arctic zone. In high-temperature regions, preserved skins are desiccated, brittle, and often impregnated with soluble and insoluble salts. In arid arctic regions, skin materials are well preserved by natural freeze-drying. These materials are not chemically stable but normally are not immediately threatened by excavation if they are maintained in a dry state.

Frozen conditions are found in circumpolar regions and are characterized by permanently frozen soil. Skin materials are well preserved in a frozen state, often with minimal chemical alteration of their preburied state, but they are unstable on recovery and subsequent uncontrolled thawing.

Terrestrial wet environments are varied. Those favorable to the preservation of skin materials are acidic-anaerobic. These are common in temperate climatic zones and arctic regions where the subsoil is frozen during the winter, but waterlogged during the summer, as a result of simultaneous thaw and precipitation. Archaeological skin objects found in these environments are well preserved but have undergone a chemical restructuring that renders them unstable immediately upon excavation.

Traditionally, excavated archaeological skin objects presented no particular conservation problems. Most excavations took place in the arid regions of the Near East, where skin objects were scarce (compared to the plentiful inorganic materials) and presented no immediate or important stability problems.

Occasionally, an isolated body was found in the bog environments of northern Europe. The deterioration that followed the excavation of many of these bodies resulted in a reevaluation of the previously assumed stability of recovered finds of skin, and indicated a lack of understanding of the processes involved.

It was not until the advent of large-scale urban excavations in the terrestrial wet areas of northern Europe and major marine archaeological recoveries (both of which yielded vast quantities of water-degraded archaeological leather) that tanned leather became a material of concern to conservators. However, unlike practitioners of waterlogged-wood conservation, who sought scientific support for their practices, leather conservators simply adopted traditional leather treatments without questioning whether they were appropriate.

The complexity of the situation is now compounded with archaeological investigations in arctic regions, where an entirely different range of archaeological skin materials—native skins, both untanned and semitanned—are being recovered (Hansen et al. 1985).

In general, skin materials do not lend themselves to in situ conservation; they must be removed. Those preserved by burial in a dry site require no treatment, or treatments less complicated than those needed for skins recovered from a frozen or terrestrial wet site.

Documentation

Except in the case of a special archaeological find, skin materials are not extensively documented; they are often too common to merit the expenditure of time. The texture and color of skin often closely resemble the embedding soil matrix, and thus photographs are frequently unable to capture detailed information, especially on terrestrial wet sites. Even when the finds are cleaned, many subtleties are not revealed to the camera. Archaeological drawing remains the most satisfactory method of pre-excavation visual documentation.

The most common are 1:1 scale drawings of washed excavated objects. Skin materials are recovered surrounded by their soil matrix and are washed on site for further documentation or storage. The value of drawings must be conveyed to those who wash the finds so that they exercise appropriate care in handling the objects. When drawing must replace photography as the method of documentation, the time required is greatly increased. In addition, when available, X-radiography of suspicious composites that include skin materials can provide valuable information on composition and construction.

Cleaning

The cleaning of archaeological skin objects is divided into two categories: physical and chemical. Physical cleaning on site is routinely undertaken (1) to document the object prior to recovery; (2) to document the object after its removal; (3) to remove abrasive adhering/embedded crystalline constituents; (4) to reduce the overall volume of the object/soil-matrix unit, especially when the object will be or has been lifted; or (5) to free the object/soil-matrix unit from the surrounding layers.

On arid sites, physical cleaning involves gentle, dry brushing; on wet sites gentle brushing is accompanied by water and ultrasonic agitation (when available); and on frozen arid sites, gentle brushing, nudging, and manipulation are accompanied by water and warm circulating air (Segal 1985:19–20; Madsen 1980:136–144). Frozen sites often yield multiorganic composites and physical cleaning, although the same task, is a major operation.

Physical cleaning can be effective, but it does remove information about the surrounding burial environment and can cause considerable damage to the objects if not done with care. The extent to which objects should be physically cleaned varies. It is determined by site policy, which defines the eventual storage, treatment, and function of the object.

Although chemical cleaning should not be attempted on site, it often is. Chemical cleaning is undertaken (1) to impart a more natural color to the skin material; (2) to remove contaminants (such as rancid fats and iron salts) that are thought likely to further degrade the skin material; or (3) to remove insoluble embedded crystalline constituents, which are abrasive and inhibit flexibility.

The literature reveals an alarming list of chemical agents used to clean archaeological skin materials (fig. 1). In general, these are used indiscriminately and their effect is not evaluated scientifically. For example, nonionic surfactants, which are only tested for compatibility with untanned arctic skins, are used routinely in the wet cleaning of vegetable tanned leathers, where they form precipitates with vegetable tannins that can never be completely removed (Segal and MacDonald 1984). In such cases, ethanol would be a more suitable surfactant.

The application of traditional recipes for cleaning contemporary or historical leather are based on the subjective evaluation of visual results. Archaeological skins are so varied that the effects of chemical cleaning can never be consistent from one object to another. The short- and long-term effects of these treatments are largely unknown.

Consolidation and lifting

Some skin objects will not be structurally sound enough to be removed simply. It is preferable to lift the object embedded in and supported by its surrounding soil-matrix. If this is not possible, or if the object is too fragile, it may be strengthened by the application of a temporary consolidant. The consolidant is sprayed or brushed on the cleaned surface, using the minimum amount to provide sufficient support.

A variety of adhesive systems are reported in the literature. Polyvinyl acetate solutions and emulsions, polyvinyl alcohol, soluble nylon, polyvinylbutural, paraffin wax, and acrylic resin solutions and emulsions should *not* be used. Molten polyethylene glycol (PEG) MW 1,000 (melting point 37°-40°C), 10%-20% aqueous solutions of PEGs of higher molecular weight, and a mixture of low-molecular-weight PEGs

*Figure 1. Cleaning Agents Reported in the Literature.**

Chemical	Terrestrial Wet Frozen		Arid	
	Tanned	Other	Tanned	Other
a) *Acids*				
(i) Ascorbic	x			
(ii) Citric	x			
(iii) Hydrochloric	x			
(iv) Oxalic	x			
(v) Phosphoric	x			
b) *Salts*				
(i) Ammonium citrate	x			
(ii) Ethylene diamine tetra-acetic acid (EDTA)				
- 2Na		x	x	
- 4Na	x			
(iii) Potassium tartrate	x			
(iv) Potassium oxalate	x			
(v) Potassium permanganate	x			
(vi) Sodium bicarbonate	x			
(vii) Sodium dodecylsulphate		x		
(viii) Sodium hydrosulphite	x			
(ix) Sodium metaphosphate	x			
(x) Sodium sesquicarbonate	x			
c) *Surfactants*				
(i) Methanol	x			
(ii) Ethanol	x	x	x	x
(iii) Non-ionics				
- Lissipol N	x	x	x	x
- Synperonic N	x	x		

*Information abstracted from Segal and MacDonald 1984; Sturge 1975; van Dienst 1985:86–92; van Soest et al. 1984:21–31; Stambolov 1969; Rector 1975:33–37; Collins 1983:5–6; Jenssen 1983b:15–16; Calnan 1976; Tilbrooke 1978; Segal 1983:5–7.

(MW less than 1,000 and ethyl hydroxyethyl cellulose [EHEC]) are water soluble after curing and suitable for water-degraded skins (Fenn 1984:138–140). Various mixtures and examples of the latter have been reported (Brown 1974:469–473; Geijer et al. 1961; Morris and Seifert 1978:33–43; Spriggs 1980:19–24; Ringgaard 1986).

No particular reference was found for the application of a consolidant to a desiccated skin object, but it must be recognized that any adhesive applied to a desiccated skin object will discolor its fabric.

Alternatively, a wet object can be consolidated by freezing in situ with the application of carbon dioxide snow, and then excavated in the frozen condition (Madsen 1980:136–144).

An alternative to consolidating the object is to apply a facing fabric impregnated with, for example, acrylic emulsion (Logan 1983:3–5; Segal 1985:19–19). The facing adheres to the surface without penetrating the material, although this napbonding may not be successful with very degraded skin material. Block lifting can be performed following normal procedures: isolating the surface with polyethylene film or aluminum foil and then filling with expandable polyurethane foam. Plaster of Paris is very heavy and its use therefore is not recommended.

Packing, storage, and transportation

Packing, storage, and transportation guidelines for archaeological leather recovered from terrestrial wet and dry sites are presented in various field-conservation and first-aid manuals and leaflets (Dowman 1970; Leigh 1972; Joukowsky 1980:244–275; Fortidsminderådet 1981; Archaeology Section, UKIC 1982 similar; Seese 1984:31–50;

Tubb 1985:17–20; Museum of London n.d.). No guidelines are available for frozen and remote terrestrial sites or for nonleather archaeological skin; nor have guidelines been published for handling larger composites or bodies.

Available guidelines stress that degraded objects made of skin are extremely vulnerable to mechanical damage through handling and movement. The procedures presented constitute preventive measures, not a final treatment. Naturally, materials often at hand, such as matchboxes, toilet paper, and paper bags, are not suitable as packing materials.

Dry finds can be packed and supported in crumpled tissue paper if they are kept in dry surroundings. If ambient surroundings are not dry, the finds can be packed in a perforated Ziploc® polyethylene bag and stored in a sealed polystyrene box containing some silica gel and a relative-humidity-indicator card. This procedure should eliminate the need for a biocide.

Wet finds can be packed wet in either the supporting soil-matrix or, gently washed, in several Ziploc polyethylene bags. Alternatively, using one bag per find, several may be stored at a time in a sealed polystyrene box. Skin objects thus packed can be refrigerated for several weeks or frozen indefinitely. Otherwise, the addition of a biocide is strongly urged.

These guidelines have been developed for archaeological leather. Guidelines for wet finds may not be appropriate for other skin materials, which, because skins have not been tanned, would subject the collagen to further deterioration if stored in water or with a biocide of inappropriate pH.

Another alternative to packing and storage is that of reburial. This has been tried on terrestrial wet and frozen sites. It is important that the pre-excavation microenvironment be maintained by, for example, covering the site with microfoam and turf.

Stabilization

The removal of water from wet archaeological skin materials, and ice from frozen ones is the most critical stage of the conservation process and the most difficult to accomplish on site, especially in remote areas where workers may have limited expertise and facilities. When the material in question is tanned leather, its proper treatment, whether carried out on site or in an established conservation laboratory, is the subject of considerable controversy.

Practitioners and theorists disagree widely concerning the function of archaeological leather objects, there being two contrasting schools of thought. One view is based on the traditional methods of leatherwork and the practices of the modern leather industry, where the aim is to return the object to a near fascimile of its preburied state. The other philosophy sees the artifact as an archaeological object, that is, as an information source that should be altered as little as possible to achieve stabilization.

Existing methods of water removal reflect these conflicting philosophies (Carlsson 1980:165–204). Therefore, the method selected depends not simply on its suitability to site conditions, but also on the function of the object being conserved. There are three principle methods of water removal: (1) replacing the water by immersing the object in baths of a water-miscible solvent, followed by evaporation and lubrication with a dressing; (2) impregnating the object with a water-miscible compound, followed by slow drying under controlled air humidity; or (3) plasticizing the object with a polyhydric alcohol derivative followed by freeze-drying (Jackman 1982).

The first two methods require the lubrication or impregnation of the leather with a variety of natural, mineral, or synthetic oils, fats, or waxes. Many of these discolor the leather, leave an oily or waxy surface, collect dust, and are chemically unstable and subject to microbic deterioration. Formulations and applications vary considerably and in many instances require elevated temperatures; most also require the addition of a biocide.

All these methods, whether or not they are otherwise appropriate, are hazardous in use. Moreover, it is questionable whether they are acceptable for site use. Some are cited in field conservation literature before the advent of freeze-drying (Dowman 1970; Sturge 1975), but several archaeological collections treated with these methods have required further conservation within five years (Peacock 1984). Furthermore, any method requiring an elevated temperature should no longer be used since degraded skin can shrink in this range.

The third method, freeze-drying, is the latest technique to be applied. It involves the sublimation of water at a temperature of -20°C to -30°C. Most commonly, this is carried out in a vacuum using specially designed equipment. It can also be accomplished "naturally," using a low-temperature freezer, or by exploiting a suitable winter climate, with careful monitoring. The latter process is, of course, only possible in countries with low temperatures of long duration.

Freeze-drying, with or without pretreatment and using a humectant, is the recommended technique for removing water from excavated frozen skin materials. It is the preferred method of a large number of conservators for water-degraded skin materials, especially leather. For leather, the cleaned skin is pretreated with a humectant (the percentage solution and duration of impregnation vary), followed by prefreezing and then freeze-drying.

Freeze-drying is difficult to accomplish on site. Freeze-drying from low concentrations of polyethylene glycols (MW 400-1000), glycerol, or Bavon ASAK could be replaced by solvent drying with acetone, ethanol, or other dewatering fluids. The danger of using these solvents must be weighed critically. Positive results have been achieved using ethanol (Sturge 1975; Peacock 1984).

Biocides

Ideally, biocides should be avoided. Water-degraded or frozen skin materials can be safely stored frozen for extended periods of time, or refrigerated for approximately one month. These materials must be well sealed to guard against natural freeze-drying. Water-degraded or frozen tanned materials can be stored for extended periods of time in containers of gently circulating water (pH 4-6 preferably).

Nonetheless, some conditions require the addition of a biocidal agent. A survey of the conservation literature reveals that a variety of biocides have been used in the storage and conservation of archaeological skin materials (fig. 2). This data is fraught with inaccuracies and inconsistencies and documents the lack of understanding of the function of biocides.

Dawson, in his survey of biocides used in the storage of waterlogged wood (1982:269–277), observes that these chemicals should be compatible with (1) the materials composing the artifact, (2) the materials employed for storage, and (3) all possible substances used in analyses or treatments. In addition, they should be effective, resistant to deterioration, and nontoxic during application and subsequent contact. He recommends orthophenyl phenol and some of the quaternary ammonium compounds for use in the storage of waterlogged wood. A survey made by Sjøgren (1978) of fungicides used specifically with leather, recommends orthophenyl phenol and salicylanilides. Dawson rejects the latter because of its toxicity to humans.

A study by Sturge (1975) addresses the pH values of various percentages of orthophenyl phenol and its sodium salt. He concludes that the sodium salt has a pH incompatible with leather. He did not explore its compatibility with other archaeological skin materials, which might not be adversely affected by a neutral pH. This biocide is widely used because, among other things, it is easily soluble in water.

*Figure 2. Biocides Reported in the Literature for Use with Skin Materials.**

Chemical Name	Trade Name or Common Name
a) *Phenol* (carbolic acid)	
b) *Chlorophenols*	
(i) Dichlorophen	Panacide BD 157
(ii) Trichlorophenate	Dowicide B
(iii) Pentachlorophenol laurate	Mystox (Mystox LPL 1007)
(iv) Sodium trichlorophenate	Preventol L
(v) Sodium pentachlorophenate	Santobrite
(vi) Sodium pentachlorophenate isocyanate	Criptogil Na
c) *p-nitro phenol*	
d) *Phenylphenols*	
(i) Orthophenyl phenol	Dowicide 1 (= I), Topane S, Preventol O extra
(ii) Sodium orthophenyl phenol	Dowicide A, Topane WS, Preventol ON extra, Ketol (Konservierugsmittels #1473)
e) *Thymol*	
Thymol, 3 hydroxy-p-cymenol	Thymol
f) *Tego*	
(i) Alcalyl-(aminoethyl)-glycine	Tego 51
(ii) Alcalyl-(aminoethyl)-glycine hydrochloride	Tego 518
g) *Salicylic acid compounds*	
(i) Methylsalicylic acid	
(ii) Propylsalicylic acid	
(iii) Sodium salicylanilide tetrahydrate	Shirlan Plus
h) *Quarternary ammonium compounds*	
Specific formula not available	Dodigen (= Desogen ?) 226
i) *Boric acid: sodium borate (Borax)* 7:3	
j) *Formaldehyde Solution*	Formalin
k) *p-dichlorobenzene (PDB)*	Mothballs
l) *Acids*	
(i) Acetic	
(ii) Sorbic	
m) *Alcohols*	
(i) Denatured ethanol	
(ii) Isopropanol	
n) *Organo tin compounds*	
(i) Tributyltin naphtenate	
(ii) Tributyltin oxide	
o) *Other*	
(i) p-chloro-m-cresol	
(ii) Sodium fluoride	
(iii) Mercuric chloride	
(iv) Dichloro-diphenyl-trichloroethane (DDT)	
(v) Specific formula not available	Sphagnum moss

*Adapted from Dawson 1982:269–277 with additions from Brown 1974:469–473; Spriggs 1980:19–24; Dowman 1970; Carlsson 1980:165–204; Sturge 1975; Sjøgren 1978; National Institute for the Conservation of Cultural Property 1984b; Stambolov 1969; Rector 1975; Bayer Materialschutz 1979; Baynes-Cope 1975:31–33; BP Chemicals 1980; Brownstein 1982:279–285; Fogel 1985; Clydesdale 1982; Ganiaris et al. 1982:12–23; Mason 1981; Muhlethaler 1973 and 1979; Nagin and McCann 1982; Werner 1968:265–290.

Research

Several recent studies point out the need for research in archaeological leather conservation (NIC 1984a; LCN 1984; Jenssen 1983b). Only one paper makes specific mention of the unique treatments appropriate for archaeological skin materials. What follows is a list of needed information and practices:

1. The exchange of information regarding methods currently used in the field (these are rarely published) so that these can be appropriately tested and either recommended or proscribed
2. A determination of which properties, such as pH and moisture content, are critical indicators of stability, and recommendations regarding safe ranges for the various archaeological skin materials, based upon studies of archaeological material
3. The development of simple and accurate methods for measuring the properties mentioned above
4. An evaluation of biocides compatible for use with the diverse skin materials during storage both before and after conservation
5. An evaluation of chemicals used in cleaning and recommendations for chemically neutral cleaning systems
6. The study of archaeological leather previously freeze-dried from solutions of humectants, to evaluate the volatility of these materials and the effect of subsequent dehydration of treated leather over time
7. The development of examination techniques and reference systems to enable the materials of archaeological skin objects to be identified
8. An evaluation of the effects of conservation treatments on the results of potential analyses, such as shrinkage temperatures
9. Research on materials for use as consolidants in the field with the diverse archaeological skin materials

References

Archaeology Section, UKIC
1982-1984 *Archaeological Artifacts Conservation Guidelines* 1-3. London.

Bayer Materialschutz
1979 *Preventol Anwendungsgebiete*. Bayer AG, Leverkusen.

Baynes-Cope, A.D.
1975 "Fungicides and the preservation of waterlogged wood: Problems of the conservation of waterlogged wood." *Maritime Monographs and Reports* 16. London.

BP Chemicals
1980 Breox polyethylene glycols. *Data Sheet PEGDS* 1. Belgium.

Brown, M.K.
1974 "A preservative compound for archaeological materials." *American Antiquity* 39 (3).

Brownstein, A.
1982 "The chemistry of polyethylene glycol." *Proceedings of the ICOM Waterlogged Wood Working Group Conference*. Ottawa.

Calnan, C.N.
1976 *An Investigation into the Use of Chelating Agents in the Treatment of Iron Stains on Waterlogged Leather*. Unpublished manuscript, Queen's University. Canada.

Carlsson, M.
1980 Konservering af vådt arkæologisk læder—forskellige metoder og principper." *Konservering og Restaurering af Læder, Skind og Pergament*. Konservatorskolen, Copenhagen.

Collins, C.
1983 "Research, projects and techniques—The Manitoba Museum of Man and Nature." *Leather Conservation News* 2.

Clydesdale, A.
1982 *Chemicals in Conservation: A Guide to Possible Hazards and Safe Use*. Edinburgh.

Dawson, J.
1982 "Some considerations in choosing a biocide." *Proceedings of the ICOM Waterlogged Wood Working Group Conference*. Ottawa.

Dowman, E.A.
1970 *Conservation in Field Archaeology*. London.

Fenn, J.
1984 "Some practical aspects in the choice of synthetic resins for the repair of ethnographic skin and gut." *Adhesives and Consolidants*. London.

Fogel, S.
1985 *Recent Advances in Leather Conservation*. Washington, D.C.

Fogel, S.and T. Raphael
1984 "Leather conservation terminology." *Leather Conservation News* 4.

Fortidsminderådet
1981 *Arkæologisk Felthåndbog*. Museumstjenesten, Viborg.

Ganiaris, H., S. Keene, and K. Starling
1982 "A comparison of some treatments for excavated leather." *The Conservator* 6.

Geijer, A., H. Boström, and H. Axelson
1961 "Modocoll som prepareringsmedel för textiler." *Svenska Museer* 1.

Hansen, J.P.H., J. Meldgaard, and J. Nordqvist
1985 *Qilakitsog—De Grønlandske Mumier fra 1400-Tallet*. Grønlands Landsmuseum, Christian Ejlers' Forlag. Copenhagen.

Jackman, J.
1982 *Leather Conservation: A Current Survey*. The Leather Conservation Centre. London.

Jenssen, V.
1983a "Conservation of waterlogged organic materials: a review." *Banff Conference 1983*. Ottawa.
1983b "Water-degraded organic materials: skeletons in our closets?" *Museum* 137.

Joukowsky, M.
1980 *Field Conservation, A Complete Manual of Field Archaeology: Tools and Techniques of Fieldwork for Archaeologists*. Englewood Cliffs, New Jersey.

Leather Conservation News
1984 "Research, projects & techniques: immediate and long-term needs 4.

Leigh, D.
1972 *First Aid for Finds*. RESCUE 1. Southampton.

Logan, J.
1983 "Red Bay 1982—textile discovery." *Textile Conservation Newsletter*.

Logan, J. and M. Segal
1985 "A 16th century costume." *Textile Conservation Newsletter*.

Madsen, H. B.
1980 "Feltkonservering af læder." *Konservering og Restaurering af Læder Skind og Pergament*. Copenhagen.

Mason, J.
1981 *Report on Waterlogged Leather Experiment*. CCI Project 398. Unpublished manuscript. Ottawa.

Morris, K. and B.L. Seifert
1978 "Conservation of leather and textiles from the Defence." *Journal of the American Institute for Conservation* 18.

Muhlethaler, B.
1973 *Conservation of Waterlogged Wood and Wet Leather*. Paris.
1979 *Kleines Handbuch der Konserverungstechnik*. Bern.

Museum of London
n.d. *Post-Excavation Care and Storage of Finds*. London.

Nagin, D. and M. McCann
1982 *Thymol and o-Phenyl Phenol: Safe Work Practices*. New York.

National Institute for the Conservation of Cultural Property
1984a *Ethnographic and Archaeological Conservation in the United States*. Washington, D.C.
1984b *Proposed Priorities for Scientific Research in Support of Museum Conservation*. Washington, D.C.

Peacock, E.E.
1984 "Mass reconservation of archaeological leather: a case study. *ICOM Committee for Conservation, 7th Triennial Meeting, Copenhagen*. Paris.

Rector, W.K.
1975 "The treatment of waterlogged leather from Blackfriars, City of London." *Transactions of the Museums Assistants' Group* 12.

Reed, R.
1972 *Ancient Skins, Parchments and Leathers*. London.

Ringgaard, M.
1986 *Imprægneringsmidler Anvendt på Jordfundne Tekstiler*. Unpublished manuscript. Copenhagen.

Seese, C.
1984 "First aid treatment for excavated finds." *Conservation on Archaeological Excavations*. Rome.

Segal, M.
1983 "Treatment of the textiles from Red Bay, Labrador." *Textile Conservation Newsletter*.
1985 "Photographic print washers." *Textile Conservation Newsletter*.

Segal, M. and M. MacDonald
1984 "Current research into the conservation of frozen skins from the arctic." *ICOM Committee for Conservation, 7th Triennial Meeting, Copenhagen*. Paris.

Sjøgren, A.
1978 *Fordele og Ulemper ved at Bruge Fungicider på Lader*. Unpublished manuscript, Copenhagen.

Spriggs, J.
1980 "The recovery and storage of materials from waterlogged deposits at York." *The Conservator* 4.

Stambolov, T.
1969 *Manufacture, Deterioration and Preservation of Leather: A Literature Survey of Theoretical Aspects and Ancient Techniques*. ICOM Committee for Conservation. Amsterdam.

Sturge, T.
1975 *The Conservation of Wet Leather*. Unpublished manuscript, Institute of Archaeology. London.

Tilbrooke, D.R.W.
1978 "Removal of iron contaminants from wet leather using complexing agents in dipolar aromatic solvents." *ICOM Committee for Conservation, 5th Triennial Meeting, Zagreb*. Paris.

Tubb, K.W.
1985 "Preparation for field conservation in the Near East." *The Conservator* 9.

van Dienst, E.
1985 "Some remarks on the conservation of wet archaeological leather." *Studies in Conservation* 30.

van Soest, H.A.B., T. Stambolov, and P.B. Hallebeek
1984 "Conservation of leather." *Studies in Conservation* 29.

Werner, A.E.
1968 "The conservation of leather, wood, bone and ivory, and archival materials." *The Conservation of Cultural Property*. Paris.

Lars-Uno Johansson

BONE AND RELATED MATERIALS

Lars-Uno Johansson studied archaeology at the University of Gothenburg. He is presently Conservator/Curator at the Central Board of National Antiquities, the Technical Institute, Stockholm, Sweden where he has specialized in the conservation of organic materials from archaeological excavations.

Much like other organic materials, bone is highly perishable. It is therefore imperative that proper procedures be observed when bone objects are found in archaeological excavations. Since, to my knowledge, objects of bone are rarely preserved in situ, this paper will address the entire process of recovery through conservation. Materials related to bone, such as horn and ivory, are dealt with peripherally; their structural similarity suggests they can be treated like bone in the excavation process.

The most common finds of bone objects in archaeological digs are artifacts, refuse from the manufacture of objects, food waste, and skeletal material. My experience is primarily with artifacts and skeletal material.

Structure

Fresh bone consists of both organic and inorganic material. The percentage of different constituents in its chemical composition varies depending on the animal from which the bone derives, as well as on other factors, such as the sex and age of the organism and its location in the body (Chaplin 1971). The organic portion consists of a network of collagen fibers.[1] Bone also contains trace elements and insoluble organic substances that can act as catalysts or inhibitors of chemical reactions. The individual structure of each bone is important in determining how its permeability will be affected in the future. Bone has a hard, impervious exterior and a permeable, porous, and spongy interior.

1. Collagen is a protein surrounded by the principal inorganic constituent, the bone salt calcium hydroxyapatite $[Ca_{10}(PO_4)_6(OH)_2]$.

Decomposition

The degree of decomposition depends both on the state of the bone when it was buried and on the chemical composition of the burial environment. Bone remains from cooking, for example, can be very different, depending on whether they were grilled, boiled, etc. Bone can be highly decomposed as far as its organic portion is concerned: It can be brittle, porous, or twisted. Another factor to consider is that animals may have chewed on bones left on the ground, while others could have landed in a waste pit that was quickly covered over. The same variables apply to artifact material, which behaves differently depending on the method of manufacture and other factors. Once bone is buried in the earth, decomposition continues in other ways, reflecting the ambient environment.

Concerning conditions of preservation, it may generally be stated that bone is preserved well in alkaline water; alkaline soils; Precambrian lime and shellfish banks; dry sand where drying prevents bacterial decomposition; fine-grained soils, such as tightly compacted soils and ooze, where aerobic bacteria are unable to live; when burnt; and together with other bones, resulting in a locally elevated concentration of calcium. Bone is preserved poorly in permeable, airy soils and in acid soils.[2] Large bones are preserved better than small ones, since the area under attack is smaller in relation to volume.

2. In acid water, however, certain proteins—such as the keratin present in horn, hooves, and hair—can be preserved.

Different types of mechanical pressure on objects while they are lying in the ground also affect their state of preservation. Seasonal climatic changes (fluctuations of temperature, humidity, etc.) are also of some importance in determining how the object is broken down. The entire decomposition problem is thus a question of the condition of the object when it landed in the soil, the type of soil in which it has lain, and how it has been affected mechanically (by pressure, vegetation, climate, etc.) and chemically (a matter of the chemical and "structural" composition of the soil; Chaplin 1971).

How to proceed in the field

When, in the course of an archaeological excavation, one is faced with the sudden appearance of bone objects in the ground, the following questions should be asked in order to proceed correctly: What is the nature of the object? Does it need to be measured for osteological purposes? Is it a unique object? Is it to be dated by the C-14 method? Is it brittle? Is it stable? How should it be conserved, if at all?

The question of how to proceed must be considered in relation to the damage that can result from an incorrect choice of stabilizing method, as well as in relation to considerations of the final conservation/stabilization method that is employed. Does the object need to be stabilized prior to recovery? Moist or wet material can be stabilized by the application of a colloidal acrylic dispersion (Primal® WS-24, 2%) before it is recovered. A dry object can be stabilized with an acrylic resin (Paraloid® B-72) dissolved in toluene or acetone (5%-10%; Koob 1984).[3]

3. For the chemical composition of this and other proprietary materials, see the Appendix.

On a grave that was recovered in its entirety, we used cellulose lacquer with favorable results after having dried out the ground with acetate. All impregnation methods entail an alteration of the material that might distort later research and analysis, such as C-14 dating. It is therefore important to take samples for analysis prior to any impregnation treatment. Do not stabilize unnecessarily! If in doubt, refrain from stabilizing! Contact an expert! Details such as paint traces or decorative carving can also be affected by consolidants. Dry objects of ivory must not come into contact with water. No stabilizer may be used if the ivory is dry—contact a conservator (Plenderleith and Werner 1971).

Complications can arise when bone is mixed with or joined to other organic or inorganic materials (such as bone and wood, or bone and metal). The bone, for example, might be waterlogged and joined to iron, which ordinarily should be kept dry. Always try to preserve the state of the most sensitive material, which is usually the bone (Johansson 1983b:109–113; 1983c:104–108).

Sometimes it may be important for purely technical reasons that the bone object be recovered for examination in the laboratory together with some of the surrounding ground. Since it may sometimes be impossible to recover an object completely in the field, the work should instead be done in the laboratory. Large preparations may have to be recovered under special circumstances, such as when a unique object or assemblage is to be examined or exhibited together with its milieu. Small specimens can be recovered by melting paraffin wax over them. The best method, however, is to dig around the object and raise it by sliding a lifting plate under it. Another well-proved method is to wrap

4. These procedures are discussed further by Louise Bacon, this volume.

the specimen in plaster bandages before lifting.[4]

A few years ago, an entire double grave in Skateholm, in southern Sweden, was raised so that it could be displayed in a museum. The grave, from about 5000 B.C., lay in gravel. A box-shaped iron framework was buried around the grave and an iron plate was inserted into the bottom of it by means of jacks. The grave was then lifted up by a crane. Prior to this procedure, the grave and the skeletons were dried out with acetone and then stabilized with Zapon® lacquer (figs. 1–3). Another example is a grave in Osthammar in central Sweden, where the skeleton lay in a partially preserved wood casket. Both skeleton and casket were badly decomposed, moist but not waterlogged. The entire casket was exposed and deep-frozen with dry ice. It was then lifted up in a case for transport to the laboratory.

When dealing with such objects the following rules should be observed:

1. During transport, the material must be firmly anchored in its transport case or box. The case must be sturdy but not too large. It should be packed with styrofoam or paper.
2. It is important to keep the object in the same state as it was when recovered until it reaches the laboratory (that is, dry if it was recovered dry; moist, if it was recovered moist).
3. In the case of large objects, cover with plastic and seal tightly, if necessary. Small objects can be kept in double polyethylene bags.
4. Do not allow moist material to remain sealed too long at room temperature because there is the risk of mold formation if fungicide is not added. For long-term storage, refrigeration is necessary (4°-5°C).

During examination and conservation in the laboratory, the principles outlined above apply here as well:

1. It is important during examination to maintain the object in the state in which it was recovered. The laboratory provides better conditions than the field for accurate documentation because of controlled light, temperature, and humidity, as well as better technical resources.
2. Be prepared for surprises. In examining several skeleton graves from Nödinge in western Sweden, investigators found remains of stomach contents as well as large portions of a brain in a cranium (Johansson 1983d:98–103).
3. The conservation method must always be chosen on the basis of the condition and state of the material.
4. If the bone is wet but stable, dehydration with alcohol is often sufficient. If the bone is wet but fragile, there are a number of methods to choose from.

In conserving bone material from the Stone Age found in Tännäs in northern Sweden, we impregnated the smallest objects (up to about 10 to 15 cm long and 1 or 2 cm thick) in a 2% solution of Primal WS-24 for a day or so. The larger objects were impregnated with a mixture of PEG 400 (2%) and PEG 1000 (18%) in water for several months. Dry objects can be stabilized with Paraloid B-72 (5%-10% in toluene or acetone) or with polyvinyl acetate (2, 5, and 6).

The casket mentioned above that was recovered by means of deep-freezing, was impregnated while slightly moist by spraying with polyvinyl alcohol (the bottom was left untreated). After eight years, casket and skeleton are still in good shape.

We have also experimented with freeze-drying waterlogged bone objects. The results are mixed. Some objects have been preserved

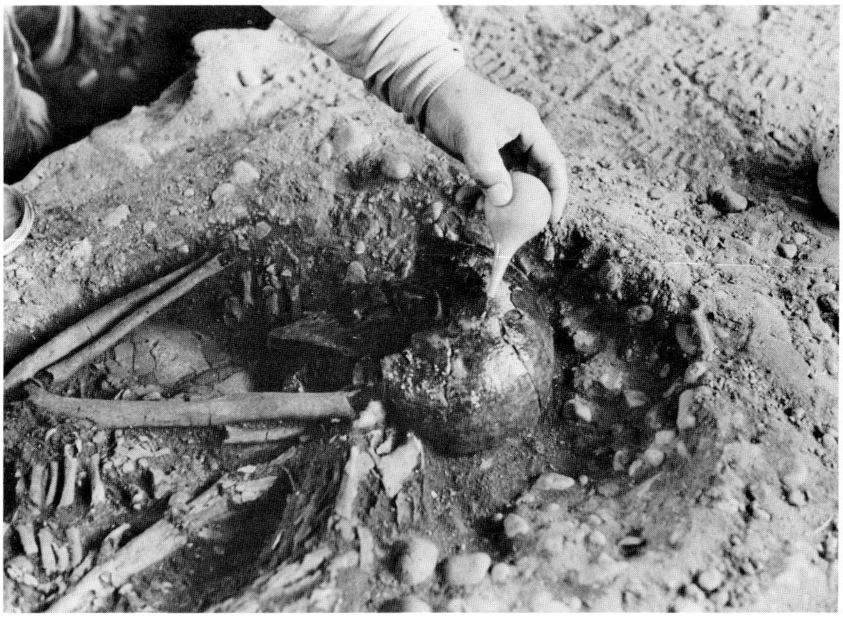

Figure 1. Stablization of the skeleton.

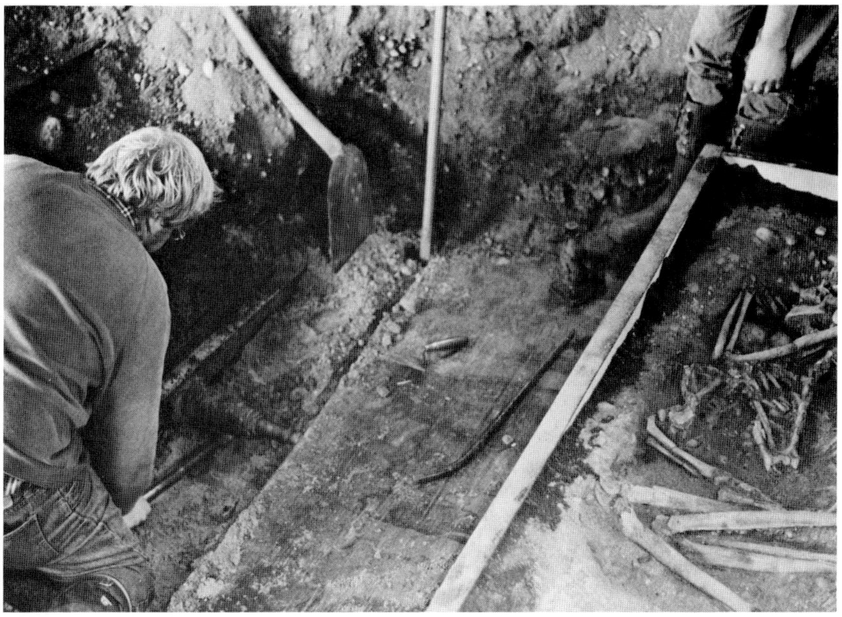

Figure 2. Inserting an iron plate under the grave prior to lifting.

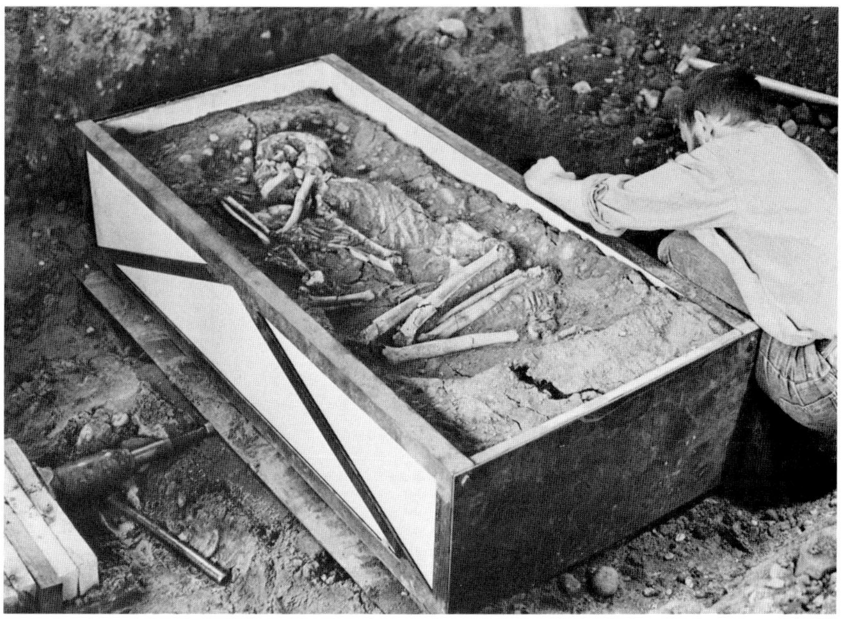

Figure 3. The lifting of the grave.

perfectly, while others have been ruined. This is probably due to the fact that the bone structure in certain objects is too dense and impermeable to water vapor in the sublimation phase and therefore ruptures during the drying process.

Summary

Bone, horn, and ivory are, like other organic materials, highly degradable. Because of the nature of their chemical composition, they are well preserved under certain environmental conditions and decompose quickly in others. The degree of preservation is higher in alkaline and chalky soils than in acidic, water-permeable soils. Artifacts made of bone are affected in much the same way as skeletal material.

In order to make it possible to lift an object in one piece, consolidation in situ is necessary in many cases. The method chosen must depend on the condition of the bone, its state of preservation, and on the subsequent treatment. Vital information might be destroyed if an incorrect method is used. The object's burial environment is clearly of great importance in determining the appropriate method.

There are also situations that require objects to be lifted together with the surrounding soil for further work in the laboratory, where conditions are more favorable for appropriate documentation and conservation. In such cases, it is important to maintain the original environmental condition.

After excavation and examination, conservation treatment is chosen according to the object's state of preservation. If the object is wet but stable, dehydration under controlled conditions is often sufficient. Brittle, waterlogged objects may have to be stabilized through the application of a bulking agent, such as a wax. Many dry objects do not need treatment by chemical means—a mechanical cleaning may suffice.

In this paper, I have presented the methods that have been employed during the last ten years at the Technical Institute.

Acknowledgments

I would like to thank Elisabet Iregren, University of Lund, Historical Museum, for valuable assistance with parts of this paper.

The grave recovery was carried out by personnel from the National Historical Museum and the Technical Institute under the leadership of conservator Björn Tägtström, University of Lund, Historical Museum.

References

Chaplin, R.E.
1971 *The Study of Animal Bones from Archaeological Sites*. London.

Johansson, L.U.
1983a "En benmejsel från Alvastra pålbyggnad" (A bone chisel from the Alvastra Pilworks). *Konserveringstekniska studier* 1. Stockholm.

1983b "Ett resesolur från Regalskeppet Stora Kronan" (A traveler's sundial from the royal warship Stora Kronan). *Konserveringstekniska studier* 1. Stockholm.

1983c "Ett resesolur funnet i Uppsala" (A traveler's sundial found in Uppsala). *Konserveringstekniska studier* 1. Stockholm.

1983d "Konservering av vattendränkt organiskt material (trä, läder, ben)" (Conservation of waterlogged organic material [wood, leather, bone]). *Konserveringstekniska studier* 1. Stockholm.

Koob, S.P.
1984 "The consolidation of archaeological bone." *Adhesives and Consolidants*. Paris.

Leigh, D.
1972 *First Aid for Finds*. 2nd ed. RESCUE. Hertford, England.

Plenderleith, H.J. and A.E. Werner
1971 *The Conservation of Antiquities and Works of Art*. 2nd ed. London.

Louise Bacon

THE CARE AND PROTECTION OF COPPER ALLOY, SILVER, AND GOLD OBJECTS ON SITE

Louise Bacon studied archaeological conservation at the Institute of Archaeology, University of London and is presently Senior Conservation Officer, Horniman Museum and Library, London. She has worked as a UNESCO consultant in Peru and the Dominican Republic, and as a British Council consultant in Seoul, South Korea. In 1973–74 she was Head of the Metals Section, Restoration Center, INAH, Mexico.

Depending on prevailing soil conditions, buried objects of copper alloy or silver are generally partially or totally corroded. High gold alloy objects survive with little change, but when alloyed with large quantities of copper or silver, gold is affected by the corrosion of those metals.

While it is not possible here to adequately address the corrosion process, it should be noted that only the collaboration of archaeologists, conservators, and corrosion scientists yields complete information on such factors as the humidity of the soil, its acidity, and the concentration of chloride ions. For instance, it is important to know the proportion of tin and copper ions leached out of bronze objects into the surrounding soil during burial. This type of information can help improve our understanding of the mechanisms of corrosion (Chase 1979: 73–109; and personal communication 1986).

While objects remain buried, the processes of decay slow down almost to a halt. But upon excavation, the environment surrounding the objects changes radically, and the processes of corrosion are activated. With copper alloy objects this process is usually evidenced by the appearance of bright green, powdery spots on the surface of the object, an effect often called "bronze disease." After excavation, copper alloy, silver, and gold objects ought to be removed from the ground as soon as possible and steps should be taken immediately to protect them from the ambient environment, as well as to ensure their security.

Lifting of objects

Usually, individual copper alloy, silver, and gold objects can be lifted safely, put into a tray for support, and taken to the processing area. In some cases—as with broken objects or when several pieces are found together in the ground—special lifting may be required (fig. 1). Before lifting, the object should be photographed with a scale, or it should be drawn in order to record information that may prove vital to later restoration.

Often an object may be lifted in a block of soil and the whole can be placed in a suitably sized box or tray, wedging the soil in tightly with packing material (fig. 2). If the soil is very friable, a surrounding wall of cardboard or wood, or a gauze bandage around the item, may be necessary in order to hold it together. A board may be slid underneath the object before it is lifted and placed in a box or on a tray.

Two other methods have been used to lift metal objects: One involves the surface application of polyvinyl acetate emulsion and a bandage; the second uses plaster of Paris and bandages as supporting

1. Soil blocks
    - Gauze bandages
    - Strips of cardboard, thin wood, or metal sheet
    - Boards or metal sheets to slide beneath blocks
    - Plastic trays or boxes for transporting blocks
    - Scissors, saws, knives, and brushes
2. Polyvinyl acetate emulsion and bandage blocks
    - PVAc emulsion
    - Containers for diluting the emulsion
    - Water for dilution and for washing brushes
    - All the materials needed for soil blocks (see above)
3. Plaster of Paris and bandage blocks
    - Plaster of Paris in airtight containers
    - Bowls for mixing plaster
    - Water to mix plaster and wash tools
    - Spatulas
    - Aluminum or plastic barrier film
    - All the materials needed for soil blocks (see above)

Figure 1 (left). Materials and equipment for lifting fragile metal artifacts.

Figure 2. Block-lifting fragile metal objects.

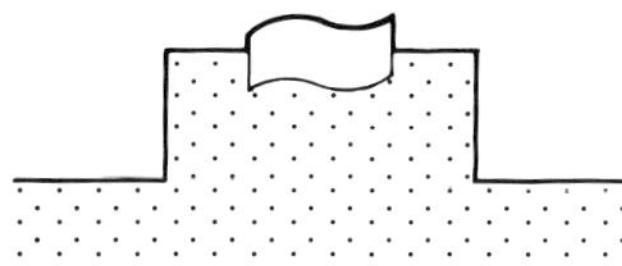

a. The fragile metal object is isolated on a block of soil.

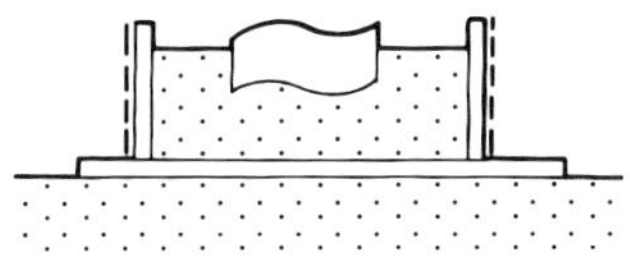

b. Soil block lift. The block is surrounded by boards and bound with gauze bandages. The lifting sheet is passed under the block.

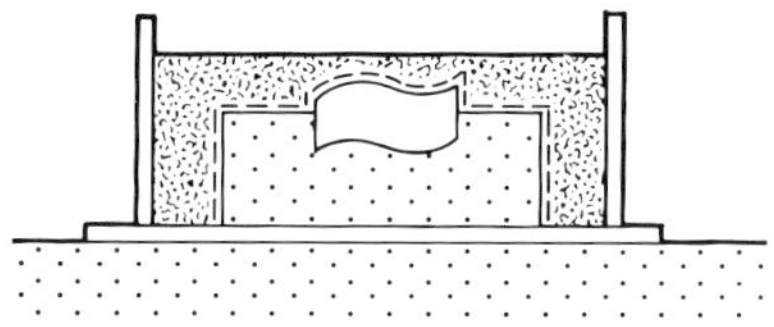

c. Plaster of Paris block lift. The plaster is separated from the object by foil or plastic sheeting.

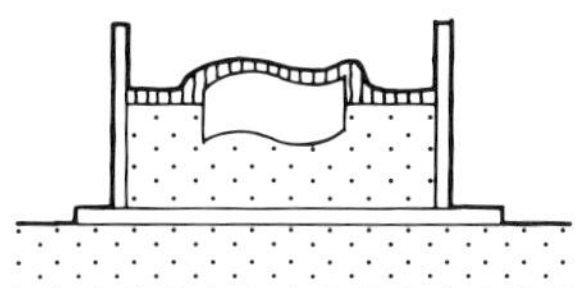

d. PVA and bandage lift. PVA emulsion and gauze bandages are applied to the surface of the object and the block.

materials (Leigh 1978; Stanley Price 1984). It should be noted that these two methods are rarely used, and then only when conservation support is assured. Objects that have been lifted with PVAc (polyvinyl acetate) emulsion and bandages must not be kept too long before they are handed over to a conservator, as the support can shrink, causing serious damage to the object. Nor should objects lifted with plaster of Paris and bandages be kept for any length of time, as moisture may be trapped between the object and the barrier film, causing deterioration.

With either method, excess soil is gently brushed off first to expose as much of the object as possible without endangering it. No attempt should be made to remove any soil that may be clinging to the metal.

If PVAc emulsion is used, it is painted directly onto the metal, and then strips of bandage are applied to cover the object and surrounding soil. Next, more PVAc is brushed onto the bandage, which is then pressed gently but firmly around the object and soil, covering both completely.

A similar technique is used to apply plaster of Paris and bandages; the difference is that the mixture is not applied directly to the metal. A thick barrier of polyethylene sheeting, metal foil, or sandwich wrap (sometimes called "cling film") is first laid over the object to act as a release agent between the object and the plaster. Then the plaster of Paris and bandages are applied, covering the object and the sides of the block of soil to be lifted. A wall of cardboard or wood is usually necessary to contain the plaster of Paris around the block.

After consolidation with either of these methods, and once it is certain that the block is fully set, a support board can be slid underneath the object, enabling it to be lifted. The block should be taken immediately to a conservator.

Packaging

All excavated metal objects should be stored in an environment that ensures the prevention of further corrosion and damage. In particular, it is important to control humidity levels (Watkinson 1987).

Acid-free tissue paper and certain plastic foams (such as polyethylene foam) are safe packaging materials (fig. 3). Cotton wool, newspaper, cheap foams, colored tissue, and brown paper should be avoided. Cotton wool tends to catch on projections, which may be broken off when the artifact is unwrapped. Inexpensive foams and paper products frequently emit harmful acidic vapors that may cause rapid corrosion. The object should be placed on a safe bed of foam or tissue; a similar layer should cover it. Objects should never be wrapped around with tissue paper or foam like Egyptian mummies; this can cause great difficulty and damage when the object is being unwrapped. Bags and boxes that fit the size of the object, inhibit its movement during transit, thus resulting in the least damage.

Figure 3. Packing materials for metal artifacts.

- Acid-free tissue paper
- Polyethylene foam
- Polyethylene bags (self-sealing or with ties)
- Polyethylene sheeting
- Polyethylene seal-tight boxes (or cake or biscuit tins)
- Adhesive tape
- Self-indicating silica gel (or plain silica gel)
- Relative humidity indicator cards
- Cotton bags for silica gel
- Waterproof labels and markers
- If available, small transparent polystyrene boxes with hinged lids
- Scissors

Containers

Copper alloy objects that are obviously not fragile can be transported in polyethylene bags, preferably the self-sealing variety that cannot open accidentally. Nonsealable polyethylene bags require some means, such as ties, to keep them tightly closed. Polyethylene bags must be perforated to allow for the circulation of air, or they must contain a desiccating agent to prevent condensation, which would cause further deterioration of the object.

Delicate objects made of copper alloy (and all silver and gold objects) should be packed in individual transparent polystyrene boxes, preferably with hinged lids, which allow clear and easy viewing of the objects inside.

These small bags and boxes should then be placed in a large polyethylene box, such as the type of container used for storing sandwiches and cakes, with a lid that can form an airtight seal.

If airtight boxes are not available, small packages can be placed in an acid-free cardboard box, which may be wrapped in many layers of polyethylene sheeting and sealed with tape. Alternatively, a biscuit tin, well-taped around the lid, may serve equally well. The area where objects are stored should not be damp. But if this is unavoidable, a constant watch must be kept on cardboard boxes, which can absorb moisture and become unsafe, and on biscuit tins, which may corrode.

Silver objects need to be protected not only against physical damage during transportation to conservation facilities, but also against sulfur pollution. Sealing them in plastic bags or boxes as described helps to achieve this.

Prior to transportation, all voids in large polyethylene boxes, and between small packages, should be filled with packing material to prevent excessive vibration in transit.

Moisture control

Copper alloy objects require a dry environment with a relative humidity of 40 percent or less. Since silver and gold objects are usually alloyed with copper, they, too, require the same conscientious storage.

A perforated bag of self-indicating silica gel should be placed in the polyethylene box, along with a relative humidity indicator card. For a large box (e.g., 30 x 30 x 15 cm), 300 to 500 grams of silica gel may be adequate, although very humid conditions may require much more. Loose-weave cotton bags can be used to contain the silica gel, but it should never be sprinkled loose into a container, nor should it come into contact with the surface of an object.

The relative humidity indicator cards are small rectangles with strips or spots of cobalt-impregnated paper, calibrated to indicate a wide range of humidities. They permit a quick and easy assessment of relative humidity, and they are visible through polyethylene. These cards should be inspected at regular intervals. Any rise in relative humidity above 40 percent should be dealt with immediately, by either increasing or replacing the silica gel in the box. (Self-indicating silica gel is blue when dry and pink when moist. It can be dried out in a domestic oven set at 110°C and reused.) The bags of silica gel in sealed polyethylene boxes should be replaced quickly, without completely lifting off the lid.

These general rules do not apply to copper alloy objects with enamel inlay decoration. Keeping such objects in a very dry environment may cause the enamel to deteriorate. They should be packed in perforated polyethylene bags or transparent polystyrene boxes and immediately taken to a conservator.

Conservation

The preceding text emphasizes the need to consider the conservation of objects when planning an excavation. Suitable containers and packaging materials must be provided to ensure the protection of metal objects. It is rarely the case, even if a conservator is present, that treatment on copper alloy, silver, and gold objects can be carried out on the site. Facilities in the field are never ideal; often there is no electricity or water. Indiscriminate cleaning of the surface of objects can remove much technological evidence. Most excavators and field assistants are aware of the inadvisability of cleaning objects of this type without proper means. Untrained and unsupervised staff, however, may do great damage.

What follows is a list of some of the categories of information that can be destroyed by injudicious handling and cleaning of metal objects in the field:

1. Copper objects often have the mineralized remains of wood, bone, or textiles adhering to their surfaces. Some information about these materials can nearly always be obtained, providing that they are not damaged. These remains are very fragile and can easily be destroyed by cleaning in the field.
2. Heavily fractured metal objects should never be lifted piecemeal. It may be difficult to reestablish the relationship of the fragments at a later date, leaving the conservator with a tedious, if not insoluble, jigsaw puzzle.
3. It sometimes happens that, beneath the outer crust of corrosion on a bronze object, a replica of the original surface of the metal lies buried. This surface may carry many kinds of information, including inscriptions, decoration, or tool marks. Removing the outer crust is a difficult and painstaking job, and even when the crust appears to be flaking off on its own accord, the replica surface should not be picked at, as it is often soft and easily damaged.

4. In the past, it has often been the practice in some areas to clean copper and silver coins on site—usually to establish the date of a stratum—using acid solutions of one kind or another. As mentioned above, the legend is often more clearly seen in the layers of corrosion on the surface of the metal, and acids may well dissolve the pertinent evidence.
5. Green products of corrosion do not necessarily mean that the underlying metal is copper or one of its alloys—it may be silver or gold. Attempts to remove corrosion by chemical or mechanical means may disintegrate a silver or gold object (Scott 1983).
6. Silver is usually very soft. Cleaning objects made of this metal with inappropriate tools, such as toothbrushes, will undoubtedly scratch the surface.
7. Enamels may be so heavily decayed that they appear to be little more than a fine powder. Even so, that powder will contain some of the elements used in the manufacture of the enamel and these may be detectable by chemical analysis; but if the powder is carelessly removed, even that possibility is lost.
8. Inlays of one metal over another are often only loosely held in place by the products of corrosion; they are easily dislodged by improper cleaning.

Unfortunately, this list is not hypothetical; examples of errors in judgment in all of these cases are common, and all could be avoided. However, if objects are lifted and packed in the appropriate manner, as described here, they can be preserved for some months until laboratory facilities are available for their treatment.

Once in a laboratory, the objects may be radiographed and cleaned under magnification with tools or appropriate chemical means. Organic matter found in the layers of corrosion can be examined, technological details can be identified, and metals can be analyzed. By these means, a great range of information about the occupants of a site may be obtained.

References

Bacon, L. and K. Janposri
1977 "The Thai bronze project." *Studies in Conservation* 22 (4). London.

Chase, W.T.
1975 *Bronze Disease and its Treatment*. Bangkok National Museum, Department of Fine Arts, Bangkok, Thailand.

1979 "Solid samples from metallic antiquities and their examination." *International Symposium on the Conservation and Restoration of Cultural Property: Cultural Property and Analytic Chemistry: November 27–30, 1978, Tokyo and Tsukuba, Japan*. National Research Institute of Cultural Properties. Tokyo.

Dowman, E.
1970 *Conservation in Field Archaeology*. London.

Leigh, D. (ed.)
1978 *First Aid For Finds*. RESCUE 1. Hertford, England.

Leigh, D.
1982 "The selection, conservation and storage of archaeological finds." *Museums Journal* 82 (2). London.

Madsen, H.B.
1967 "A preliminary note on the use of BTA for stabilizing bronze objects." *Studies in Conservation* 12. London.

1971 "Further remarks on the use of BTA for stabilizing bronze objects." *Studies in Conservation* 16. London.

Scott, D.A.
1983 "The deterioration of gold alloys and some aspects of their conservation." *Studies in Conservation* 28 (4). London.

Sease, C.
1978 "Benzotriazole: A review for conservators." *Studies in Conservation* 23. London.

Stanley Price, N.P. (ed.)
1984 *Conservation of Archaeological Excavations with Particular Reference to the Mediterranean Area*. ICCROM. Rome.

United Kingdom Institute for Conservation
1984 *Environmental Standards for the Permanent Storage of Excavated Material from Archaeological Sites*. Archaeology Section. Conservation Guidelines.

1983 *Packaging and Storage of Freshly-excavated Artifacts from Archaeological Sites*. Archaeology Section. Conservation Guidelines 2.

Watkinson, D. (ed.)
1987 *First Aid For Finds*. RESCUE 15A. Hertford, England. In press.

White, A.J. and Partington-Omar (eds.)
1981 *Archaeological Storage*. Society of Museum Archaeologists and Yorkshire and Humberside Federation of Museums and Art Galleries, England.

Acknowledgments

I would like to thank Janey Cronyn and Martin Read who both read this manuscript in draft and gave much good advice and made many useful suggestions. I would also like to thank Miss Louise Millard, County Museum Officer, who gave me permission to work on this paper while I was employed by the Kent County Museum Service, England.

Henry W.M. Hodges

THE CONSERVATION TREATMENT OF CERAMICS IN THE FIELD

Henry W.M. Hodges is presently Professor of Artifact and Archaeological Conservation at Queen's University, Canada and was previously Lecturer in Archaeological Technology at the Institute of Archaeology, University of London. The author of standard works on the history of technology, he is a Council member of the International Institute for Conservation.

A review of *Art and Archaeology Technical Abstracts (AATA)* over the past fifteen years shows that very little has been published on the subject of the preservation of ceramic artifacts, and nothing of any great moment has been said about their treatment in situ or during archaeological excavation. This silence might be taken to imply that no outstanding problems exist—that this is the one area in which conservators have devised a methodology that is entirely satisfactory. Or the silence could be interpreted to mean that the preservation of ceramic materials is not one of our most pressing problems, that we have by trial and error arrived at apparent solutions, and thus it is not absolutely essential to conduct further research on this subject. The second of these interpretations seems to lie nearer to the truth. It is, therefore, my intention to examine briefly some of the ills to which pottery is normally heir and to determine to what extent our solutions are fully understood. The findings, as I hope to show, have implications far beyond the narrow scope of this paper.

The removal of water-soluble salts from ceramics

Without dwelling upon the disastrous results that may accrue from a failure to remove salts from infested pottery, it can be said that two major systems are currently in use for eliminating salts from ceramics: The pottery may be soaked in water to dissolve, and thereby remove, the salts; or, where the pottery is too friable to allow this, wet poultices may be applied to the surfaces to draw out the contaminants. The first of these methods is widely used, but when one inquires more deeply to discover what system of soaking is employed, how long the soaking must take place, and how effective is the process, one finds no answers that are supported by data—merely the supposition that each process advocated appears to give satisfactory results.

The diversity of methods employed and the total lack of data to support their use attracted the attention of one of our students at Queen's University, who decided to make this the topic of a research project (Last 1979). Beginning with porous test sherds of modern origin, he impregnated them thoroughly and uniformly with sodium chloride. He then proceeded to wash out the salt using various techniques, measuring the rate at which the salt was removed and continuing until the salt was eliminated (fig. 1). A sufficiently large sample of sherds had to be used to ensure the validity of his results, and this made the project far more time-consuming than my brief description might suggest.

Looking at these results, one must bear in mind that the figures are not absolute; that is, while the sherds used in this project were free of salt after five hours of cold ultrasonic cleaning, this does not apply to all pottery. Nevertheless, the experiment does indicate the relative efficacy of the methods tested. The most common and time-honored methods (Nos. 6, 7, and 8) are least efficient; the rate of leeching is slow and hence an extravagant amount of water is needed, which often can be ill-afforded.

While ultrasonic cleaning (Nos. 1 and 2) seems to offer great advantages, its practicality is questionable. Many archaeological ceramics are certainly not strong enough to withstand such treatment, and the use of this equipment demands a supply of electricity not always available in the field.

Certainly, the most unexpected result was achieved using changes of water with agitation and heating (No. 3). I say unexpected because the solubility of sodium chloride is not greatly enhanced by higher temperatures. One suspects that thermal currents and the expansion and contraction of the ceramics during the hourly changes of water must account for the differences between this treatment and that carried out in ambient temperatures (No. 6). What can be said with some certainty is that method Number 6 will take roughly four times as long as method Number 3, that it will require four times the quantity of water, and that the use of the more efficient process could provide important economies in time and water in the field.

Method of washing	Time needed to remove all salt	
1. Ultrasonic cleaning at 60° C	4.0	hours
2. Ultrasonic cleaning; ambient temperatures	5.0	hours
3. Changes of water with agitation and heat (60° C)	6.5	hours
4. Electroionophoresis at 20 volts	8.0	hours
5. Water agitated and circulated through a deionizing resin	19.5	hours
6. Changes of water with agitation; ambient temperatures	23.0	hours
7. Soaking in running water; ambient temperatures	23.0	hours
8. Soaking in static water; ambient temperatures	30.0+	hours

Figure 1. Comparison of methods used for removal of soluble salts.

I am unaware of any comparable studies that have been made on the use of poultices—a deficiency we hope to correct in the near future. The process is very simple to describe: A compact of wet, porous material is applied to the surface of the pottery; the water in the poultice dissolves the salt; and as the compact dries, the salts in solution are drawn into it and so removed. There is, however, a long list of factors about which we know very little: (1) what poultice material to use (kaolinite, fuller's earth, cotton linters, paper, pulp, and many others have all been suggested); (2) the optimum moisture content; (3) whether the pottery should be presoaked in water before applying the poultice; (4) the optimum rate of drying; and (5) the effects of heating the poultice. Considering the frequency with which poultices are used for salt removal, we seem to have shown remarkably little curiosity about what we are doing.

The impregnation of friable pottery

We probably all agree that some pottery is so friable when excavated that it must be consolidated if it is to survive. We need not concern ourselves here with the actual selection of a suitable consolidant; a far greater concern is how to get any consolidant into the pottery where it can do the most good.

The three traditional methods of applying consolidants in solution—painting, soaking, and soaking under reduced air pressure—were examined by Fox (1978). Carefully prepared friable sherds of red earthenware were impregnated and then tested to determine to what degree their strength improved and the extent to which the consolidant penetrated the bodies of the ware. No matter what strength solution was employed, it was found that painting and soaking gave very inadequate results. The consolidant seldom penetrated more than a few millimeters. By contrast, total impregnation was easily achieved when the ware was soaked under reduced air pressure.

A fourth method of impregnation was also tested—suspending the sherd above the impregnating solution so that only the bottom edge of the sherd was in contact with the liquid. Some of the solution was drawn up into the sherd. When the capillary rise stopped, the sherd was lowered into the solution to that level, enabling it to absorb even more liquid. The process was continued until the sherd was fully immersed. The results achieved in this way were by no means inferior to those achieved by soaking the sherd under reduced air pressure. One cannot avoid the conclusion that impregnation by capillary rise might well be used in the field, thereby eliminating the need to provide a vacuum pump and a chamber capable of withstanding reduced air pressure.

Lime-encrusted pottery

Normally the removal of lime or calcite incrustations from the surface of pottery presents few problems. One simply uses a few drops of one's favorite acid (as often as not hydrochloric acid), washes the pottery thoroughly, and all is well. However, when the body of the ware in question contains a filler of calcite, chalk, or shell, the removal of lime incrustations is not so simple a matter. This was the problem that Leskard (1981) encountered in the field, and she elected to make a study of the phenomenon as a research project. Her methodology was quite simple: Calcite-filled earthenware sherds were prepared, a number of different treatments were applied, and the sherds were sectioned and examined under the microscope for possible damage.

In the literature, two different approaches were advocated, neither of which seemed to have been given sufficient scrutiny to justify much confidence. The first suggested attacking the offending incrustation with a fairly concentrated acid, followed immediately by washing. The other insisted that the sherds should be soaked in water prior to treatment—the idea being that should any acid penetrate the body, it would be so diluted that its effect on the inclusions of lime would be minimal. In either case the sherds should be given a prolonged and thorough washing afterward to remove the products of solution. Of these two methods, the latter (the one calling for a prior soaking) proved to do less damage to the lime inclusions, although that is not to say that the body of the wares was unaffected. In the untreated control samples, the lime particles were totally surrounded by clay matter. In the treated samples, a considerable gap had developed between the clay matter and the lime particles—a gap that seemed excessive if accounted for only by the solution of the particles. Indeed, it seemed an inescapable conclusion that some of the clay matter had been removed. What had happened became painfully clear when the prepared sections were examined under a microscope during acid treatment. A small amount of lime will generate a great deal of carbon dioxide, and the violent effervescence was quite literally hammering into sludge the clay matter that surrounded the lime fragments. Indeed, there were far greater losses due to the evolution of carbon dioxide than could be attributed to the slight solution of calcite particles.

It now becomes clear why one must have reservations about the use of acids to remove calcite incrustations from the surface of normal ceramics—those in which the body contains no lime filler. If lime has entered the external pores of the ware during burial (as, indeed, is probable), the use of acids to remove the incrustation may very well disrupt the surface, a risk one should endeavor to avoid. This was a problem anticipated by Plenderleith and Werner (1971:335). Clearly, a process that does not involve the evolution of carbon dioxide is preferable.

A number of alternatives for the removal of lime incrustations are possible. Calgon® (sodium hexametaphosphate) is one, but because this reagent is known to affect glazes adversely, its use would have to be limited to low-fired, unglazed ceramics. Zeolite® (sodium aluminium silicate), the active ingredient in some water-softening beads, applied in a poultice is another possibility, and cationic deionizing resins, used in the same way, would probably be even more effective. In a somewhat similar situation, Podany of the J. Paul Getty Museum has had considerable success removing lime incrustations from marble using deionizing resins (personal communication 1986).

In brief, this is yet another area where we have little comparative data upon which to work. Until more information becomes available, it is impossible to advise on a suitable methodology for lime removal without running the risk of damaging the underlying ceramics.

Stained pottery

For conservators working with archaeologists whose interests lie in more recent periods, the staining of white earthenwares during burial can present real problems. The same is true, of course, of white porous stones such as marble. Usually the stains have entered pores or cracks in the material, and their removal is seldom easy. Because the exact nature of the stain is uncertain, the situation is often made even more complex. Broadly speaking, however, one can presume that the majority of these stains derive from either organic matter (which may include food residues absorbed during use) or ferruginous materials in the burial environment. In some instances these stains emanate from both sources, as when ceramics absorb iron compounds as well as tannic acids from a burial deposit. If one were dealing only with organic stains the problem would hardly exist, since any oxidizing agent, such as hydrogen peroxide or sodium hypochlorite, would suffice to remove them. But some iron staining is almost invariably present with organic matter, and such stains cannot be removed by an oxidizing agent alone. Indeed, the latter reagent could make removal even more difficult, since any iron present could be oxidized to the ferric form.

As it happens, iron is far more readily removed in the ferrous form because such compounds are generally more reactive and more readily soluble in water than equivalent compounds in the ferric state. This being so, one essential component when removing iron stains appears to be a reducing agent. For this purpose sodium dithionite (also known as sodium hydrosulfite) or even sodium hyposulfite, appears to be ideal. It is in fact the active component of dye-strippers and some commercial mixtures designed to clean toilet appliances. For the removal of iron compounds, once they have been reduced, a range of sequestering agents could be employed. But the research done by Rabin (1983) and Prytulak (1985) has shown that trisodium phosphate, a major ingredient of many commercial water softeners and paint-cleaning mixtures, is the most effective reagent, although sodium citrate comes in a close second. In either case, it is important to note that the solutions should be freshly prepared immediately before use; once in solution, sodium dithionite ceases to be reactive after about twelve hours. In practice, the most satisfactory method of application is to soak either very heavy filter paper or white blotting paper in the solution, immediately place the paper on the surface to be cleaned, and then cover the paper with polyethylene film to prevent rapid evaporation. After twelve hours, the film should be removed and the paper allowed to air-dry. It should be added, parenthetically, that the same method has been used satisfactorily to remove iron stains (including dried blood) from textiles and paper objects, in which cases the operative pH of approximately 7.0 may be seen as a great advantage.

This very brief review suggests that we still have a great deal to learn about the conservation of ceramics, despite the fact that we may deem the opposite to be true. It is high time, I believe, that as conservators we abandon the empirical approach to conservation methods and settle down to the more rational, and often more tedious, business of experimentation and critical appraisal. If this is true of the conservation of ceramics, an area in which we already feel fairly confident, is it not, perhaps, equally true of all areas of our endeavors?

References

Fox, Louise
1978 "A Preliminary Study of the Impregnation of Ceramics." M.A.C. Project Report. Queen's University, Kingston, Ontario. Unpublished manuscript.

Last, Joseph H.
1979 "The Desalinization of Ceramics." M.A.C. Project Report. Queen's University, Kingston, Ontario. Unpublished manuscript.

Leskard, Marta
1981 "The Effects of Hydrochloric Acid on Archaeological Ceramics." M.A.C. Project Report. Queen's University, Kingston, Ontario. Unpublished manuscript.

Plenderleith, H. and A.E. Werner
1971 *The Conservation of Antiquities and Works of Art.* London.

Prytulak, G.
1985 "Optimization of Iron Stain Removal Using Sodium Dithionite and Sodium Tripolyphosphate." M.A.C. Project Report. Queen's University, Kingston, Ontario. Unpublished manuscript.

Rabin, Paul
1983 "A Study of Iron Stain Removal from Marble." M.A.C. Project Report. Queen's University, Kingston, Ontario. Unpublished manuscript.

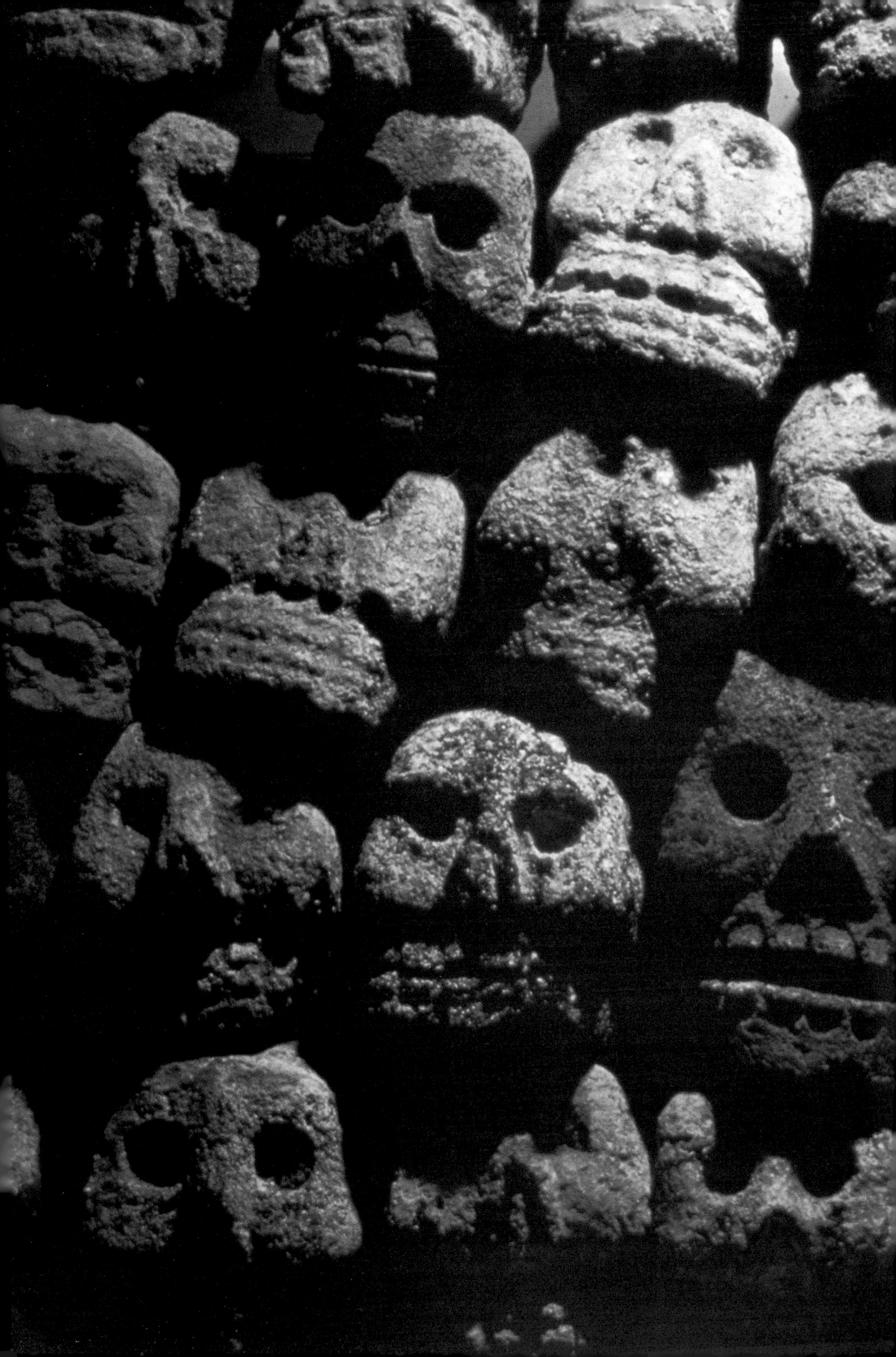

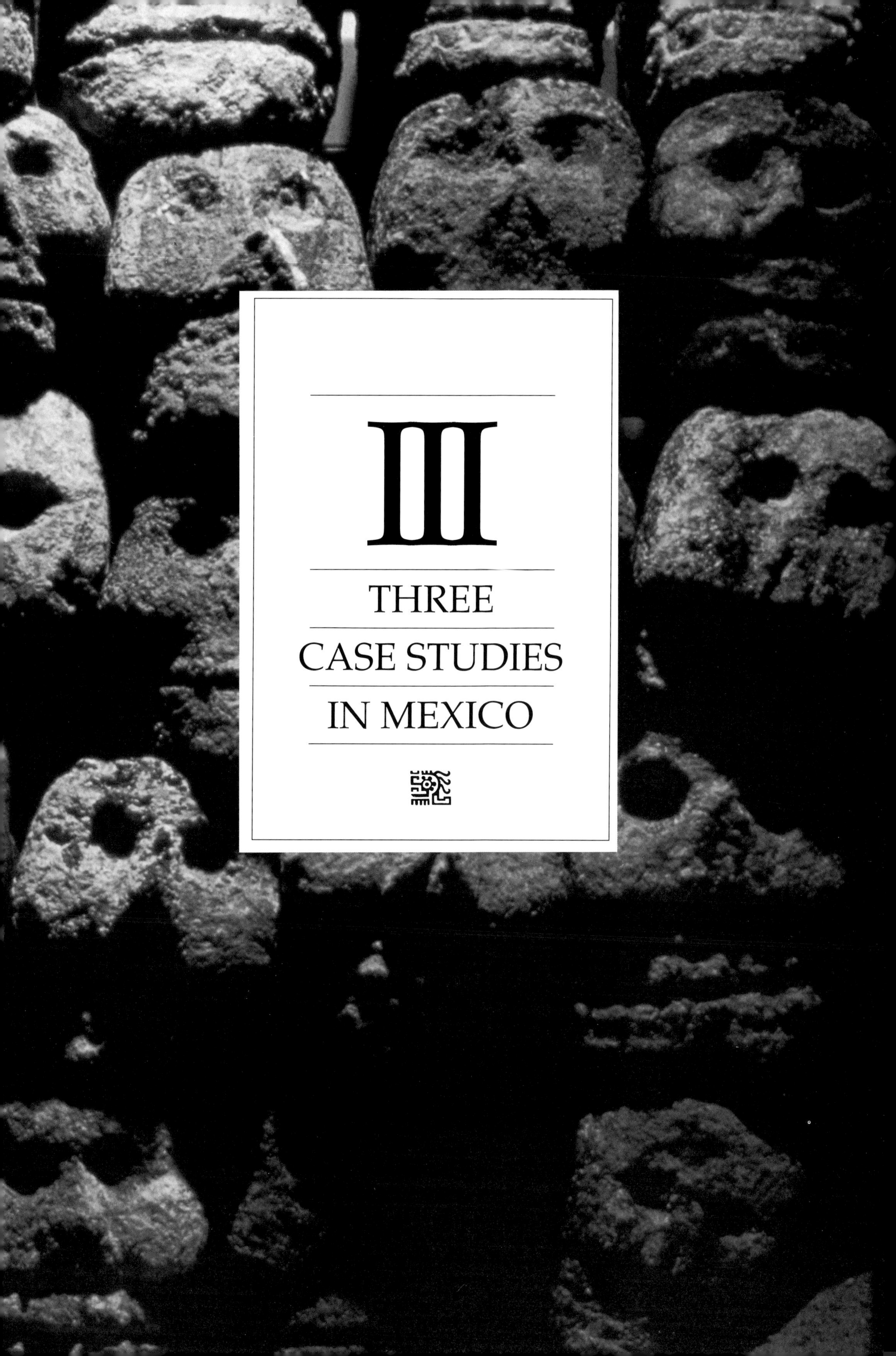

III

THREE CASE STUDIES IN MEXICO

THREE CASE STUDIES IN MEXICO

Three archaeological sites—the Templo Mayor, Teotihuacan, and Cacaxtla—were chosen for visits by delegates to the conference. The choice was a particularly fortunate one, for it allowed the visitors to see for themselves in situ conservation as it is practiced in Mexico today, and also, in the case of Teotihuacan, to have a glimpse of conservation and restoration methods in use over the past century.

Before each visit, members of the conference attended introductory lectures explaining the salient features of in situ conservation to be seen at each site. One of the many advantages of these talks was that they were presented by those who were actually responsible for the work. They thus provided both a background and a discussion of the objectives that resulted in the measures that were adopted. These lectures were, indeed, a most valuable contribution to the conference, and are therefore published in the following pages.

The Templo Mayor is situated in the very heart of Mexico City, immediately to the east of the Catholic Metropolitan Cathedral. The temple pyramid and other associated structures were razed shortly after the Hispanic conquest, and the area remained built over until quite recently, when the site became available for archaeological excavations. These began in 1978, and the remains of the temple are now on display to the public.

Teotihuacan, an ancient religious and administrative city, lies 48 kilometers northeast of the center of Mexico City. The site occupies a broad valley crossed by the Juan River, and is largely encompassed by mountains. Its great size—it occupies roughly 18 square kilometers—makes it the largest pre-Columbian city in the Americas. Once believed to be of Toltec foundations (ca. A.D. 600–1200), it is now recognized that the foundations belong to a far older civilization dating from circa 200 B.C., and is thus Teotihuacan. Moreover, the Teotihuacan architecture clearly influenced that of later conquerors—Toltec, Maya, and Aztec—as did its sculpture and other decorative features.

Cacaxtla, the site of a considerable pre-Hispanic settlement, lies about 70 kilometers due east of the center of Mexico City, close to the town of San Miguel, although the traveling distance is far greater, since the road perforce winds through mountains for much of the way. A large part of the site has already been exposed and conserved, and is available to the public. Excavations at the site continue.

Opposite: Sunrise over the Pyramid of the Sun, Teotihuacan.

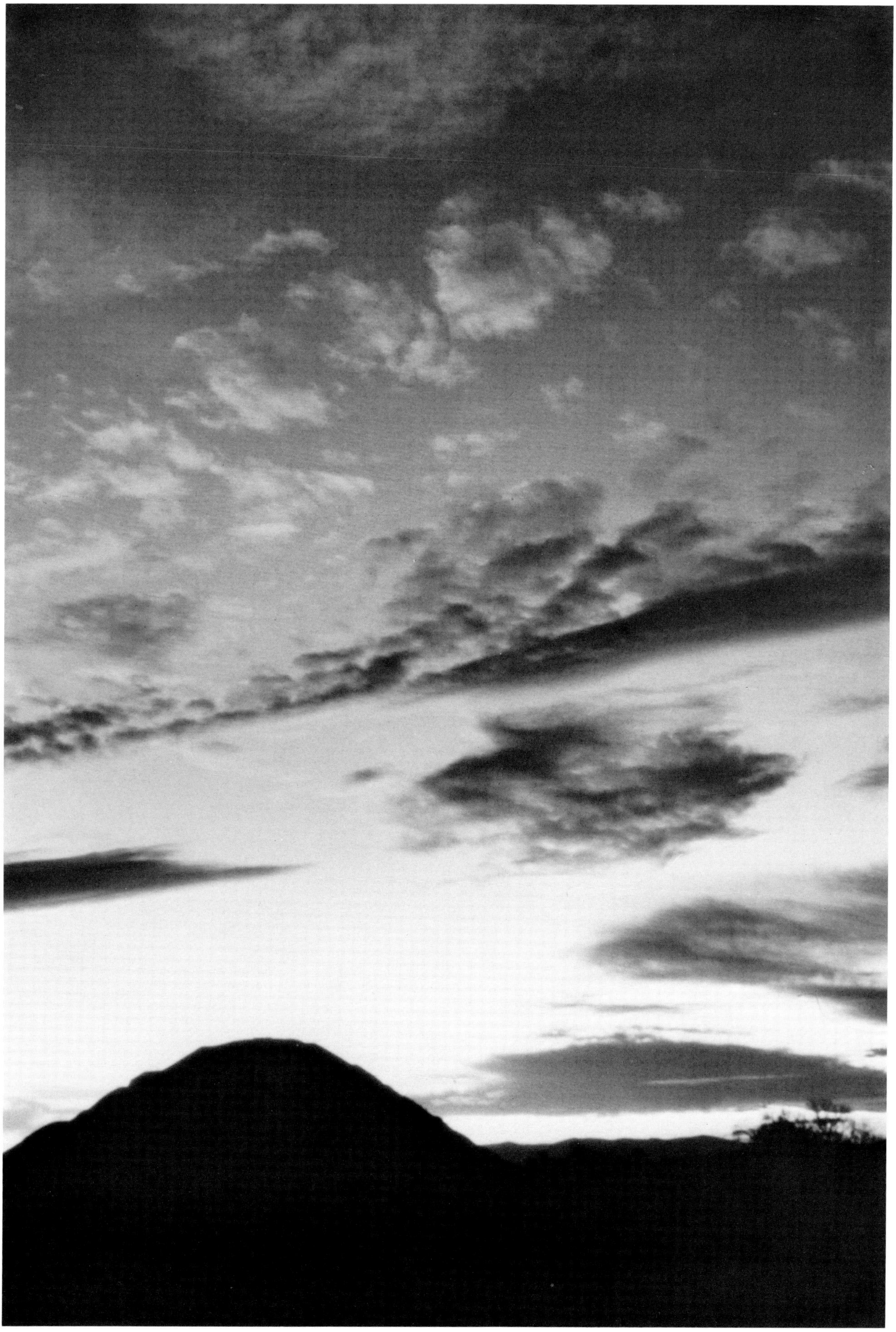

Eduardo Matos Moctezuma

ARCHAEOLOGY AND CONSERVATION IN THE GREAT TEMPLE OF TENOCHTITLAN

Eduardo Matos Moctezuma, Director General of CIESAS, Mexico, directed the work at the Templo Mayor. He was Director of Prehispanic Monuments, President of the Archaeological Council, and Director of the National School of Anthropology.

On the evening of 21 February 1978, workers from the Light and Power Company were digging at the corner of Guatemala and Argentina streets, in the heart of Mexico City. After breaking through the thick concrete pavement of the street and penetrating approximately 2 meters below street level, they encountered a hard stone that blocked their way. After removing the mud on its surface, they saw that the stone was covered with a series of reliefs. They stopped working and the next day called the office of Archaeological Preservation of the Instituto Nacional de Antropología e Historia (INAH). A team of archaeologists came to investigate. It took two days to verify that what they had found was indeed part of a sculpture on which could be seen the profile of a face with ornaments on its head. The salvage work, directed by the archaeologists, continued until 27 February and revealed an enormous monolith, 3.25 meters in diameter. The upper part was carved in relief to represent a naked woman, decapitated, with the arms and legs separated from the trunk. Without a doubt, it depicted Huitzilopochtli's sister, Coyolxauhqui, the moon goddess who, according to myth, was killed by her brother in combat on the hill of Coatepec.

The newly discovered carving of Coyolxauhqui formed part of the ancient Great Temple of the Mexicas (or Aztecs), so as a consequence of the find, there was a rush of interest in excavating the site (fig. 1). Parts of it were already known, such as the southwest corner, excavated in 1913–1914 by Manuel Gamio, as well as other elements from later excavations, such as the one carried out by Hugo Moedano and Elma Estrada in 1948. Also known was the 1933 work of Emilio Cuevas who located part of the wall and a small part of the stairway to the platform upon which rested one of the last construction stages of the Great Temple. Attempting a complete excavation of the Coyolxauhqui carving required a strategy that took into account the urban environment in which the temple was located. A plan of action was established and the problems to be solved were determined; all this effort was based on specific theory. That was how we came to propose three phases for the research project, as detailed below.

First Phase: Case study and statement of objectives

This phase consisted of gathering all available information from the archaeological literature and from historical sources. We obtained the data from previous work done in the temple or near it—from 1790 when the Aztec Sun Stone (Piedra del Sol) and the monumental carving of Coatlicue were discovered, up to those excavations of the Great Temple already mentioned. Regarding the latter, we have the sixteenth-century accounts of Bernal Diaz del Castillo and Hernan Cortes, who were in

Figure 1 (opposite page). Plan of the Great Temple of Tenochtitlan.

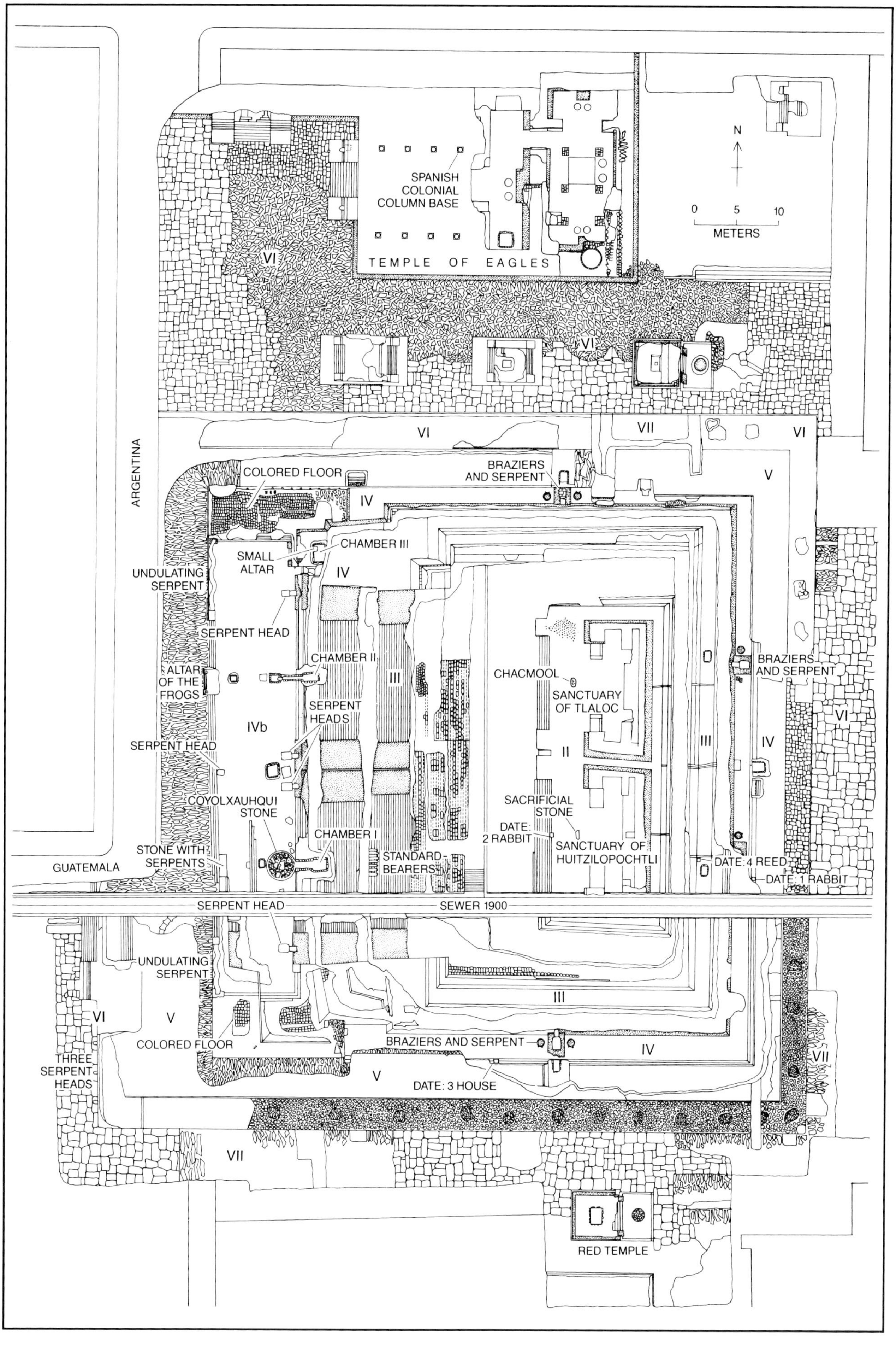
N
0 5 10
METERS
SPANISH COLONIAL COLUMN BASE
TEMPLE OF EAGLES
VI
VI
VI
VII
VI
ARGENTINA
COLORED FLOOR
BRAZIERS AND SERPENT
IV
V
CHAMBER III
SMALL ALTAR
UNDULATING SERPENT
IV
SERPENT HEAD
CHAMBER II
ALTAR OF THE FROGS
III
CHACMOOL
SANCTUARY OF TLALOC
BRAZIERS AND SERPENT
VI
SERPENT HEADS
IVb
SERPENT HEAD
II
III
IV
COYOLXAUHQUI STONE
SACRIFICIAL STONE
CHAMBER I
DATE: 2 RABBIT
STONE WITH SERPENTS
GUATEMALA
STANDARD-BEARERS
SANCTUARY OF HUITZILOPOCHTLI
DATE: 4 REED
DATE: 1 RABBIT
SERPENT HEAD
SEWER 1900
UNDULATING SERPENT
VI
V
III
COLORED FLOOR
BRAZIERS AND SERPENT
IV
THREE SERPENT HEADS
V
DATE: 3 HOUSE
VII
VII
RED TEMPLE

the Great Temple, as well as those of Bernardino de Sahagun, Diego Duran, Hernando Alvarado Tezozomoc, and others. All this rich and varied information was useful in sustaining the hypotheses postulated on the basis of the temple's characteristics. According to these data, the Great Temple faced west and had two stairways to the upper part. At the top were two sanctuaries: one dedicated to Tlaloc, god of rain, water, and fertility; and one dedicated to Huitzilopochtli, god of the sun and of war. The presence of these gods may have been indicative of the basic needs of the Mexica. Their economy seemed to be based on agricultural production—thereby making water and rain very important—and tribute obtained from other nations through military conquest—thus making war important. If these assumptions are true, then all the elements associated with the Great Temple—such as votive objects and sculpture—would in some way relate to this basic sociocultural pattern. To determine if this was indeed the case, it was necessary to use the appropriate excavation technique, which also had to address the surrounding urban environment. This consideration determined the plan of action in the second phase.

Second Phase: Excavation

First, all existing structures on the land designated for excavation were evaluated, and an assessment of their historical importance was made. The Advisory Board on Monuments considered each case individually before deciding to raze a house for the sake of continuing the excavation. Of all the houses removed, only two contained elements from the Colonial period. These features were photographed and registered and then were removed to the Department of Monuments.

Because most pre-Hispanic buildings were destroyed at the time of the Conquest and their stones were used to build new houses and colonial buildings, we realized that we would find only remnants of these structures. In the case of the Great Temple, we could determine that the last stage of construction was leveled to the foundations, leaving only traces of the building on the flagstone floor of the great square of the ceremonial precinct. Only 1 meter on the north side of the wall of the platform was preserved. The earlier stages of construction, however, were in better condition. This could be attributed to the level of the water table. At present, the water table is 4 to 5 meters below street level.

The first stage of construction of the Great Temple (designated Stage II) dates from about the year A.D. 1390. It was found to be almost complete, but only the upper part (i.e., the last of the four or five structures that constitute it and the two sanctuaries) could be excavated. The constructions below could not be excavated without lowering the water table, which could affect adversely the stability of the temple and the colonial and modern buildings in the vicinity. This constraint is not entirely negative. In a few years, when the level of subsurface water can be controlled, the building can be excavated using more sophisticated techniques. The results will either confirm or disprove the assumptions that rest on our current knowledge and techniques.

We began our work by marking off 2-meter squares, which we labeled with letters and numbers. We also used a depth-tracking system starting from specific markers, with the main one fixed in place. A support team provided the expertise indispensable to archaeological field work: photography, drafting, inventory control, and conservation. The involvement of the conservator was essential since it was expected that we would unearth a great diversity of archaeological materials, both movable and immovable, organic and inorganic. Thus we planned to have a team of curators and the necessary facilities for adequate treatment of the materials as they were found. Moreover, it was agreed by Project Coordination that throughout the excavation process the archaeologist and the conservator would always work together. Only in this way could we guarantee that preservation efforts would begin from the precise moment the archaeological object was found, taking into consideration the place where it was found, the conditions under which it was found, and the special treatment it might require. While the archaeologist was responsible for the work in general, the conservator could intervene when an object had to be given special care. The mutual respect of both specialists and their common interest in archaeological objects bore fruit, and because of it, many of the objects retained their characteristics, which now facilitates their study. The installation of a laboratory-workshop in the excavation area made possible the proper treatment of the object once it was removed from its original location. In addition, there was the support of other specialists, including biologists, chemists, and geologists from the Department of Prehistory of INAH. An advisory panel of soil engineers was also needed, since it was dangerous to work close to the adjoining buildings with excavation probes (there was also the previously mentioned problem with the water level).

Third Phase: Interpretation of the material

After five years of uninterrupted work, the first two phases were completed. The third phase comprises the study, analysis, and interpretation of the excavated materials. This aspect of the research is ongoing and will take some time to complete.

Evaluation of the excavated complex

At present the excavated zone is open to the public. The interest generated by the work over the last five years has been enormous. Nevertheless, before the zone was opened, various aspects had to be studied and they are listed below.

Setting

Some years ago, I had occasion to do a study of the different surroundings in archaeological zones; I arrived at definitions for three types. The first I call "natural surroundings," that is, those archaeological sites in which the buildings are adapted in some way to the natural environment. In these cases, it is necessary to protect the natural landscape in order to avoid altering the unity of the monument and the site. An example is Teotihuacan, where the hills are in harmony with the temples, and we can see how the Pyramids of the Sun and the Moon reflect their shapes. Obviously, changing the natural setting in any way would radically alter the integration of the entire complex.

The second type is the "urban setting," in which sites are surrounded by cities. This is the case in Tenayuca, Tlatelolco, and the Great Temple of Tenochtitlan (where other problems also come into play, as we have already seen).

The Great Temple is a typical example of the problems of urban settings. The excavation demanded that several buildings be removed, and as a consequence the backs of the remaining buildings that surround the Temple were exposed to view. This created a deplorable setting, which required immediate resolution. A new museum was being built behind the excavated area to house the artifacts. It was decided that the height of the new building would reflect the height of the surrounding buildings, in order to preserve the harmony of the urban landscape. The facades and backs of the present-day buildings were covered with the same smooth gray plaster that was used to surface the new museum building. Some of the walls of the contemporary buildings were used to display descriptions of chroniclers such as Cortes, Montolinia, and Bernal Diaz del Castillo.

The third type is what I term "megaspace." It refers to archaeological sites and monuments placed on high ground in open spaces where they blend with the sky. This is the situation in Monte Alban and

Zochicalco. This kind of setting is affected by (among other things) airplanes passing overhead, intruding visually and aurally.

Roofs

The ways in which to ensure adequate protection of the murals, sculptures, and other elements were studied in detail. It was finally decided, after several meetings in which both archaeologists and restorers took part, to put modern roofs over some of these elements. An alternative would have been to detach these archaeological elements and move them to the museum, together with pieces such as the Chac-mool. To determine the feasibility of this option, some tests were made on the walls, with negative results. Furthermore, the total ensemble would have been affected had this procedure been followed.

Interior walkways

We decided to install metal walkways over the site that permitted viewing below. Visitors who might be tempted to take away "souvenirs" were thereby prevented from doing so, while still being given an opportunity to see the site. The metal walkway went through a sewer main, 2 meters in diameter, under Guatemala Street. This sewer main was built near the end of the last century beneath Santa Teresa Street (later renamed Guatemala) and cut through all the construction phases of the Temple, destroying them in a 2-meter-wide swath.

A special characteristic of the walkways is that they are removable. They rest on supports over the archaeological structures without altering or destroying them and may be moved. The view from the interior walkways is complemented by the aspect seen from a walkway around the perimeter, which runs alongside the present-day buildings and makes use of the space between the buildings and the area of excavation.

Adjacent streets

Because of the great amount of pollution in the area caused by motor vehicles, a request was made to the Transit Authority of the city to reserve the streets next to the Great Temple for pedestrians only. It was finally agreed that the streets of Seminario, Guatemala, and Lic. Verdad, and the first block of Argentina would remain closed to vehicular traffic. This measure reduced the very high level of pollution, even though the urban surroundings in which the Great Temple is located cannot be completely protected from the atmospheric environment of Mexico City.

The zone was opened to the people of Mexico on 25 November 1982 and is now visited by about four or five thousand people daily. In the meantime, studies on the findings continue, and this will take many more years.

Summary of the results of the excavations

What follows is a general summary of the major discoveries during our five years of excavation (1978–1982), of the Great Temple.

Architecture

Until a few years ago the accounts of the Great Temple left to us by the chroniclers of the sixteenth century were our main source of information about its characteristics. Now, thanks to archaeologists, we have the Temple in front of us, and we can see that the descriptions of the chroniclers were really very close to what they saw or what had been described to them by the natives. Archaeology has also allowed us to discover the very ancient stages of the Great Temple, which were not known even to the last generations of the Mexica.

The Temple was oriented with its principal facade facing west and was set on a platform on a large pedestal composed of four levels. Two stairways led to the top where there were two sanctuaries. The one on the south side was dedicated to the tutelary god, Huitzilopochtli, the god of war; and the one on the north side to Tlaloc, the god of water, rain, and fertility.

Over the years the Temple was enlarged for different reasons. The city of Tenochtitlan was subject to flooding, making it necessary to raise the level of the architectonic complexes. Moreover, the settling and cave-ins of the unstable land aggravated the structural flaws of the buildings. Also, historical sources reveal that rulers superimposed new temples over existing ones. We now know that the Temple was enlarged on all four sides, on at least seven occasions, and there were also four additions to just the main facade.

The following is a general description of each of these construction stages, with the exception of Stage I, which was found inside Stage II in such a deteriorated state that any description would be useless and unnecessary. The successive Roman numerals indicate full augmentation on all four sides, and when the Roman numeral is followed by a letter, the additions are only to the main facade.

Stage II. This building is characterized by the fact that its top is very complete. On it we see the remains of two sanctuaries of stone and some of the stucco (a mixture of limestone and sand) that covered them. In front of the entrance to this sanctuary of Huitzilopochtli, we find a sacrificial stone (fig. 2). On the last step, on an axis with the sacrificial stone, we see the face of a personage with a glyph above (2 rabbit), which indicates the year A.D. 1390. In the interior of the sanctuary, there is a low platform running north to south; in the middle of it is a small altar on which the god was placed.

On the side of Tlaloc (north), there is a polychrome stone statue that represents Chac-mool, the divine messenger in charge of bringing offerings to the sanctuary. The piers framing the entrance to the sanctuary are covered with their original painting, black and white circles—possibly the eyes of the god— over blue and red bands. Below this are black-and-white vertical stripes. This stage dates to before the year 1427, when the Mexica freed themselves from Atzcapotzalco.

Stage III. This stage is characterized by well-built stairways and vertical

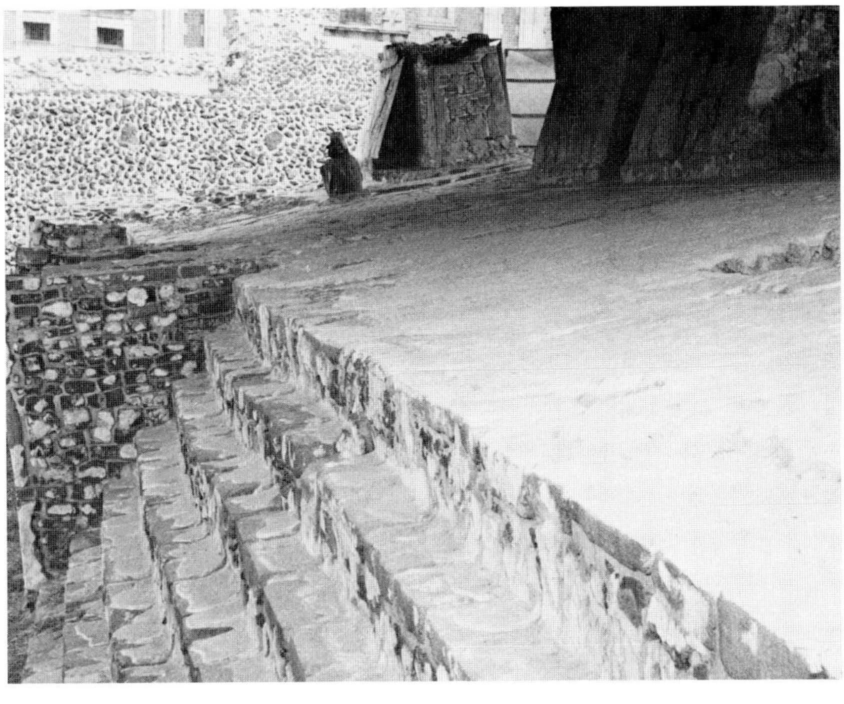

Figure 2. The sacrificial stone (right) at the entrance to the sanctuary of Huitzilopochtli. In the background is the Chac-mool at the entrance to the sanctuary of Tlaloc.

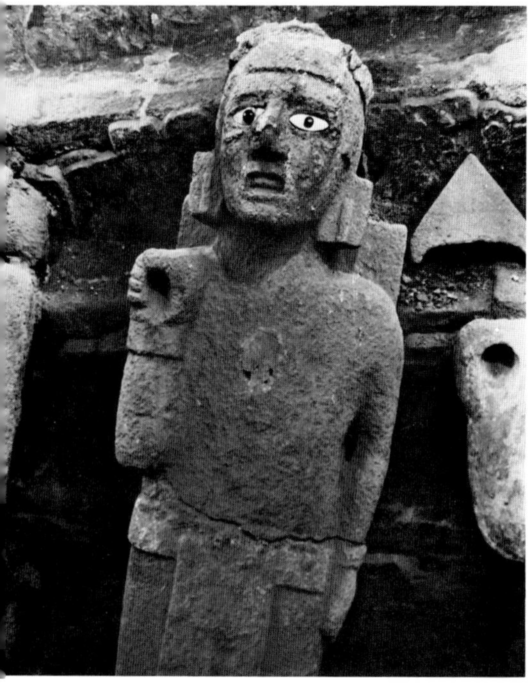

Figure 3. Detail of one of the eight sculptures reclining on the stairway of the Huizilopochtli side of Stage III.

facings on the base. On the stairway on the Huitzilopochtli side, eight statues representing standard bearers recline on the steps (fig. 3). Originally, these figures probably decorated the building, but when the next construction stage started, they may have been collected and placed ceremonially on the stairway where we found them.

At the rear of this building, on the platform wall on the Huitzilopochtli side, there is the "4 reed" glyph, which is equivalent to the year A.D. 1431, the reign of Itzcoatl.

Stage IV. This stage and its additions is among the richest in elements. The common platform is decorated with braziers and serpent heads on all four sides (fig. 4). The braziers on the Tlaloc side (the one at the rear) bear the likeness of this god, while the ones on the other side are equipped with bows, the attributes of Huitzilopochtli. At the foot of these braziers and serpents are various offerings still in their cists.

Stage IVb is an addition to the main facade (facing west), which has a multitude of elements. This addition is the large platform on which the Great Temple rests. There is one flight of stairs with enormous snakes at both ends; their undulating bodies are capped with large heads that still retain some of the original pigment that once covered them. The platform stairway is interrupted by a small altar with two frogs on its sides. This element is on an axis with the part of the stairway that leads to the upper part of the temple sacred to Tlaloc. On the Huitzilopochtli side, in front of the flight of stairs leading to the sanctuary of this god, is a stone slab, 2-meters long, decorated with serpents, which forms part of the fourth step of the platform.

The foundation formed by the remains of the two stairways leading to the upper part consists of four serpent heads: two on the outer balustrades and two on the balustrades in the middle, which mark the joining of the two buildings. In the middle of the platform on the Huitzilopochtli side, one can see a monumental stone sculpture of Huitzilopochtli's sister, Coyolxauhqui, with whom he fought on the hill of Coatepec.

On this platform various offerings were found, some around the statue of Coyolxauhqui, others between the two serpent heads, and still others midway on the stairway of Tlaloc. All of them were under

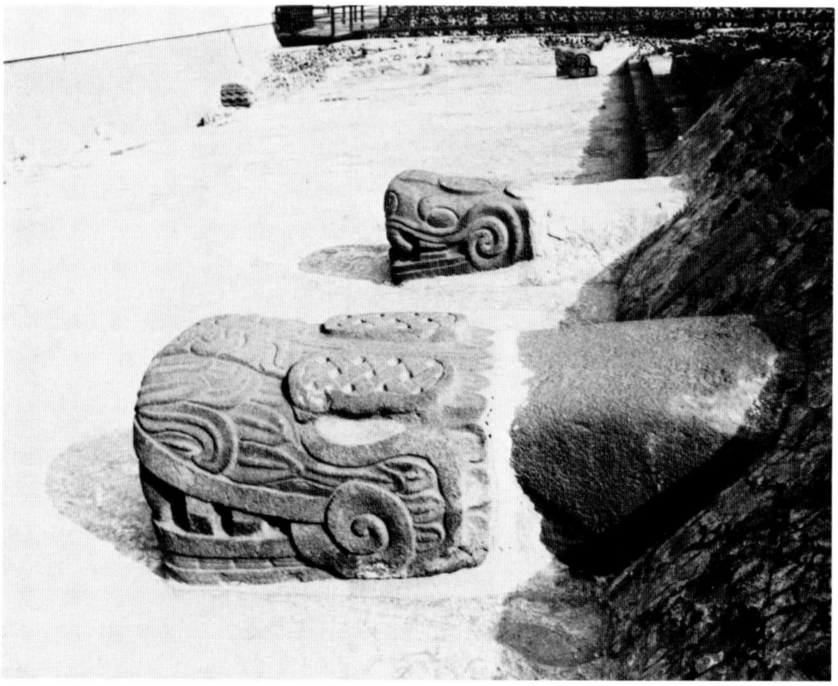

Figure 4. Serpent heads on the Stage IV platform. The braziers bear the images of Tlaloc and Huitzilopochtli.

the floor of the platform, and only Chambers I and II were found behind the steps at the exact midpoint of each of the buildings. Rooms with floors of colored marble were found at the north and south ends. Additionally, on the north side there was a small altar with a short flight of steps. Inside it, two offerings were found, one with more than forty skulls and bones of children, below which was a space we have called Chamber III. Both offerings were dedicated to Tlaloc.

We think that, chronologically, Stage IV corresponds in general to the reign of Moctezuma I, since there is a numeral on the back part of the platform on the Huitzilopochtli side with the glyph, "1 rabbit," equivalent to the year A.D. 1454. (All these dates are subject to revision.) The addition of Coyolxauhqui and the serpents could well belong to the Axayacatl period, because of another glyph on the south side of the building with the symbol "3 house," which in numbers would be equivalent to the year A.D. 1469, when this emperor *(tlatoani)* began his reign.

Stage V. Of this stage, the only vestiges that have been found are the common platform covered with a layer of stucco and part of the flagstone and stucco floor of the great ceremonial precinct.

Stage VI. This stage is the penultimate, and we can see part of the great platform that remains. On its principal facade there is a wall with three serpent heads facing west, as well as a decorated balustrade and part of the stairway, which was destroyed by the sewer main built in 1900. This balustrade was the one found by the archaeologist Emilio Cuevas in 1933. The temples on the north side and the base of the Temple of Eagles, as well as the great flagstone square where the sanctuaries are located, correspond to this construction stage. On the south side of the Great Temple there is another temple, which we have named the Red Temple; it is similar to a corresponding temple on the north side.

Given their importance, each of the buildings on the north side deserve a brief description.

Sanctuary A. This building is set on a small platform and has two stairways that give access to the rear part of it, one oriented to the west and the other to the east. The walls have no decoration.

Sanctuary B. The stairs face west and the other three sides are decorated with skull-covered panels (figs. 5, 6). It is what we call a *tzompantli* (altar). The skulls, numbering around two hundred forty, show several layers of stucco with which they were covered.

Sanctuary C. The main facade faces east and comprises a vestibule in the center of which are the remains of a circular altar. At both sides it has two walls decorated with circles of stone painted red, which symbolize *chalchihuites* (a kind of rough emerald) or precious stones. At the back of the vestibule is the stairway. The building is totally decorated and retains the *talud-tablero* (sloped panel-vertical panel) motif, a reminder of the typical architectural order that prevailed many centuries earlier in Teotihuacan.

Base and Pedestal of the Eagles. To the north of this grouping of three sanctuaries stand a large base and pedestal facing west, whose beams are decorated with an eagle's head that bears traces of the original pigment. This facade is complemented by one facing south, forming a configuration we have called the Courtyard of the Eagles. The floor on which all these buildings are set out shows several superimpositions, and was found in magnificent condition.

Rear Court. The rear court of the Great Temple dates from this same phase; like the one on the north side it is built of flagstones joined with stucco. The eastern edge of this court formed the boundary of the great ceremonial precinct containing the seventy-eight buildings of which Sahagun spoke—the sacred area of the Mexica. It is made up of

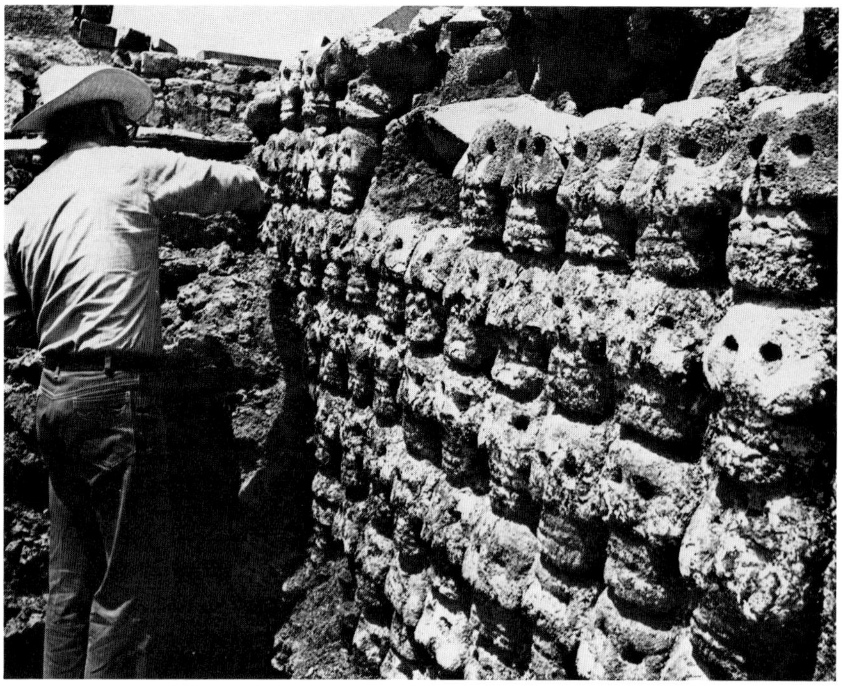

Figure 5. The excavation of Sanctuary B in the patio north of the Major Temple, which is decorated with skull-covered panels.

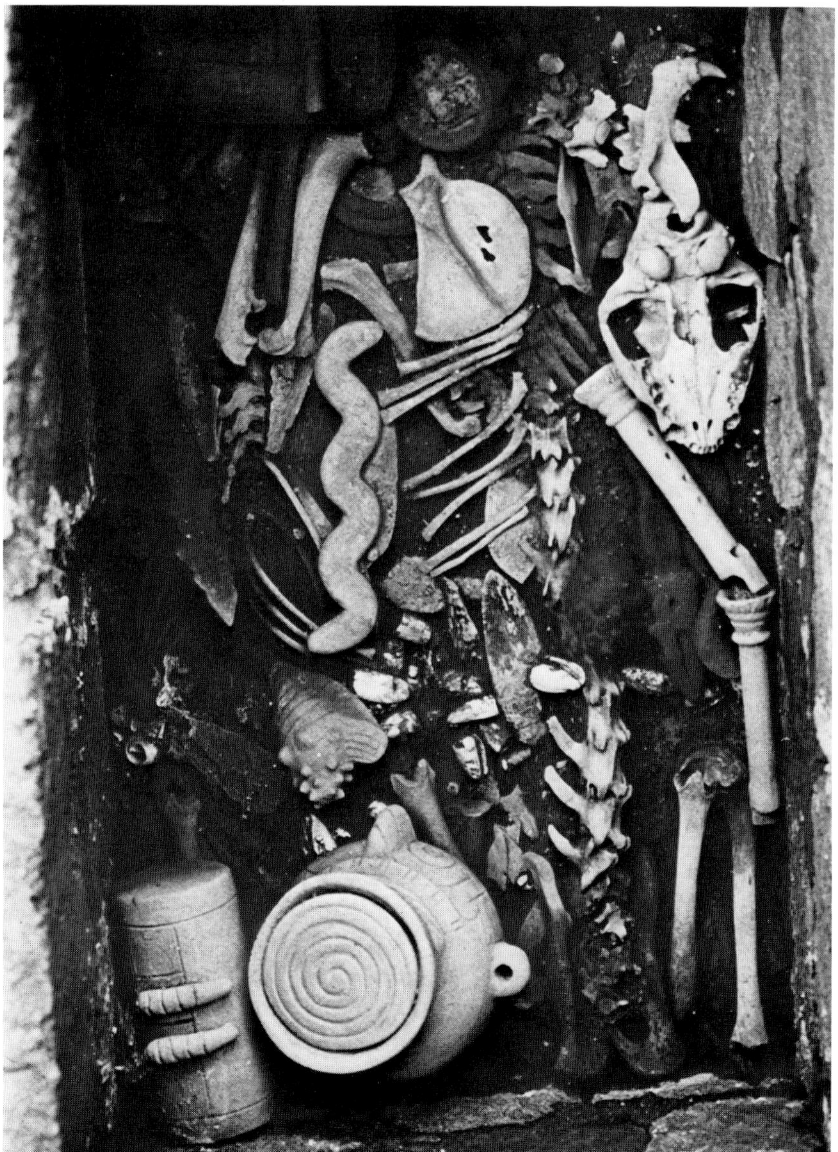

Figure 6. An example of a burial deposit in a cist found in Sanctuary B.

flights of stairs alternating with balustrades, resembling the configurations at Tlateloloco, a neighboring city of Tenochtitlan. Therefore, it is not, as some chroniclers have called it, a *coatepantli* (wall of serpents), which is characterized by the small serpent heads such as the ones we see in the walls of the Temple platforms.

Stage VII. This is the last phase of construction of the Great Temple, which is the one the Spaniards saw. About all that remains of it is part of the flagstone floor of the ceremonial precinct and traces indicating where the temple had been situated. On the north side, however, one can see part of the platform, corresponding to the platform of the preceding stage, except that on filling and covering the previously mentioned buildings (Sanctuaries A, B, C, and the Base and Platform of the Eagles), and constructing the floor of the last stage on top of the buildings, they made use of the same platform; that is, the Great Temple was not extended further, but rather the already-existing platform was used.

Offerings

More than one hundred offerings were found in the excavation of the Great Temple. Some of these were out of the context of the Temple; that is, they were found as offerings in sanctuaries near it. Here we shall describe some of the general characteristics of the finds directly associated with the Great Temple.

To start with, all the offerings were found to have been deposited in one of three ways: (1) they had been placed in cists whose walls and floors were of stone with stucco remains; (2) they had been placed inside movable stone boxes with lids of the same material; or (3) they were placed directly in the fill or nucleus of stone and mud that covered a construction stage.

In general, the offerings were placed on set axes. In the main facade there are three principal axes: The first two are under the floor of the platform in front of the middle of each of the stairways, another is in front of where the two buildings join (Tlaloc and Huitzilopochtli). On the corners of the Temple they were also placed the same as on the north-south axis, approximately in the middle of the building. In the back we also have three axes: One in the midpoint of each of the buildings and one where they join. Some offerings were placed around the base and pedestal, equidistant from one another.

Some of the finds were funeral urns. Although in our nomenclature they were called "offerings," they actually had a function related to funerary rites.

The placement of objects within the offering was not haphazard, that is, the objects occupy a place conforming to some symbolism that needs to be deciphered. In other words, the objects and their placement are a kind of a language. Thus, for example, there are materials that in some offerings are always found in the lower part, on the floor, while there are others that are always found in the upper part. We have also observed that the materials are oriented in a certain way. The offerings on the west side (main facade) and those at the back of the temple are oriented toward the west, where the sun sets, while the offerings found at the middle of the building in its north and south facades are oriented toward those directions. Another interesting aspect is that their placement within the offering sometimes was intentional. In Offerings 7 and 61 (the former, located in the middle of the building in its south facade, and the latter, on the north side), the materials were distributed in the same way: in the lower part, *strombus* snails were oriented north to south; over them were crocodiles; and, in the upper part were the seated gods we have named Xiuhtecutli, since they are represented as old people. On the right side of the seated gods there was a sea coral and on their left side, a clay container with the effigy of the god Tlaloc

Figure 7. Polychrome vessel with the image of Tlaloc, the god of war.

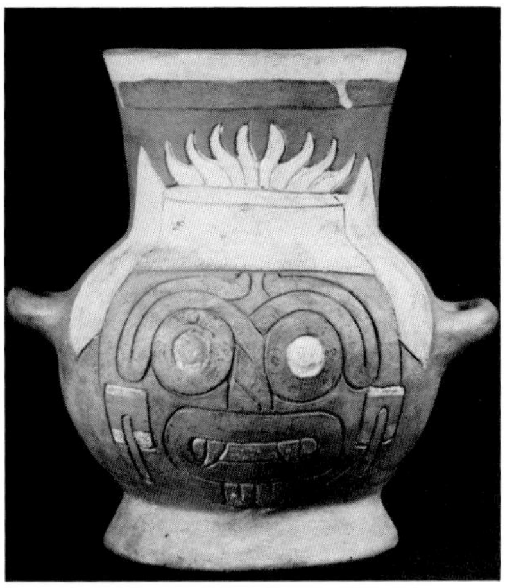

Figure 8. Sculpture of a snail shell, symbolizing fertility. 84 cms.

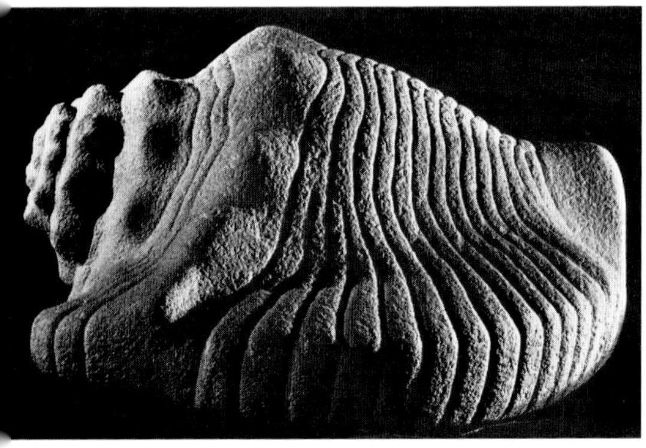

Figure 9. Mask of green stone from the present state of Guerrero, which was under the control of the Aztecs.

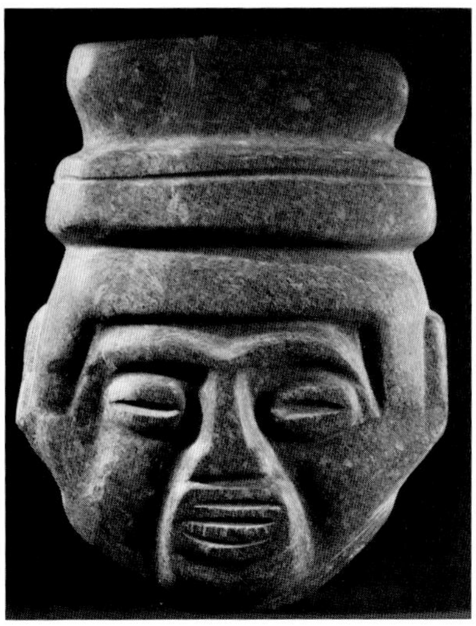

(fig. 7). It could be that this distribution signifies that the snails represent the sea; the crocodiles, the land; and Xiuhtecutli and Tlaloc, the sky. The same positioning occurs with Offerings 11 and 17, the former located in the main facade between the two serpent heads that mark the union of the temples of Tlaloc and Huitzilopochtli, and the latter at the back, where both buildings join. The two offerings are located in cists with stone sides, and their objects resemble each other and are in similar positions.

The material obtained from the eighty-six offerings associated with the Great Temple was abundant and varied. There were more than six thousand objects. Some of them are actually of Mexica origin and others must have come from tributary areas, mostly from the present states of Guerrero, Oaxaca, the Gulf Coast, and Puebla. There was not a single object from the Tarascan culture (which, as we know, was not under Mexica control), nor any from the Maya. Of the Mexica material, the most abundant are the sculptures of seated old people wearing only their loincloths *(maxtlatl)* and a headdress characterized by two protuberances. It has been thought that they represent the god Xiuhteculti; other authors, such as H.B. Nicholson, think they could well represent Tepeyolotl. We are inclined to favor the first hypothesis, given that the carving shows an old person and the headdress could correspond to the two pieces of wood used to make a fire. We know of the importance of Xiuhteculti as the center of the universe, the home, and as the father of the gods.

Other representations of the Mexica are: effigies of the god Tlaloc carved out of volcanic rock *(tezontle)* and other stones; some coiled serpents; representations of rattlesnake heads made of obsidian; stone braziers with knotted bows; and of course, the magnificent representations of snails (fig. 8), which are true works of art, although they do not appear among the offerings. Coyolxauhqui and the serpent heads which decorate the facades of the temple are included in this category.

Also found were interesting pieces from the tributary areas, such as a large quantity of masks and figures of different types and sizes from Mezcala and pieces in alabaster from the region of Puebla, such as deer heads and arrowheads, along with seated deities. From the Gulf Coast we have two magnificent covered funerary urns of orange ceramic. Inside the urns were found the remains of burned bones, necklaces, and other materials. The great variety of snails and conches, fish, and rostral cartilage from sawfish and coral came partly from the Gulf Coast and partly from the Pacific. The same is true of the crocodiles and jaguars which may be from Veracruz, Tabasco, or Chiapas.

Another group of objects clearly belongs to societies that existed long before the Mexica. Teotihuacan and Olmec masks are examples of this, with the provenance of the latter, according to petroglyphic analysis, being the region comprising the present-day states of Puebla, Oaxaca, and Guerrero (figs. 9, 10).

There is not much more to report, as all the material is still being studied. Nevertheless, we can say that the majority of materials including all the objects of marine origin, canoes, and other items represent Tlaloc or symbols associated with him. In addition, we have objects associated with Huitzilopochtli—not representations of the god, but symbols, such as braziers with the bow or knot, decapitated skulls, sacrificial knives called *tecpatl* with eyes and teeth of conch—and objects that came from the tributary areas, as a result of military conquest.

What has been described serves to confirm our hypothesis of an agricultural and military nation maintaining itself by both agriculture and tribute, water and war, life and death, all of which are represented in the Great Temple of Tenochtitlan. Moreover, the site is of fundamental importance, for it represents the concentration of Mexica power, the

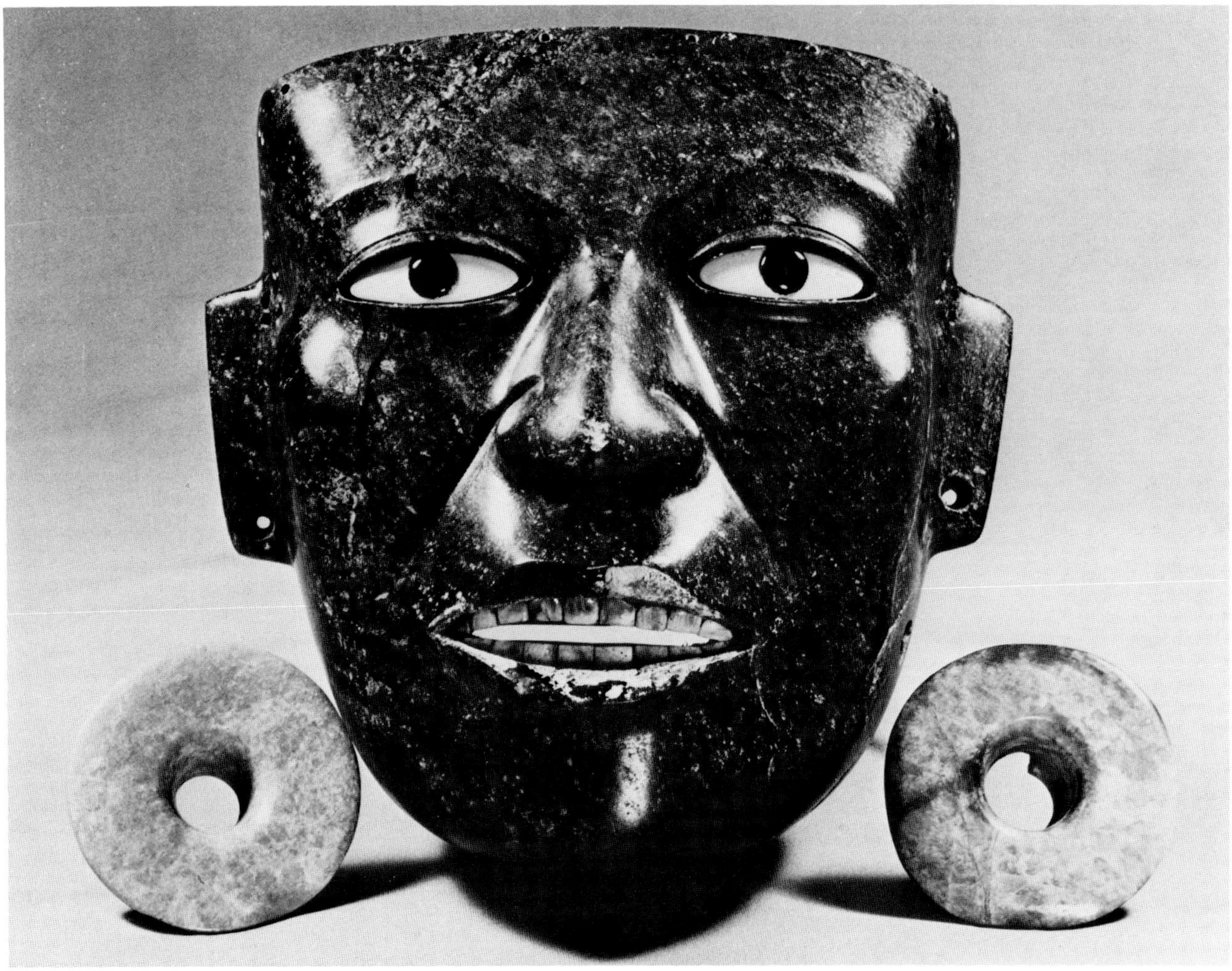

Figure 10. A mask from Teotihuacan with incrustations of conch and obsidian.

appropriation of the destiny of the conquered nations, just as Huitzilopochtli did with his brothers and sisters, conquering them and taking control of their destiny *(anecuyotl)*. On a symbolic level this is important, since in this way the Mexica carried on the mission started by their tutelary god. But not only did the Mexica take over the "destiny" of the conquered nations, the mighty civilization also took control of their production, agriculture as well as other activities, including the making of cloaks, war, shields, furs, and skins.

Maria Luisa Franco

CONSERVATION AT THE TEMPLO MAYOR OF TENOCHTITLAN

Maria Luisa Franco is a restorer at INAH and was in charge of the restoration section at the Templo Mayor. She has been responsible for Mexican archaeological materials exhibited in museums in Paris, Rome, New York, and Italy.

This paper will provide a general overview of the conservation work carried out at the archaeological site at the Templo Mayor, on the architecture, mural paintings, related sculpture, and various materials found in ritual deposits.

Building materials

The Templo Mayor, in each stage of construction, and the adjoining altars were built using very porous volcanic rock (*tezontle*) for the core and walls, fine-grained volcanic rock (andesite) for the staircases and patio floors, and fine-grained metamorphosed rock (slate) for the patio floors. The original mortar was composed of mud, or clay and sand, in some cases; lime mortar was also used. All of the surfaces were coated with stucco plaster of varying thickness.

Mural paintings

Mural paintings are found on three substrata: stucco plaster, clay plaster, and stone. The black and blue pigments are organic; the red, orange, and all other pigments are of mineral origin. According to the analysis of Ing. Huerta of the Directorate of Restoration, the technique is tempera, and although the medium has disappeared, the layers of paint and plaster remain perfectly distinct. These paints are found in Stage II, the oldest stage of construction, on the Enclosure of the Eagle Knights (Recinto de los Caballeros Aguila), on the Red Temples (Templos Rojos), and on the southeast corner of the rear patio of the Templo Mayor.

Sculptures

Some of the carvings associated with the buildings are attached, while others are freestanding. The attached sculptures are of Coyolxauhqui and Chac-mool and nine sculptures of serpents and two of frogs, all of which are in polychrome. A second representation of Coyolxauhqui, six braziers, a human face, and a sacrificial stone are in monochrome.

Of the freestanding sculptures, the nine standard bearers in stone are in polychrome, while the two clay representations of Mitlantecuhtli and the two clay figures of Eagle Knights are uncolored. There is also a sculpture in green stone.

The attached carvings were left in situ, hence, the restorer worked on them where they were found. The freestanding carvings were removed from their original positions and were taken to the conservation workshop for treatment.

Ritual deposits

The materials found in ritual deposits include mud, bone, stone (basalt, green stone, and flint, turquoise), calcareous materials (conch, shell, and coral), obsidian, metal (gold, silver, copper, and tombac), copal, wood, and animal materials (tortoise shell, fish and crocodile scales, and cartilage from the beaks of sawfish).

Environmental conditions

Among other agents that cause the deterioration of archaeological materials, the environmental conditions at the Templo Mayor are characterized by two such factors in particular—namely, atmospheric contamination and high moisture levels in the subsoil—which threaten the survival of materials at the site.

The Templo Mayor is located in downtown Mexico City, a highly contaminated urban area. The atmosphere contains solid and gaseous substances from mineral-oil combustion, gasoline and hydrocarbons in general, as well as dust, soot, carbon dioxide, sulfur dioxide, and nitrogen oxides.

Solid particles are deposited on all surfaces, forming a porous bed that absorbs and retains gases and atmospheric moisture. Carbon dioxide increases the acidity of rainwater and dissolves the constituents of the limestone. Even more important, sulfur dioxide reacts with atmospheric moisture or with rain to form sulfuric acid, which reacts with calcium or sodium compounds in the stones, forming soluble salts and weakening their structure. Or it may dissolve calcium compounds in the stuccowork, causing the loss of both adhesion and cohesion.

At one time five lakes could be found in the valley of Mexico: Chalco, Xochimilco, Texcoco, Xaltocan, and Zumpango. These lakes were interconnected by a manmade network of deep canals and ditches. Little by little these lakes have been drying up, but the subsoil of downtown Mexico City (Old Tenochtitlan) remains marshlike with a rather high water table, since rainwater adds to the subsoil through filtration. In the Templo Mayor the highest water-table level is found as close 15 centimeters from the pre-Hispanic floor.

Other deteriorative factors that have been noted in the Templo Mayor are the composition of the soil, excessive humidity, temperature, light, biological forces, wind, seismic movements, and the activities of human beings.

Soils

The soil in the Templo Mayor is mildly acidic as a result of decomposing organic materials that produce humic acids. This has caused some of the stucco layers to lose their strength and adhesiveness and to acquire a "spongelike" texture. Materials found in offerings have been likewise affected. We find, for example, that carbonates and phosphates present in bone dissolve readily in acid, and that humic acid seriously damages metal and decomposes calcium found in ceramics. Objects have also undergone serious damage and distortion as a result of the weight and pressure of compacted soil.

Humidity

Humidity results from capillary action (i.e., rising damp) or filtration. Groundwater rises to a certain height in walls and floors through capillary spaces. As this happens, the water extracts and dissolves the soluble salts (chlorides and sulfates) from the subsoil and the construction materials and usually carries them up to the surface. When the water evaporates, the salts absorb carbon dioxide from the air, crystallize, and form a deposit on the surface (efflorescence). Upon crystallizing, the salts exert pressure on the capillary walls around them, causing fissures. However, some salts crystallize beneath the surface (subefflorescence), depending on their solubility and the rate of water evaporation. For example, chlorides can break down crystalline structures, dissolve mortars, and ultimately cause exfoliation or pulverization of an entire structure. Thus, alternating periods of wet and dry weather are accompanied by dissolving and crystallization; the result is the serious deterioration of structural materials.

Rainwater enters the walls and floors of the structures through pores, gaps, and cracks. Adding to the soluble salt dissolved from the building materials, this water contains dissolved atmospheric gases, which it deposits on the structure. Among these, sulfur trioxide is particularly harmful since, upon contact with rain, it forms sulfuric acid, which in turn reacts with any calcium or sodium compounds present in stones and forms soluble salts; this gradually weakens the structure of the building. Any gaps in the plaster caused by humidity, destroying the adhesion to the support, are susceptible to the action of crystallizing salts (introduced by water filtration) that will push against the plaster until it falls away. In the case of stucco, the corrosion of the material is caused by a solution of the calcium compounds and the acids present in water. This causes the loss of adhesion and cohesion and produces friability and separation. Wind can aid the penetration of rainwater into the plaster, or its force can cause a difference of air pressure on the plaster.

Temperature

Although in Mexico City the temperature extremes of tropical or desert regions are not found (8°-26°C on the average), the alternation of sunshine and rain causes the expansion and contraction of construction materials and thus the cracking and pulverization of the stones. Temperature acts as an agent of deterioration by altering the moisture content in the pores of the material and by affecting the solubilities of salts and gases. It also accelerates chemical reactions and hastens hydrolysis.

Humidity and temperature are closely related. Conditions of very high temperature and humidity, along with the presence of light, encourage the growth of organisms, such as moss.

Light

Natural and artificial light cause materials to deteriorate, especially those of organic origin, because light aids in chemical reactions (Espinosa 1974). Ultraviolet rays discolor pigments that do not have the ability to reflect light at this wave length; the energy is thereby absorbed and this can break chemical bonds in materials and cause them to change. Ultraviolet light attacks synthetic resins and varnishes in the same fashion. Sometimes, however, ultraviolet light can have beneficial effects; at high intensities, this light can impede the development of fungi and lichens. Natural light is related to biological processes of degradation and may hasten them.

Biological damage

Biological damage is caused primarily by the effects of moss, plants, birds, or rodents. Moss poses the greatest and most constant problem since it proliferates on mud, stucco, mortar, or stone, causing structural damage through the penetration of rhizoids. Moss survives by absorbing nutritive substances (principally calcium, potassium, and sodium compounds) from the host material, and its cells store water through absorption. As it feeds and grows, the rhizoids cause the materials to crack. The greatest damage from moss occurs in the dry season, when the rhizoids and the materials to which they are attached lose water; the moss shrinks and the materials crack. Plants attack the substrata, causing fissures with their roots, which, in turn, produce humic acids (see above). Bird excrement produces organic acids that can cause corrosion.

Wind

Windborne particles erode surfaces of buildings. They are especially harmful when they are large and become embedded in the pores of the material, where they can absorb moisture and dissolved gases.

Seismic movements

Seismic movements can cause serious shifts in the loads, resulting in fissures and cracks in structures and surfaces. The cataclysmic Mexico City earthquake on 19 September 1985 caused the water table to fall 1 meter below its normal level, 10 centimeters beneath the floor of the site. Buildings of all periods were affected, but after a month and a half, the subsoil readjusted, and the water table returned to its normal level. Thus the damage to pre-Hispanic, colonial, and modern buildings that would have resulted from settling following the drying out of the subsoil was avoided.

The destructive effects of human activity

By far the greatest damage to the Templo Mayor was caused by man. The Spanish conquest of Tenochtitlan in 1521 destroyed the site in its last stage of construction. The monolithic carvings and the materials from the destroyed temples were used to build a cathedral and other structures.

Today, we hear of events that, if not the same, are indeed similar. Modern buildings are erected on top of pre-Hispanic sites; buildings, murals, and pre-Hispanic structures are damaged; and archaeological elements are vandalized and mural fragments or entire paintings are stolen. As a result of these activities, it is necessary to protect objects and structures from direct contact with people; visitors must be confined to walkways that keep them at a distance. As for the votive offerings, these were found settled in a humid environment after having undergone various kinds of defacement.

Conservation treatments

The choice of conservation treatments was based on the relative degree of deterioration of the materials. Each object and each material were treated according to the specific problems they posed.

Stonework and stucco

In general, the stone structures of the Templo Mayor and the adjoining temples are very sound, although the material tends to lose strength when the marshy subsoil dries and shrinks. Large and deep cracks were noted in all the foundations of the walls as well as in the staircases and floors. The solution chosen for the reinforcement of the foundations was to extract the earth and mud from between the stones and to introduce a mortar consisting of one-half part powdered lime and one-half part cement to three parts coarse sand, and crushed volcanic rock as needed.

As for the cracks, the stones were arranged perpendicular to the fissures in order to prevent them from opening up again at this spot. Near the surface, on the cracks as well as in the nuclei of the stones, the same mortar was used (but without stones) as a filler paste to give a finished touch to the interventions. The proportions were: one-half part powdered lime, three parts coarse sand, one-quarter part cement, and water as needed. This was placed directly onto the subsurface layer, which left it at a slightly lower level—1 centimeter—than the original surface. It was given a light, furrowed texture using a small brush so it could be easily differentiated from the original. This entire process was also carried out on the steps, rafters, walls, and floors.

The riser (vertical wall) of each stairway was reinforced with a retaining wall to prevent it from collapsing. First, the riser was covered with plaster (using a wood frame) to maintain humidity. Hollow rods were then inserted into the plaster and wood frame into which cement was poured to create a wall of concrete. After it dried, the frame was removed. Another coat of plaster was placed over the concrete and imbedded with basalt and *tezontle* so that the appearance would blend with the rest of the building. Two untouched areas were left as samples of the original construction.

The stucco was saturated with water and covered with mud. It also had structural cracks, fissures, exfoliations, and air pockets. In places, it was powdery as a result of attack by contaminants, salts, and moss. The fissures and cracks were retouched, first with coarse and then with fine retouching paste. The coarse paste was composed of powdered lime, coarse sand, cement, and water; the fine paste was made of lime paste, fine sand, and polyvinyl acetate emulsion (Mowilith® DM4 and water).[1] In areas where there were gaps, the exposed edges of the stucco were strengthened with Curasol® AR in distilled water or acrylic copolymer emulsion (Primal® AC-33) in distilled water. Subsequently, these edges were treated with fine retouching paste. Detached layers or flakes of stucco were put back into place with polyvinyl alcohol (Curasol AR) in distilled water or Primal AC-33 in distilled water. Air pockets were eliminated and the stucco was reattached to the stone support by mechanical pressure, using Curasol AR in lime water or distilled water as the adhesive.

1. This and other proprietary materials are listed in the Appendix.

The oily and porous crust on the stucco caused by deposits of solid particles was eliminated mechanically (in some parts only partially) using a solution of ammonia in distilled water. To inhibit such deposits from reforming and to prevent sulfuric acid attack, the stucco was soaked after being reinforced with Wacker's® impregnant 190L or silicone resin. Salts were removed mechanically. Moss was killed with a nontoxic biocide (Diuron® and Bromacil® in water) and then it was removed mechanically.

The fact that mural paintings were found on three different bases meant that three different treatments were required. In cases where they were found in good condition, they were protected with a permanent roof.

There were several overpaintings on stuccoed plaster. They were given a mechanical cleaning, and then the plaster fragments and the pictorial layer were joined to the original base with a polyvinyl alcohol adhesive, Endurol®; the painting was fixed with Paraloid® B-72 in xylol. Paint flakes were reinforced with Mowital® B60H in ethyl acetate and alcohol; the stucco was strengthened with a mixture of lime paste, lime water, and Mowilith DM4. Finally, the fissures and gap edges were treated with a fine retouching paste.

Clay plaster has a very irregular surface and several overpaintings were found on this support. Since clay plaster dries more quickly than stucco and shrinks, it had to be strengthened immediately if the pictorial layer was to be reapplied. Curasol AR in distilled water was used as the strengthening agent, and air pockets were filled with the same substance mixed with fine soil. Plaster fragments were joined to their original base with a semiliquid paste of Curasol AR in water, fine soil, fine sand, and a polyvinyl acetate emulsion, Mowilith DMIH. Fissures and cracks were retouched with fine sand mixed with fine soil and Mowilith DM4. The cleaning of the pictorial layer was mechanical and the pigments were applied with Paraloid B-72 in xylol.

Murals painted directly on stone posed a serious problem. This was the case in the Corridor of the Warriors at the Enclosure of the Eagle Knights. The stones are placed directly on the ground only 20 centimeters above the water table. Water rises by capillarity into the stones and salts, causing exfoliation and the crumbling of the paint layer and the stone.

Owing to the constant humidity, in many cases the salt has not crystallized; but at higher levels, there may be salt deposits covering the paint. The treatment of the pictorial layer has been merely mechanical cleaning, followed by a light application of Primal AC-33 in low concentration to prevent crumbling. However, because the effect of humidity creates a continual problem, the Directorate of Restoration of INAH is working on a more satisfactory solution.

Sculptures

Each of the previously mentioned sculptures presented a different problem, and each was treated according to its particular condition and characteristics. Because this work was extensive, only two attached and one-free standing carvings are discussed here as representative cases, Coyolxauhqui, Chac-mool, and the Eagle Knight.

Coyolxauhqui. This fractured sculpture was found on a circular bed of stone slabs (fig. 1). The central portion of this foundation, which had been filled with *tezontle* and clay, probably "upon sinking and settling, created an open hole beneath the stone, which weakened its resistance, causing the crack." Adding to this possibility was the sheer weight of the superimposed materials and the movement and settling of the subsoil. Furthermore, fragments in the area of the crack had broken off and in the rest of the piece some exfoliated and eroded areas, accumulations of earth in the pores of the statue, and loss of pigment were also discovered. The condition of the stone was quite good, and as a primary protection a canvas cover was provided. Later, an aluminum roof was installed, covering only the stone.

The proper protection of the stone and its base necessitated the design and installation of a support.[2] Two triangular iron plates, separated from the stone by a neoprene layer, were assembled under the stone, supported by beams with supports. In the center were three mechanical jacks on sand supports "to avoid the bending (or flexing) of the beams and to ensure the safety of the stone and the personnel working under the beams."[3] Finally, to keep the stone from sliding, five iron stops were installed on the edges, each separated from the stone with neoprene.

Two types of treatment were carried out: one preventive and one for consolidation. The preventive treatment, which was done before

2. This structure was designed and installed by engineer Julio Chan INAH's Office of the Restoration of Cultural Property with the collaboration of engineer Samuel Ruiz of DDF and the agreement of Prof. Eduardo Matos, Coordinator, Templo Mayor Project.

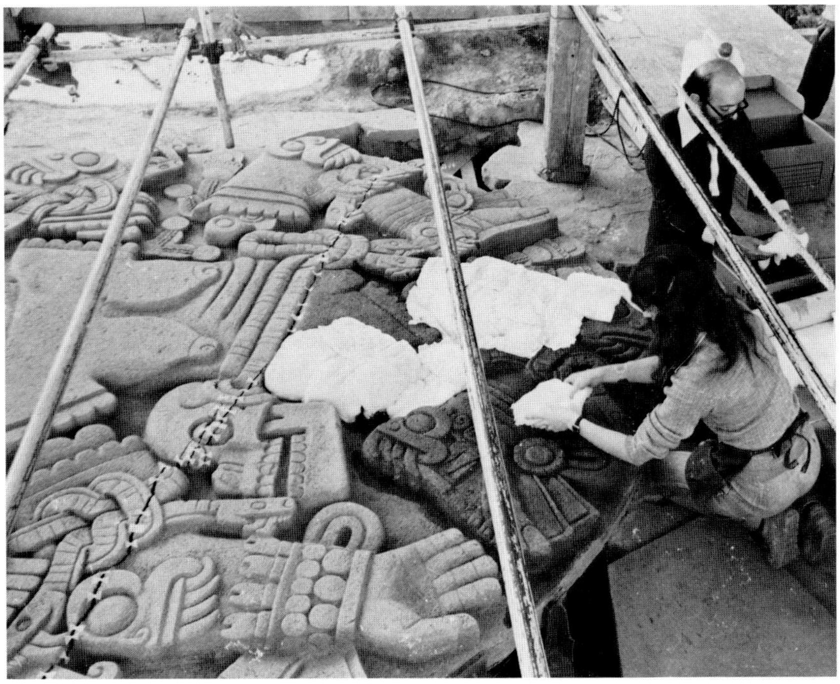

Figure 1. Treatment of the Coyolxauhqui carving.

3. Office of Engineer Julio Chan, 25 May 1978.

our arrival at the project, consisted of mechanical cleaning with water and alcohol. The application of pigments was done with soluble nylon in ethanol. The edges of fractures were covered with Japanese paper and Mowital, and the joining of most of the fragments in the fracture was accomplished with an epoxy resin (Araldite®) and its respective hardener; some fragments were joined with Mowital B60H.

After testing and the deep cleaning of the stone, the consolidation treatment was done with Wacker's OH strengthening agent (an ethyl-silicate-based adhesive) in order to restore cohesion to the stone. This treatment causes a double chemical reaction in the stone, first between the components of the consolidant, and then between the products of this reaction and the crystals of the stone itself, providing a chemical bond. The agent was diluted in naphtha (50%) and was applied with cotton compresses directly on the stone; the area being strengthened was sealed with polyethylene sheeting and weights in order to keep the material from gelling upon contact with air. Consolidant was added daily to the previously applied compresses in order to obtain maximum penetration, until each section was saturated and the vapors had dissipated. A total of 371 liters were used (Franco 1982).

Chac-mool. This sculpture was protected by a thick layer of mud covered by basaltic stones, *tezontle,* and andesite. It was necessary to partially remove this mud before removing all stones that were on top of it so that when the stone was raised it would bring with it the mud which would, in turn, bring with it the pigment that covered the Chac-mool. In other cases, it was necessary to fix the pigment or consolidate the preparative base before removing the stones. A large piece of stone (20 cm x 7 cm) was attached to the headdress, and it was difficult to remove without also removing part of the stucco to which it was attached. The stucco fragment was removed to the on-site workshop to be detached and subsequently re-attached to its original place. Only the left foot had fractures—a total of five, two of which were quite large. There were also some small exfoliated areas, fissures, and lacunae.

This sculpture, made of andesitic vitreous tufa, was found in a very good state of conservation, as was the paint layer. Its treatment consisted of mechanical cleaning and the application of pigments with

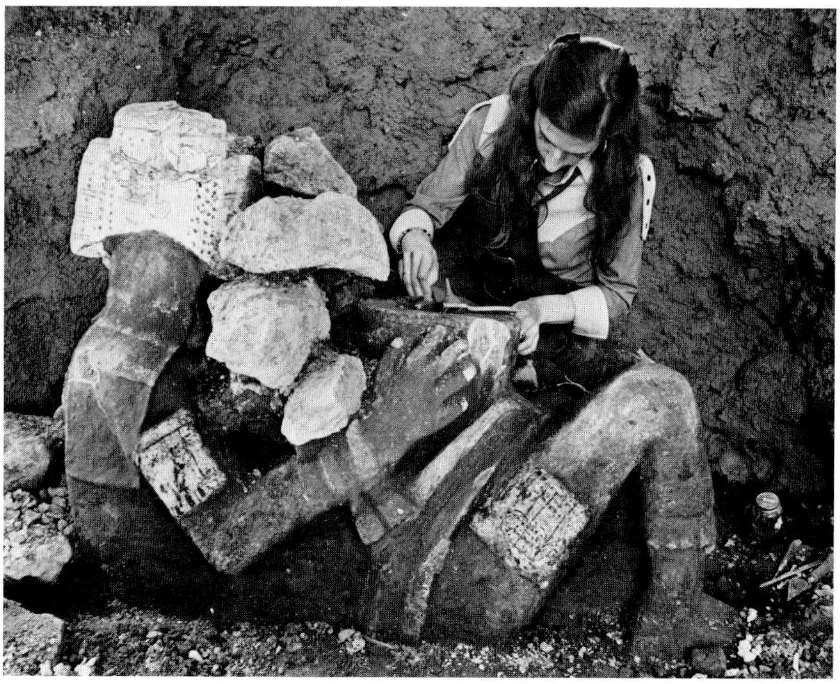

Figure 2. Chac-mool, a sculpture from Stage II (A.D. 1390). Note the stones that covered the sculpture.

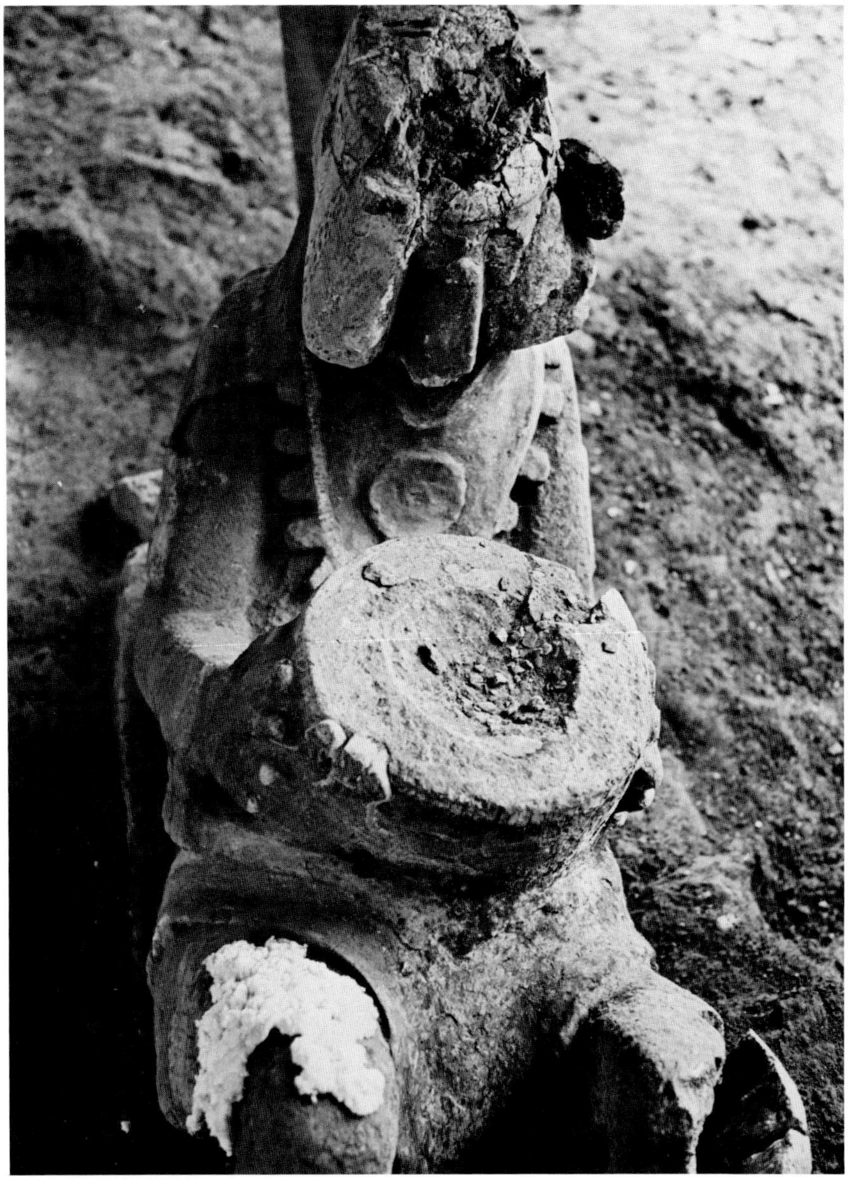

Figure 3. Frontal view of Chac-mool showing the affected part of the stucco of the head and the broken left leg.

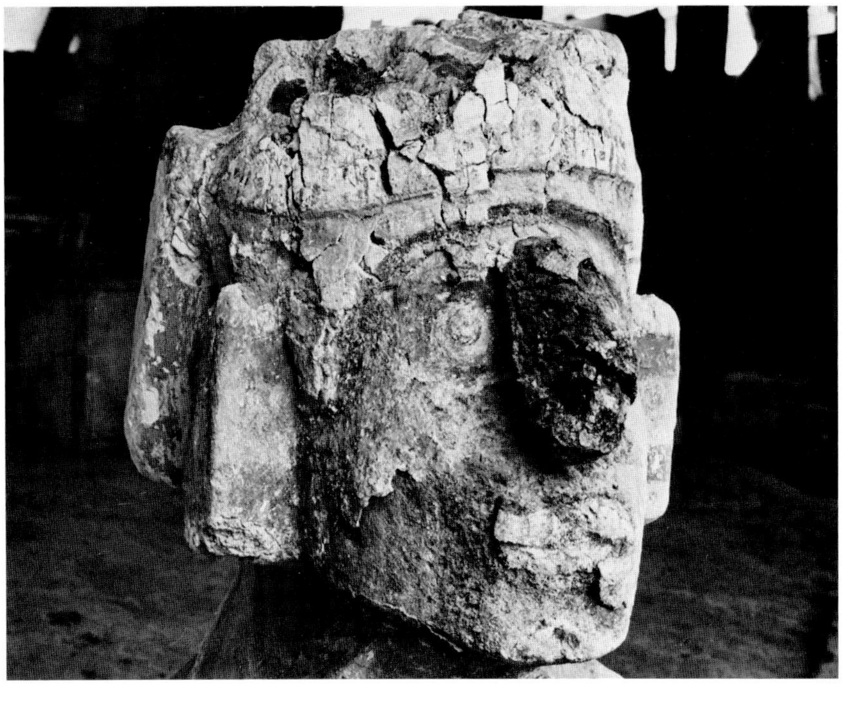

Figure 4. Chac-mool's head before restoration.

Figure 5. Detail of Chac-mool's left leg after conservation treatment.

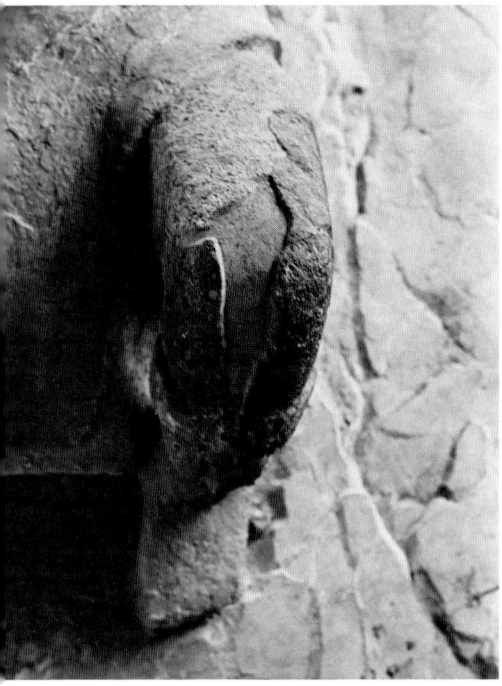

Paraloid B-72 in xylol (figs. 2-4). Stucco fragments were joined either with a mixture of fine sand, lime paste, cement, lime water, and Mowilith DM4 or with Endurol in distilled water. The Trinidad asphalt of the nose was joined with Mowilith DM4 and DMlH. The left-leg fragments were joined with Colmadur® B-V liquid epoxy resin and its hardener. Finally, the retouching was done with fine retouching paste.

The headdress fragment that was treated in the workshop was composed of stucco and small *tezontle* stones. It was detached from the stone with 5% Mowilith DM4 in lime water, added drop by drop, with a preliminary application of alcohol. This hardened the stucco to such a degree that, once dry, it loosened the mud that was attached to the stone making it possible to completely remove the stone from the headdress fragment. At the same time, the stucco from the Chac-mool headdress was consolidated in the same way, by means of drops and injections, and the fragment remained in place.

Small headdress fragments were treated with 10% Endurol in distilled water. Headdress exfoliations were consolidated with 50% Mowital B60H in ethyl acetate and alcohol or, better, 5% Mowilith in acetone. The fragment joints of the left leg were treated with a fine filler containing a small amount of earth from the vicinity (instead of sand) to hide the filler (fig. 5).

The Eagle Knight. This is a clay figure almost 2 meters high that represents a man adorned as an eagle, made in four interlocking parts. It was found on one of the walkways of the Enclosure of the Eagle Knights next to a similar figure that had a stucco covering in the form of feathers. It was removed from the place where it was originally found, and care was taken to record each fragment so that the figure could be assembled correctly in the restoration workshop. The pieces were cleaned mechanically and by washing with distilled water. The fungal stains were eliminated with hydrogen peroxide. All of the fragments were strengthened by immersion in Paraloid B-72 in xylol for three days, after which they were left to dry in an airtight container. The fragments were joined with Mowital B60H in ethyl alcohol and ethyl acetate. The joints were retouched using a ceramic paste made of ceramic fibers, calcium carbonate, kaolin, Mowilith 30, and acetone.

Missing parts were replaced using a skeleton of AJK dough, and ceramic paste was applied over this framework (Gedye 1968). All the retouched areas were restored with natural earth pigments applied with Paraloid B-72 in xylol. The joining of the figure's arms, because of their weight and size, had to be done in the following way. Each arm was filled with a mixture of polystyrene capsules and Mowilith DM4, and into this mixture, at shoulder height, two stainless-steel notched rods were set. Once the rods were attached, Mowital B60H in ethyl acetate and ethyl alcohol were applied to the surfaces that were to be joined. Once both arms were fixed, the joints were filled with ceramic paste and were retouched with natural earth pigments.

Deposits on objects in offerings

The treatment of objects in offerings began in situ. The objects were found covered in mud, some broken, some having retained their colors. Upon excavation they were exposed to air and light, thus it was necessary to control their drying immediately. Here, the restorer's in situ participation was merely preventive; that is to say, the processes of cleaning, fixing pigments, and reinforcing the material were done while the archaeologist was recording data and making drawings of each object. When the finds were brought to the studio, the final conservation treatment could be carried out. This treatment included gradual drying, deep cleaning, complete strengthening, the joining of fragments, the fixing of pigments, and the elimination of fungi.

As a typical example of an offering we have taken number 52, which consists of thirty-three polychromed ceremonial knives worked in flint, with conch and obsidian incrustations fixed to a copal base. These elements formed a facial profile on one side of some knives. The pigments are of mineral origin. In addition to the knives, many objects of obsidian, mother-of-pearl, conch, copal, copper, coral, sea-urchin spines, and fish-scales were also found in the offering. I shall, however, discuss only the treatment of the knives.

The knives were found on a bed of ocean sand covered with mud and protected by large volcanic stones. The side of each knife that was in contact with the bed of sand retained little coloring, in contrast to the side protected by stones. For their in situ treatment, the knives were carefully unearthed with wood-toothed chisels and a dissection needle. The pigments were found well attached, except the red, which had been reduced to dust. Paraloid B-72 in xylol was applied to the red areas. Copal, one of the materials most affected, suffered from a lack of cohesion. Before its removal it was necessary to strengthen it with Primal AC-33 in water. After the work of the archaeologist was finished, the knives and the rest of the materials were removed and transported to the studio. There, the knives received further cleaning, strengthening, and adhesion of the incrustations using Mowital B60H in ethyl acetate and ethyl alcohol.

As we have observed, the presence of the conservation specialist in an excavation project from the beginning is very important. The participation of such a professional on site helps to prolong the life of the materials that are found, as he can combat the immediate sources of deterioration and apply efficient methods to conserve and protect them from future decay. Interdisciplinary work involving archaeologists and other specialists (chemists, biologists, etc.) then, is essential for the conservation of our cultural heritage

References

Espinosa, Agustín
1974 "La iluminación, una causa de deterioro." *Boletin Informativo* 5. CEDOCLA. Mexico.

Franco, Maria Luisa
1982 "El tratamiento de conservación en piedra: tres casos." In *El Templo Mayor de México: Excavaciones y Estudios*. INAH. Mexico.

Gedye, I.
1968 "Pottery and Glass." In *The Conservation of Cultural Property*. Plenderleith, H.J. (ed.). UNESCO 112.

Rubén Cabrera Castro

ARCHITECTURAL RESTORATION AT TEOTIHUACAN: A HISTORICAL ANALYSIS

Rubén Cabrera Castro is Curator of the Teotihaucan archaeological site and museum and has coordinated archaeological excavations at Chicayán, Tzintzuntzan, and Cocula. He has taught courses in archaeological methods and published widely on various research projects.

Teotihuacan still retains the character of a continental metropolis—a great city whose size and architectural importance were unequaled in pre-Columbian American history. Perhaps for this reason, and the site's proximity to Mexico City, various fields of anthropological research have dedicated more time and energy to this area than to any other in Mexico. As a consequence, numerous restoration projects have taken place at this site.

Unfortunately, documentation of the restoration work is insufficient. This is not only the case at Teotihuacan, but throughout Mexico; most of the information that could have given us greater insight into the history and development of archaeological restoration in Mexico has been lost. This is due, in large part, to the lack of interest on the part of specialists in establishing principles and standards to govern restoration projects. Documents and other written materials on this theme are scarce. In the words of Molina Montes, "No studies exist that would reveal the conceptual basis on which the archaeologist may have based his restoration work" (1975). In the absence of information that might have illustrated the evolution of restoration theories and criteria that were applied to the treatment of buildings, the preparation of this paper, especially as far as architectural restoration is concerned, has been difficult. Therefore, most of the information presented is based on direct examination of the buildings that have been explored and restored at Teotihuacan over its several archaeological explorations.

Since restoration projects of pre-Hispanic buildings are closely related to archaeological excavations, it will be necessary to refer to the main interventions that have occurred at Teotihuacan in order to set a time frame and understand their development.

The first studies of this important cultural center were launched during the past century. In the beginning, the focus was on the largest and most monumental buildings—a practice that continued for quite a while. A series of minor unrelated projects were carried out by Almaráz, Charney, Batres, García Cubas, and Holmes, who did not seek a complete knowledge of the Teotihuacan culture, but rather just of some of its architectural aspects. The focus of these first projects in Teotihuacan was essentially to become familiar with the pre-Columbian culture in order to foster a sense of Mexican nationalism. With this purpose in mind, the issues relating to the restoration of pre-Hispanic buildings were not discussed, despite the fact that in European centers with great artistic traditions, such as France, it was a widely discussed theme at the time.

The most important work at the beginning of this century was carried out by Leopoldo Batres, who, by presidential decree, explored and restored the Pyramid of the Sun (pls. 21, 23), a portion of the Citadel, the Temple of Agriculture, and a portion of the Superimposed Buildings (Armillas 1950). The impetus for this vast project was ideological and political in nature, since preparations were being made for the celebration of the Mexican Independence Centennial, and it was essential that the foreign diplomats invited to these celebrations visit and admire part of Mexico's grandiose past.

According to many written sources, during the exploration of the Pyramid of the Sun, Batres dynamited part of the monument, sacrificing architectural information and detail for the sake of large-scale exposure. This action resulted in a strong negative reaction against him in subsequent years. Nevertheless, Batres (1912) himself has given us a description of his restoration work that somehow vindicates his excavations:

> It was now necessary to reinforce the wall surfaces, and to do that I proceeded to insert between each and every stone a mortar of lime, sand and cement, reinforced with fragments of volcanic rock stone, using this to fill the openings left by the clay that had been previously extracted in order to be replaced with the protective paste. This entire . . . operation was carried out without moving the stone from its original site, and thus managing to conserve . . . the old age and original aspect of the primitive construction.

This procedure is still employed today to reinforce pre-Hispanic walls at many of the country's archaeological sites. Even so, Batres's works at the Pyramid of the Sun contain serious reconstruction errors. This is evidenced by subsequent interpretive studies that have detected the source of the *taludes* (talus mounds) and their original dimensions.

Architect Ignacio Marquina, during his participation in the 1917-1922 excavations and restorations, drew up a hypothetical reconstruction of the pre-Hispanic city and its main buildings (Marquina 1979). He presented a reconstruction study of the Pyramid of the Sun, which—although it was widely accepted at first—was later refuted by new interpretations. Marquina's work adjusted the shape and the size of the pyramid indicating it was made up of five stepped bodies and making each side of its base 220 meters long. According to the same author, the fourth body of this building was formed by a *talud-tablero* style wall.

From 1945 to 1947, Rémy Bastien (1951) conducted a detailed study of this monument, also basing it upon the slope of its *taludes*. He arrived at the conclusion that the pyramid consisted of only four large bodies, with a wall at the end of the fourth body, which, according to Marquina, was a *talud-tablero* motif. He also included a cornice projecting from the face of the *talud*-style wall at the end of each body. The debate about the true shape and size of the Pyramid of the Sun began with Batres's errors.

From 1917 to 1922 an integrated anthropological research project was carried out in Teotihuacan under the direction of Manuel Gamio. Archaeologically, this project focused on the Citadel, where extensive exploration and restoration took place, although a few locations along the Street of the Dead were also examined. The project was undertaken largely to explore the "cultural manifestations, intellectual as well as material, of the population of the Valley of Teotihuacan, from pre-Hispanic times to the more recent cultures" (Gamio 1979). Regarding pre-Hispanic cultural manifestations, this project placed great importance on the architecture—an importance that is reflected in the care given to the restoration of the lavish Temple of Quetzalcoatl, as evidenced in the restoration of the great sculptured heads on its main facade (pls. 24, 25).

The work under Dr. Gamio's direction involved a labor of anastylosis because the building was covered with large interlocking sculptured blocks. In some places reinforcement was sufficient, in others it was essential to put back numerous pieces that were out of place. In some cases, such pieces were not taken down and put back in their original position, which was vertical and perfectly level; instead they were consolidated as they were found. As for large sculptural blocks that had fallen from their original positions, their reintegration was carried out with skill and moderation. The slightly fragmented stones were set in place in the same condition in which they were found, without any attempt to complete them or substitute them. The sections with some missing pieces were left as they were, properly reinforced, without the addition of new stone.

The interventions exemplified careful exploration and restoration, respect for ancient matter, and moderation in reintegration, without unnatural additions. With his efforts, Gamio brought prestige to Mexican anthropology.

Nevertheless, during the restoration of the new building of Quetzalcoatl, which is next to the Old Temple (*Templo Viejo*), the same care and principles unfortunately were not applied. Nor were they used in the restoration carried out at the other secondary temples of the Citadel compound during this period. This can be attributed to the construction method used in these buildings—irregular stones coated with thick layers of plaster and stucco—which made it impossible to use anastylosis to restore their shape. This being the case with the elements that were conserved in situ, the result was a reconstruction rather than true integration.

The present uniformity of these buildings make evident the fact that a large portion of the stone wall surfaces came apart and had to be reconstructed, and that it was necessary to cut a considerable number of stones to accomplish the restoration. Some of the materials used were completely different from the original. For example, thin pink stones were substituted for the basalt flatstones used as *istaplatetes* which, in Teotihuacan construction, functioned as fitted supports in the core for the moldings that surrounded the *tableros.*

Although in general the reconstruction is fairly reliable, in some cases doubts do exist. For example, in the fourth body of the attached pyramid where, according to Marquina (1979:55),

> some probing was done . . . perpendicular to the mound, discovering during several probings exterior structures consisting of leveled concrete walls. Upon pursuing these remains with the utmost care, the front of the building was progressively becoming uncovered . . . then it became clear that it comprised four talud-style bodies, adorned with tableros and a central staircase . . . . On the top portion, given the topography of the mound, there seems to have existed two small symmetrical structures, leaving free a central space, corresponding to the staircase.

The building presently consists of four bodies, and inasmuch as in the last of them no evidence of any original portions is to found, it is very possible that this body was added by analogy to the three previously extant restorations.

The reconstruction of the fourth body of this building and the restoration of the other secondary temples of the Citadel, such as the inside platforms, in some way contributed to the reconstructions that were subsequently done at Teotihuacan culminating with the massive reconstructions in 1962–1964.

Gamio's project also included the exploration and restoration of the area called the "Superimposed Compound," situated inside the Street of the Dead complex. This group contains several periods of construction of which two can be seen, the more recent phase situated on the upper level. The deeper structures belong to earlier periods, of which a relatively small portion has been explored, "due to the difficulties presented by the emptying and the conservation of the upper structure" (Gamio 1979). In order to conserve the upper level of structures, a roof held up by small beams and planks resting on iron columns and brick pillars was created. These are placed so close together that they obstruct a large part of the view of the lower compound. However, the Superimposed compound is the only example at Teotihuacan of portions of the lower level being exposed without destroying the upper level.

During 1932 to 1935, a section of the residential area of Teotihuacan was revealed by Linné's excavations of the compounds known as Xolalpan and Tlalmimilopan, undertaken in order to define ceramic sequences. The information that the author obtained permitted him to reconstruct the site in drawings and maps, but not the buildings themselves, since the excavation was covered up, avoiding the problem of restoring architectural elements.

In 1942–1944, archaeologist Pedro Armillas explored the residential-religious complex known as the "Viking Group." The restoration of the archaeological vestiges was achieved without complications. Walls were strengthened just to the height at which they were found, and despite further excavation to obtain information about earlier constructions, only a few earlier floors are visible, so that the visitor may better understand the sequence of development.

In 1942, when the Viking Group was being worked on, an accidental discovery was made on the site known as Tepantitla. These findings, comprising the remains of an ancient Teotihuacan palace decorated with mural paintings, shed much light on the religious ideas of the Teotihuacans. Painter Agustín Villagra restored them in such a praiseworthy manner that even today their colors remain quite vivid.

The architectural restoration of this complex was carried out by Orellana and Saenz, who cleared the rest of the murals. For the in situ protection of these paintings, the old rooms were reconstructed, with the original walls and pilasters extended in height in order to support a reinforced concrete roof. At this location there are other groups of pictures, protected under tile roofs that seem more provisional than the concrete roof.

In 1944, just east of the ceremonial center of an area designated as Tetitla, a site was identified at which, four or five years earlier, a complete mural was uncovered in clandestine excavations and removed. It was at this time that excavation began at, first by Carlos Margain and then by Pedro Armillas and, in subsequent years, by L. Séjournné. As a result of their efforts, we can now see a great number of the wall paintings that decorated the rooms and porticoes of this Teotihuacan palace (pl. 20). Unfortunately, the in situ protection of this 3800-square-meter site has been a real problem. Roofs have been erected with no regard for the shape or dimension of the original architecture. The supports and columns that hold up the protective roofs often obstruct important architectural elements. This results in the loss of the original spatial relationships and separates the painting from its intended context.

Atetelco is another of the residential complexes located in the same Tetitla area. Its exploration, which was begun in 1945 by Armillas, was continued in 1947 by Margain and Villagra. This site consists of various sections, but I will only refer to two of its patios where the restoration is fairly significant. In the so-called White Patio, because of the presence of several murals in three rooms, the rooms were completely reconstructed. The walls were raised to a height of 3.5 meters, and a concrete roof crowned with merlons was added. This was done in order to protect the pictures in situ and to integrate the great quantity of fragments recovered near the walls that had become detached when the subsequent level of occupation was built.

The reconstruction of the three temples with porticoes was based on the reconstruction on paper of the mural painting from fragments. These pieces were numbered and placed in their corresponding positions, which not only allowed one to determine the size of the porticoes, but also to study the motifs and designs represented on them. Once the buildings were reconstructed, the mural fragments were integrated on their walls. That which was authentic and that which was reconstructed were differentiated by the use of two shades of red, one bright and the other pale. This activity is still going on under the sole supervision of Santo Villa Sanchez, who collaborated with Villagra from the beginning of the reconstruction. According to Villagra, the reconstruction of these buildings in order to reintegrate the paintings offers "an approximate sample of their original appearance." Today such a practice would be questioned, but it is part of the history of the restoration of Teotihuacan.

In the White Patio section of Atetelco, carved stone blocks were found from a small altar that had been completely destroyed. With the restoration of this altar, the entire building was completed using new stones.

There were two more palace-type buildings—the Zacuala and the Yayahuala—which had been completely uncovered during the last construction period. There, ruins of wall paintings that formed a part of the architecture were also found, but because they had no protection from the elements, they have practically disappeared.

Nineteen sixty-two saw the beginning of one of the largest research projects, of utmost importance for Teotihuacan. I am referring to René Millon's Mapping Project, which basically consisted of a surface survey; the few excavation pits that were sunk did not permit analytical conservation and restoration. The floors in the Citadel and in Tepantitla that were cut up in order to create the stratigraphic pits were replaced with a stone pavement, which preserved the floor and indicated where the excavations had taken place.

Between 1962 and 1964, at Teotihuacan, INAH carried out one of its largest archaeological excavation and restoration projects, and it received international recognition. It was a project that came into being through governmental resolution. Its purpose was to present the site as evidence of the nation's glorious past and to make it a first-class tourist attraction, which would bring more foreign currency into the country.

The project was coordinated by Dr. Ignacio Bernal, who could depend on a sizable team of archaeologists and restorers and the participation of architects and other technicians. Among those who deserve special mention are the late archaeologists Jorge Acosta and Ponciano Salazar. The late restorer, Manuel Gaitan, who was in charge of the treatment of wall paintings with his team of restoration specialists, also deserves special mention.

Here, architectural restoration work was closely coordinated with the archaeological excavations focused on both sides of the northern stretch of the Street of the Dead, from the Viking Group to the Pyramid of the Moon, including the Palace of the Butterflies (pls. 26, 27). Outside the city center, it included the completion of the explorations of Tetitla, Zacuala, and Yayahuala.

This project has been singled out by numerous conservation critics and specialists (theoreticians, historians, restorers, architects, and archaeologists), as a "massive large-scale reconstruction project, accused, understandably enough, of showiness," and later used as "an equally debatable background for Light and Sound" (Gendrop 1984). This has led, unfortunately, to its underrating with respect to international standards of conservation.

In fact, this project, like many of its type that are state-sponsored with tourist-promoting, political, and nationalistic ends in mind, sought to display the monumental nature of the site and the buildings as seen from the Street of the Dead or from its main facades. Thus the facades of the buildings were restored, adorned with *tableros* and stairways, while the rear mounds were destroyed. The reconstructed sections cover structures from previous periods, which makes future exploration difficult and limits our knowledge of earlier periods.

The restoration, then, was basically achieved by a reconstruction that completed the lowest bodies of the pyramidal foundations, according to the information provided by archaeological excavation. On the upper parts, the reconstruction was done by analogy, based on the extant lower portions and on other buildings from the same archaeological area or from other places such as Tepeapulco. The buildings that make up the northern portion of the Street of the Dead, the Pyramid of the Moon, and the pyramidal foundations that make up the plaza of the same name, were all restored in this way.

To cite one example, there is Building 4, a pyramidal foundation located on the west side of the Plaza of the Moon, joined directly to the Palace of the Butterflies. The restoration of the compound was carried out by the archaeologist Jorge Acosta. According to Acosta, this building had four stepped bodies, each one with a closed *tablero* that was twice the height of its *talud*. For the restoration of the first body, the information was in situ, and for the upper bodies the restoration was based on the north side of the building, which connects with a platform with bodies that are better preserved. The restoration of the fourth body was based on the proportions of the lower bodies. Its upper portion was reinforced with a stone pavement, indicating a central area that, according to Acosta, was the space that the subgrade of the temple had occupied.

The restoration of this building was the model for the restoration of other buildings that make up the Plaza and the Street of the Dead. The method and criteria used were very similar to those used in the restoration of the pyramid attached to the temple of Quetzalcoatl in the Citadel, more than forty years earlier. International standards and criteria of restoration had not yet been set forth. But by 1962, when the Mapping Project began, these standards and principles had already been discussed (the Athens Charter dates from 1931).

The Palace of the Butterflies, which was found in a state of collapse, was missing many parts. It was totally reconstructed by extending the walls and adding a roof of concrete reinforced with small steel beams, not visible to the public. The heavy roofing was supported by capitals and columns of concrete, covered with the carved stones and fastened with strong perimeter beams.

The roof was placed at a height determined by the restoration of one of the carved pilasters that had survived nearly intact. The pilaster was restored by inserting the missing sculptural blocks. Then the eleven remaining pilasters could be assembled, integrating new carved stones when necessary. The rest of the reconstructed components were made by analogy, based on models and drawings (Acosta 1964).[1]

1. I should mention that detailed information regarding the shape and the materials that were used in this reconstruction was published by Acosta, so it is not necessary to repeat here what is explained in detail in his book.

The deep excavation that was performed in the central portion of the patio detected older constructions and remains of wall paintings. The compound found there is known as the Temple of the Plumed Snails. Even though a large part of the compound was reconstructed, the average visitor can draw a reasonably accurate impression from it. For specialists, however, it belies both authenticity and the internationally accepted standards and rules of restoration and in no way should be repeated.

As a consequence of the massive and arbitrary reconstruction

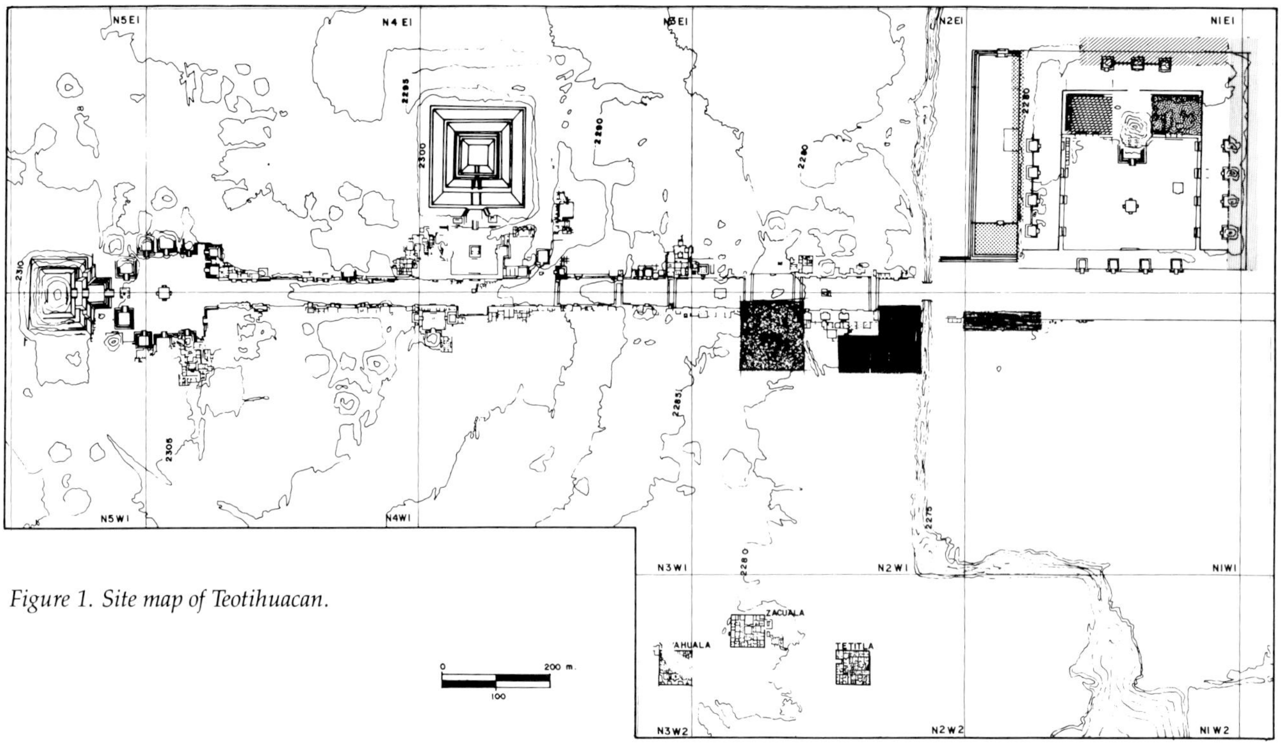

Figure 1. Site map of Teotihuacan.

performed during the 1962–1964 Archaeological Project, a wave of protest in national and foreign anthropological circles was unleashed, voicing strong disapproval of the attitude of those responsible. Even so, many archaeologists paid no attention to these protests, maintaining that international restoration standards are not applicable to pre-Hispanic Mesoamerican structures. They continued to carry out complete reconstructions at other sites.

Finally, we have the work carried out by the Proyecto Arqueológico Teotihuacán (PAT) 1980–1982, under the auspices of the Instituto Nacional de Antropología e Historia. Like the 1962–1964 Project, this one was of a political-administrative origin, endowed with ample economic and human resources and maintained within the framework of the Mexican government. This time, however, the participants resolved not to repeat the same actions as in 1962–1964, although the political conditions were very similar.

The works of the Proyecto Arqueológico Teotihuacán, which I coordinated, were based on discussions among participating archaeologists, physical anthropologists, restorers, architects, biologists, and other specialists.

As for the investigations, even though their original proposals were traditional in character because projects for exploration of the central part of the archaeological zone were still being contemplated, it was nevertheless possible to work in the periphery, planning programs that attempted to achieve an integral knowledge of the different social problems of the great metropolis.

The restoration is based on the international rules regarding conservation and on an ethical compromise to respect the original data. Some Mexican documents and publications were taken into consideration, such as the *Reunión Técnica Consultiva sobre conservación y monumentos de Zonas Arqueológicas* (Advisory Technical Session Concerning Conservation and Archaeological Zone Monuments) and Augusto Molina's work (1975) on the same theme. We could thereby profit from both the good and bad experiences in the work that preceded ours. We also welcomed a critical evaluation at any time and from any specialist in our field, especially regarding restoration and the measures that were taken to protect certain specific elements, which are included below.

The PAT excavations uncovered large areas of architectural structures. To get a better idea of the magnitude of the restoration work, the central part of the archaeological zone where different excavations were done covered a total of 96,465 square meters. This does not include the investigations that were carried out in the periphery—excavations that were by necessity backfilled, with proper steps taken to protect the architectural remains.

Another study, still in preparation, makes reference to the way in which movable materials were treated. In this volume, Esperanza Teyssier explains the problems relating to stuccowork, wall paintings, and carvings. Before I conclude, I will touch on some specific cases of architectural restoration and the preparation of the same.

Different methods were used and appropriate solutions sought depending on the nature of the problem. For example, using the terminology employed by Molina Montes (1975), the most common method used during this period was restoration by consolidation. With minimal intervention of this type, additional stability was given to the structures without altering their form; at the same time, a more realistic image of the monument in ruins was put on display (Cabrera et al. 1984:26).

Restoration by reintegration was rarely done because the majority of the buildings that were treated were not constructed of blocks of sculptured, cut, or carved stones. Instead, we had to deal with irregular stone masonry that formed fairly coarse walls. This technique was applied to the Temple of Quetzalcoatl on its south side, where the facade had been dismantled by pulling off part of the sculptures that appeared close to the building (Cabrera and Sugiyama 1982:181). Some of these were replaced and set into position according to the drawings that had been prepared or according to the main facade of the monument that was restored between 1917 and 1922.

Even though restoration by reconstruction is furthest from the theoretical concepts that were recommended, this technique was applied to some buildings in the 1980–1982 project. This work undertook a simple strengthening of buildings to lend stability to the architectural elements, indicating clearly the fractures of the building and differentiating between what was original and what was reconstructed. The technique is called *el rejoneo*[2]; it has been used in Teotihuacan since Gamio's restoration of 1917–1922 and is commonly used at other architectural sites. Nevertheless, in some cases it was necessary to carry the reconstruction a little further. This has meant extending some of the architectural elements in the areas of heaviest destruction, with the objective of giving more protection and strength to the monument—as well as for didactic purposes, to make the site more understandable to the public.

2. A term borrowed from bullfighting, it means "jabbing with a short lance."

Since one of the objectives of the 1980–1982 project was to become familiar with the complete sequences of architectural development, restoration by unearthing was the method employed most widely; this made it necessary to deepen the excavations in certain places. Some of the sequences of construction, those examples that were clearest and would be most understandable to the public, were revealed as a means of explaining the development of Teotihuacan architecture.

Previous projects have carried out this type of restoration. Such is the case of the Superimposed Compound, the restoration of the Temple of Quetzalcoatl, and Acosta's work in the Palace of the Butterflies where two periods of construction are shown (only here, as we have already seen, the upper building was completely reconstructed). Other examples that met with less success include the Agriculture Compound and the Palace of Tetitla, where several superimpositions of construction are displayed; there too, it is difficult to understand which occupation level they correspond to because they are presented in a confusing fashion.

The most important examples of restoration by unearthing carried out by the 1980-1982 project include: the building designated IB′ situated in the esplanade of the Citadel, which displays a sequence of seven superimpositions (Cabrera 1982b); Structure 87 of the northwest compound, with two superimpositions; the Plaza of the Superimposed Altars, with four superimpositions; two small superimposed portions in the central area of the Street of the Dead complex; and the stairway with serpent heads and jaguar, in building 40A of the West Plaza (Morelos 1983).

These are not the only examples where the excavation was deepened to uncover and display sequences of construction. There exist other places in the more recently excavated areas where parts of older constructions were left uncovered, but in all the cases, the excavators made sure that these samples were easy to understand. When it was not possible to present an understandable sequence, it was thought preferable to cover up the older construction, taking the necessary conservation steps.

The restoration by unearthing of the 1980–1982 project, which made the display of architectural development sequences possible, entailed as a consequence the necessity for collateral repair work and direct intervention on the monuments, in order to provide them with protection while managing at the same time to display them adequately in keeping with the context in which they were situated. To accomplish this with joint input from architects, restorers, and archaeologists, the different suggestions that were presented were analyzed and discussed to decide the best solution to the problem at hand. Even though the Project carries the responsibility, INAH authorities also participated in making decisions about this work. In this manner, appropriate roofs were designed and completed to protect the archaeological elements that had special artistic value or historical and social content, and they were integrated into the architectural compound. In each case adequate protection was provided against water, rain, sun, dust and human intervention.

One of the more outstanding examples (which are therefore more open to criticism) is the roof of building 1B′ in the Esplanade of the Citadel, constructed competely of metal to protect a compound with a sequence of seven superimpositions and the mural painting that it contains. There are also two roofs constructed of reinforced concrete supported by small steel beams—one of these protects the Superimposed Altars situated southwest of the Superimposed Compound, which also contain mural paintings; and the other protects the serpent heads in the West Plaza Compound of the Street of the Dead complex. In both cases, the roofs allow the viewer to distinguish the different stages of construction while preserving the structural relationships and original architectural finishes, as well as protecting the carvings and the color that was applied to them.

In this brief description of the history of restoration and its development, there has been no intention of either discrediting or justifying the meaning and the objectives of the projects that have taken place in Teotihuacan, for the practices they engendered cannot be understood apart from the historical situation and political conditions that existed at the time they took place. The ideas and experiments that are common and accepted in one era, erroneous as they may be, become the basis for the development of subsequent work. We should repeat what has already been stressed many times in this colloquy: In finding solutions to our problems, in this case in order to preserve cultural treasures, the participation of interdisciplinary teams is indispensable.

References

Acosta, Jorge R.
1964 *El Palacio de Quetzalpapalotl*. INAH. México.

Armillas, Pedro
1950 *Teotihuacán, Tula y los Toltecas*. RUNA 3. Buenos Aires.

Bastien, Rémy
1951 "The pyramid of the Sun at Teotihuacán: A New Interpretation." In *Memorias del XXIX International Congress of Americanists*. Chicago.

Batres, Leopoldo
1912 "Descubrimientos y Consolidación de los Monumentos Arqueológicos de Teotihuacán." In *Memorias del XVIII International Congress of Americanists*. London.

Bernal, Ignacio
1963 *Teotihuacan: Descubrimientos, Reconstrucciones*. INAH. México.

Cabrera Castro, Rubén
1982a "El Proyecto Arqueológico Teotihuacán," *Memoria del Proyecto Arquelógico Teotihuacán 80–82*, Primeros Resultados. Coordinación de Rubén Cabrera, Ignacio Rodriguez y Noel Morelos. INAH. México.

1982b "La Excavación de la Estructura 1B′ en el Interior de la Ciudadela." *Memoria del Proyecto Arqueológico Teotihuacán 80–82*. Coordinación de Rubén Cabrera, Ignacio Rodriguez y Noel Morelos. INAH. México.

Cabrera Castro, Rubén and Saburo Sugiyama
1982 "La reexploración y restauración del Templo Viejo de Quetzalcoatl." *Memorias del Proyecto Arqueológico Teotihuacán 80–82*. Coordinación de Rubén Cabrera, Ignacio Rodriguez y Noel Morelos. INAH. México.

Cabrera Castro, Rubén, Noel Morelos, and Ignacio Rodriguez
1984 "Los Problemas de Conservación Arquitectónica en Teotihuacán: Análisis y Criterios." In *Cuadernos de Arquitectura Mesoamericana* 3. División de Estudios de Postgrado-Facultad de Arquitectura. UNAM. México.

Gamio, Manuel
1979 *La Poblacion del Valle de Teotihuacan*. Edición Fascimilar. INI. México.

Gendrop, Paul
1984 "La Problemática de la Conservación." Introducción editorial in *Cuadermos de Arquitecura Mesoamericana* 3. UNAM. México.

Margain, Carlos
1966 "Sobre sistemas y Materiales de Construcción en Teotihuacán." In *Teotihuacán*. SMA. México.

Marquina, Ignacio
1979 "Arquitectura y Escultura." In *Manuel Gamio: La poblacion del valle de Teotihuacán*. Edición Fascimilar. INI. México.

Millon, René
1976 *Urbanization at Teotihuacán, Mexico. Volume One: The Teotihuacán Map*. Austin.

Molina Montes, Augusto
1975 *La Restauración Arquitectónica de Edificios Arqueológicos*. Colección Científica 21. INAH. México.

Morelos, Noel
1983 "Esculturas y arquitectura en un conjunto teotihuacano," ponencia a la *XVIII Mesa redonda de la Sociedad Mexicana de Antropologia*.

Villagra, Agustín
1951 "Las Pinturas de Atetelco en Teotihuacán." *Cuadernos Americanos* X 1. México.

Esperanza Teyssier

ARCHAEOLOGICAL RESTORATION AT TEOTIHUACAN: A HISTORICAL ANALYSIS

Esperanza Teyssier is Restorer of the archaeological zone of Teotihuacan.

A historical analysis of the restoration of the architectural and city-planning elements of the Teotihuacan archaeological site was presented by Rubén Cabrera Castro (this volume). This paper offers a historical analysis of the restoration of the stuccowork and mural paintings during the subsequent excavation periods and the results obtained to date. It will look at the wall and floor covering treatments, as well as the sculptural elements of the stuccoed stone.

For thousands of years, stuccowork has been used as an architectural finish. We find it in practically all large ancient cities, and it was an integral architectural element for the Mesoamerican civilizations. As a support for mural paintings, stucco was a medium for expressing religious and social beliefs.

Teotihuacan was not a gray city constructed of stone. Present evidence and the descriptions of chroniclers of the Colonial period tell us it was teeming with life; buildings, streets, and plazas were covered with polished stuccowork in white, red, and pink. It is vital that we not allow these polychrome elements to disappear, or the concept that future generations will have of Teotihuacan will forever be altered.

Each archaeological intervention has left a characteristic imprint on the site. Restoration work always reflects the state of technical knowledge at the time and also the criteria that guide the archaeologist in charge of the exploration.

Since we have no documentation regarding the earliest restorations, my chronological study will begin with the excavations of 1905 led by Leopoldo Batres. In his first excavations of the Pyramid of the Sun, Batres found colored stuccowork covering the walls and floors; he also mentions mural paintings in the Palace of Agriculture as well as remains of stuccoed stone (Batres 1912). He did not, however, consider restoring the stuccowork except, perhaps, the loose fragments of mural paintings and their concrete plaster.

During the 1900s, archaeological practice focused on the reinforcement of buildings, walls, and plaster, but there was no attempt to preserve the stuccowork. Not until the second decade, under the direction of Manuel Gamio, did the conservation of stuccowork and mural paintings take place. His clear sense of the architectural integrity of the site allowed him to see the stuccowork as an indispensable component of Teotihuacan. He carried out restoration work not only on the concrete plaster that acts as support for the stuccowork, but also on the stucco itself, binding the thin layers to the support along their edges.

Opposite: The Citadel, Teotihuacan.

The discovery of the Old Temple of Quetzalcoatl inside the Citadel dates to the same period. The Temple facade was covered with poly-

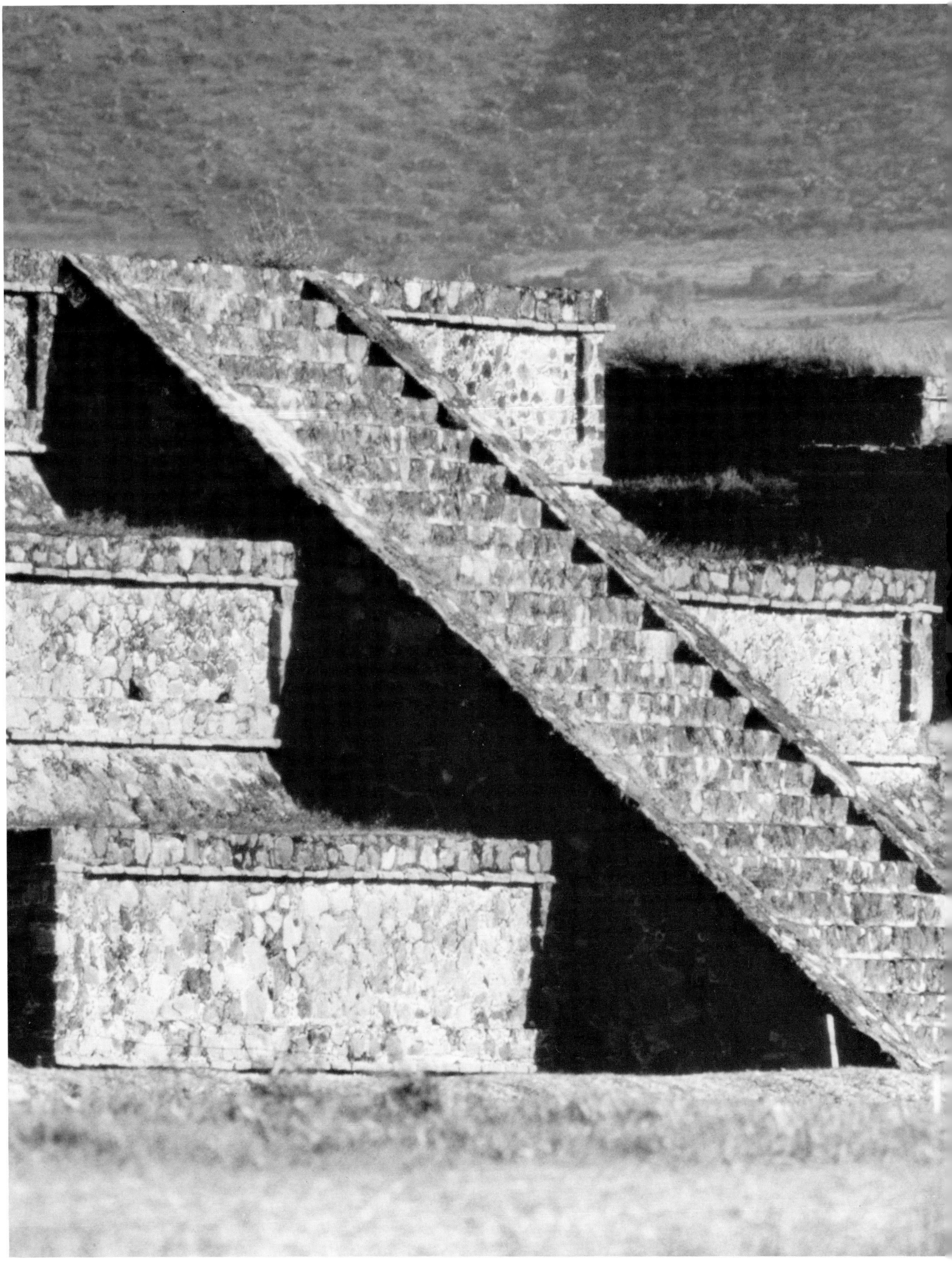

chrome carved stone and had the remains of multicolored stuccowork on the sculptures and staircase. This stuccowork is preserved to this day without the benefit of any restoration work. No restoration was performed on the mural paintings, only reintegration, and in some places reconstruction (pl. 27). Painters, rather than restorers or archaeologists, executed the work.

One fragmented mural painting was reconstructed by painters Agustín Villagra and Mateo Saldana (Marquina). Villagra and Santos Villa worked on many fragments of the mural paintings found in the excavations that were done between 1920 and 1960. Their method consisted of restoration, reinforcement, and cleaning. The reinforcement was actually done on the concrete plaster, not the pictorial layer; liquid cement was applied to the back side of the plaster fragment. In fact, Santos Villa continues to work at the Atetelco Palace today. In 1942–1943, Armillas (1950) reported having found polychrome painting on adobe, but did not mention how it was preserved.

It was not until 1962–1964, under the direction of Ignacio Bernal, that a group of restoration experts from the Index and the Directorate of Restoration of INAH collaborated with archaeologists on the Teotihuacan project. At that time, conservation of the mural paintings and the wall stuccowork was undertaken. The techniques employed included reinforcement through injection, surface protection of stuccowork and mural paintings remaining in situ, and the removal and mounting on frames of the mural paintings that would deteriorate if left in place (Bernal 1963). In later years, essentially the same techniques were used by what is now the Directorate of Restoration of INAH, although over the years the processes as well as the materials employed have changed.

The 1980–1982 Teotihuacan archaeological project was the most recent excavation. It was designed from the beginning as an interdisciplinary effort. The work was performed under the supervision and control of the senior archaeologist, together with the Resident in Restoration from the Directorate of Restoration. The close collaboration between archaeologists and restorers resulted in timely and appropriate treatments to the stuccowork. Newly excavated stuccowork on walls and floors was restored immediately. Floors that had been untreated during previous excavations received attention. There were few mural paintings revealed, but two small fragments were taken to the laboratory for mounting.

Plumed serpent heads and large mask fragments of Tlaloc, which originally decorated the north facade of the Old Temple of Quetzalcoatl, were found in the excavation of Compound 1D (where they were left in situ). Similar artifacts from the south facade were discovered in the living quarters unearthed in the 1917–1922 excavations.

In the West Plaza Complex, along the Street of the Dead, the polychrome serpent heads that line the beams and the staircase of the first construction period were restored. In the restoration laboratory, cressets of stuccoed stone or ceramic, as well as stuccoed ceramic pieces, were restored. In the 1982 season, in collaboration with the archaeologist Laurette Sejourne, conservation work was done on stuccowork and mural paintings in the Atetelco Palace. During the same season, work was done on the coloring of thick mud plaster that remained in place and on the altar of the first construction period of the Superimposed Altars Plaza and others that had collapsed in Complex 1D.

Our restoration work was not limited to treating stuccowork. On an archaeological dig, the materials that require conservation and restoration are quite varied, especially if they are to remain in place. Such was the case with the adobes sealing the accesses in the sunken patio in Structure 57B and those visible around the mud altar of the Superimposed Altars.

Another important element that needed restoration treatment was the demijohn-shaped cave in the parking area. Its walls and floors are composed of a material with a high clay content that is native to the area. The floor and the small altar, which serve as a foundation for an andesite slab, are covered with mud plaster. As it was being recovered from the material that obstructed it, the entire area was reinforced. Subsequently, when the humidity of the clayey material had changed sufficiently to allow a more permanent reinforcement treatment, it was treated once again.

Along the Street of the Dead, in the northwest corner of the Citadel, a section of stratigraphy with a volume of 60 x 60 x 70 centimeters was consolidated. This was done by the drip method to allow for a greater penetration of the reinforcing material. Unfortunately, this section is subject to inevitable deterioration because it is not adequately protected from visitors.

As previously stated, each excavation at Teotihuacan has left its imprint and the result is a sample of restoration "methods." To visit Teotihuacan is to leaf through the history of its conservation and restoration, which could be summed up in four stages. The first was characterized by the absence of any specific treatment for stuccowork; the second began with the Gamio project and lasted until 1962, when the restoration of stucco began in a free, empirical way; the third, from 1962 to 1980, was when restoration appeared as a technical discipline, partially integrated with the Teotihuacan Project; and the fourth began with the Teotihuacan Archaeological Project of 1980–1982, which viewed restoration as a discipline intrinsic to archaeological work.

First stage: From the mid-nineteenth century to Batres

There is little that can be said regarding the stuccowork unearthed in the excavation performed in the last century and the early part of this century. Batres (1912) mentions it but we know little about the conditions in which it was found, how extensive an area it covered, and when it was unearthed. An analysis of Batres's writings leads us to conclude that at excavation time, much of the stuccowork covering the walls and floors remained in place. At present, however, there are large areas with stucco missing, probably as a result of the lack of conservation and restoration; this exposed the concrete plaster, entirely altering the appearance of the place.

As for mural paintings, no great effort was made to conserve them. The early excavators simply picked up fallen fragments with their plaster support, identified the iconography, and subsequently placed the fragments in storage. At present, the adhesion to the support is good; there are no signs of chalkiness or distortion.

Second stage: From the Gamio excavations to 1962

The second period reveals significant progress regarding the conception of what the conservation of archaeological treasures should entail. The International Documents on the Conservation of Monuments and Sites (the Venice Charter), ICOMOS, and their specific guidelines on the subject were still to come, but this does not diminish the significance of Gamio's conservation and restoration efforts related to stuccowork in 1917–1922 and subsequently.

The excavations of this time revealed a great wealth of stuccowork, not only in buildings and temples of the final construction period, but also of previous periods, such as that of the Superimposed Buildings. Gamio's work involved an excavation technique that allowed the disclosure of large areas of stuccowork without damage. The finish displays a polished, bright texture that remains to this day, and the original colors are conserved as well. There has been speculation that this stuccowork finish is the result of the application of a varnish for protection or reinforcement; however, investigations have failed to reveal any trace of resin.

In general, this second period was characterized by the conservation of stuccowork by finishing the edges with a fine paste of white cement to keep it adhered to the concrete plaster that acts as the support. Large missing surfaces were covered with the same fine white paste, and the isolated pieces of original stuccowork were left embedded in it. A clear example of this technique is the Temple reconstruction in the Superimposed Buildings.

In the course of my investigations, I interviewed some of the oldest workers in the area, who took part in past archaeological projects—although it must be stated that none of them worked directly with Gamio and that their information is hearsay. My informants concurred in mentioning the use of cactus juice to enhance the adhesiveness of the aforementioned cement paste.These people attribute certain qualities to the cactus gum resin that it lacks in actuality, for we have found that it does not augment the adhesiveness of modern cement. Since pre-Hispanic days, cactus juice has been known for its medicinal applications, and it is possible it was used as an adhesive in fresco painting. Mixed with water and lime, it could have served as a medium for the pigments.

We may conclude that the result of the stuccowork restoration methods used during the second period—affixing loose pieces to the concrete plaster by finishing off the edges with a fine white cement paste—was positive since the restored stucco has endured over half a century in good condition. However, the floor stuccowork remained untreated and it is impossible to know whether we are seeing the same stucco today that was unearthed by the excavation. As for the mural paintings, the restoration work was at times complete and competent, as in the case of the Atetelco Palace.

Third stage: 1962 to 1980

The third period was characterized by great activity in the treatment of mural paintings disclosed in the Teotihuacan Project and of paintings dating from previous times, both in storage and in situ. There is less to report regarding stuccowork restoration during this period.

The 1962–1964 Teotihuacan Project revealed a large number of mural paintings, most of which were removed from the walls and transported to a restoration lab in the Old Museum for mounting onto new supports. The restoration techniques used during this period were the same as those used by the Directorate of Restoration and Index of Cultural Heritage, founded by Don Manuel Castillo Negrete, who maintained a close and fruitful relationship between Mexican restoration centers and those in Europe and the United States. The group of restorers commissioned to work on the Teotihuacan Project comprised twelve experts under the direction of Manuel Gaytan and Jesus Alvarado Lang, a chemical engineer in charge of the research laboratory.

The technique used for the removal was usually the *distacco (stacco)* method; the *strappo* method *(stacco a macello)* was mentioned only once. First, the surface was cleaned and the fragments were covered with cloth and polyvinyl acetates in emulsion and reinforced with an iron rod and gypsum structure. Then the fragments were removed from the walls and the wall plaster was eliminated. Each fragment was placed on a uniform surface, where it was encased in wood strips to frame it. A thin layer of modeling clay was applied to the fragment, 3 or

4 millimeters from the area between the fragment edges and the wood frame strips, which left the painting at a different level.

The new support was a thin aluminum sheet with a frame reinforced with T bars and aluminum corners. The adhesives used most often, synthetic resins, were applied uniformly on both the fragment and the layer of modeling clay. Then the aluminum support was positioned and pressure was applied by weights.

After forty-eight hours the mural was uncovered; the modeling clay was removed, exposing the painting fragments raised above the set resin like the printing areas of a woodblock. A mortar of lime and sand was applied to fill in the depressed background, resulting in an even surface, with the painted fragments surrounded by mortar. To finish, the carbonate layer over the pictorial surface was removed, and color was restored on the replastered walls by applying a final protective layer.

The mural paintings in situ and the wall stuccowork were reinforced and replastered. Pigment was applied, surface inlay was removed, and a protective coating was applied.

For the most part, the materials used were PVA type resins in emulsion, epoxy resins, silanes, and Fester Transparent Silicone. To set the surface and fragment, lime caseinate was also used.

Some of these paintings were put back in their original location, as in the Temple of the Plumed Snails, but most remained for years in storage at the Directorate of Restoration. From there, they were shipped to the Anthropology Museum, and in turn disbursed, as there is no adequate place there to display or store them.

When the Teotihuacan Project ended, the work of the experts of the Directorate of Restoration continued with some variation in the materials used. Also at that time, work was ordered by the Regional Zone Administrators, but as there were no restorers hired, this work had to be executed by untrained personnel.

Conclusions useful to future projects can be made about the activities during the third period. First, the restoration of both wall stuccowork and mural paintings was undertaken according to the scientific criteria of the time. For the first time, stuccowork restoration was done not by archaeologists or persons outside the conservation field, but rather by a group of experts assembled to achieve the rescue of the national archaeological heritage. Second, the efforts of these restorers were not empirical or personal. European conservation techniques were known and used, new and pre-Hispanic materials were researched, and solutions were sought within the limitations of the situation. Now that more than twenty years have passed, we can judge if the techniques and materials employed were appropriate and effective.

Fourth stage: 1980 to Present

The fourth stage covers the current period. It began in 1980 with the Teotihuacan Archaeological Project 1980–1982. The Restoration Division started conservation work shortly after the beginning of excavations. It consisted of the Resident, working for the INAH Conservation Division; an expert from the mural painting laboratory of the Directorate of Restoration; and a group of untrained workers that fluctuated in number from forty to one hundred. A group of workers was assigned to each excavation site according to the needs of the area; a field head and a foreman were appointed to carry out the supervision of the sites, which encompassed an area from the Citadel in the south to the Street of the Dead Complex in the north. Besides the field personnel, there were administrative and storage personnel working for the Division, assigned as the Project headquarters. The laborers were trained on the job in the field. Some of them reached a high degree of competence and expertise.

From the outset, the principal goals were the conservation of the greatest possible amount of wall and floor stuccowork and the salvage of all archaeological material when necessary. To reach our objectives, we took into account time and budgetary factors: the Project had set the budget, excavation, and restoration time. However, when the Project was over, further work had to cease, because there were no more funds available.

Among the project objectives was the restoration of the murals of the Manuel Gamio school. Some were quite deteriorated and others were covered with lime paint. Baltazar Trejo, of the Directorate of Restoration, was in charge of this restoration. He was assisted by four laborers, and the work was completed within the projected time.

Another issue was the treatment of the west facade of the Old Temple of Quetzalcoatl. For years the stucco-covered beams and the cut stones that are at an angle with the beams' outside surface had shown deterioration. This deterioration was caused by humidity, which stemmed from the water retained inside the wide beams the entire year. It was compounded by the capillary humidity that may result from a spring beneath the level of the first section, which is located in the area of the great staircase. Taking advantage of voids left during the reconstruction of the beams' lateral walls, we removed the damp mud inside and allowed the interior to dry out. The holes were subsequently refilled with new dry material, and the surfaces were covered with volcanic building stone and rubble stone, as had been done in the Marquina excavations.

This solved the humidity problem as far as the beams were concerned, but not that affecting the first, second and third areas where the beam joined the *tablero* (vertical panel) and *talud* (sloped panel)—the humidity there stemmed from the subsoil. We asked the Project Director to allow us to dig a pit at the level of the floor of the second body next to the beams as far as the level of the lower floor. This excavation would not damage the floor stuccowork because the area was covered by white cement paste. The request was granted and the pit was dug, revealing mud with a great degree of moisture that increased at lower depths. The pit has been covered with sheets and left open for the past three years in order to observe the change in water level through rainy and dry seasons. This open space in the rear of the *tablero* and *talud* allowed the stones to dry thus stopping deterioration.

In addition to eliminating the cause of deterioration, we reinforced the stone blocks, then treated the edges surrounding individual affected areas. The latter measure was very important to keep the exfoliated layers in place and to prevent the penetration of runoff water. As a preventive treatment, the entire stone facade was impregnated with Wacker® OH reinforcement and waterproofed with Wacker® W190L.[1]

1. See Appendix.

In Front number 3 the altar of the plaza of the Superimposed Altars was treated. It has a stone core with a thick layer of mud plaster, over which there is a white stucco layer with red and green figures. When the alter was unearthed, we could see that the plaster did not adhere well to the core and that it was coming loose at the cornice corners. The stucco layers showed good adhesion to the plaster, but we could not determine if there was surface carbonation because everything was very damp. As a preventive measure, the surface was reinforced with Curasol® AR 7% in water, and pressure was applied to the plaster areas that were separating. As it dried, we observed that the mud plaster did not have good cohesion. The situation worsened due to volume loss and the widening of the fissures. The heavy cornice corners were in danger of falling off since they had no flag stones as foundation support. Because the structure's top part had been destroyed, the core was exposed. Several stones were removed from the core and the mud plaster was reinforced from the inside. For structural reinforcement a 1/4-inch iron rod chain with mortar was placed inside; the loose plaster was affixed to it. When the piece dried, the surface was reinforced and the pigment set by using Paraloid® B-72 at 5% and 10%.

This altar, which is approximately 70 centimeters lower than the previous floor, has withstood two floods that lasted several days, and neither the stuccowork nor the mud structure deteriorated.

The Teotihuacan Archaeological Project ended in 1982. The restoration of the stuccowork on floors in the last excavation areas remained: Front 4 had yet to be unearthed, and Front 6 remained unfinished. In two subsequent short seasons, the budget allowed for floor restoration. During the first season, work was done on Front 6 and during the second one, on the *Gran Plaza* of Front 4.

We believe that the great progress in stuccowork restoration was due to the cooperation between the Archaeology Division and the Restoration Division. The work strategy allowed the development of both areas of specialization without hampering their field work.

The support of the Directorate of Restoration was very important for the Restoration Division work. The laboratories allowed us to do research on pigments with the expert help of Alejandro Huerta and research on the stone with the help of Julio Chan. In our field work, we also received the support of the Studio of Mural Painting ("Teller de Pintura Mural") through the collaboration of restorer Yolanda Santaella, who participated as a temporary coordinator.

At present, the conservation conditions of the site are critical. We lack economic and human resources for adequate upkeep. We cannot protect it against the agents of physical, chemical, and biological deterioration because of lack of funds, and we are defenseless against human beings, the site's greatest menace, without adequate and effective security.

Daniel Molina Feal

CACAXTLA

Daniel Molina Feal received his degree from the National School of Anthropology and History in Mexico. He has worked on the Cacaxtla and Yohualichan projects and carried out restoration and maintenance in archaeological areas. He is a research professor and has published numerous articles.

In 1975, an episode of looting in the archaeological site of Cacaxtla brought to light some archaeological remains that by their very nature demanded immediate protection.

The looters had exposed a polychrome mural, which was in danger of being buried by the wall of the vault dug by the looters—a situation caused in part by the tunneling and in part by the people who came after and squeezed into the tiny space to look at the "Apache" (pl. 28). The figure had been damaged by the tools used in the looting, and a clay element affixed to the north wall of the mural had been destroyed.

A plan for rescue and protection was laid out. In time, because of the importance of the discovery, this plan turned into a major project carried out in four field sessions over the course of almost five years—not including the time spent analyzing the recovered materials.

The project was entrusted to the Puebla-Tlaxcala Regional Center of the Instituto Nacional de Antropología e Historia (INAH) under the direction of Diana Lopez, with whom I collaborated for the duration of the project. The restoration of the murals and reliefs was carried out by the personnel of the current Directorate of Restoration at INAH.

From the very beginning, there were multiple options for the Cacaxtla excavation. The solution selected would determine the conservation method that would be applied: Would the substructures be excavated? Would pieces of evidence be left? Would the excavated area need to be enlarged? Should the elements be removed or left in situ? In planning the necessary restoration measures, one had to keep in mind the complexity of the process, as well as both short- and long-term maintenance, the cost, the number and kind of security personnel that would be required, and many other considerations quite separate from the first aim of the archaeologist, which is to obtain the data necessary for further study and interpretation.

Regarding the decision-making process, it must be explained to those unfamiliar with Mexico that all activities of this kind are under the control of the state through the office of INAH, whose operations are based on laws and regulations aimed at the protection of the country's cultural heritage. One branch of INAH is an agency that has the responsibility of ensuring that archaeological projects in Mexico meet certain requirements—the presence of a trained director, a theoretical assessment of the problem, appropriate methodology, planned budgets, institutional support, etc.—and also that the projects are carried out within the limitations of the permit. Finally, the agency is in charge of reviewing the body of technical information; it may approve the findings or request amplification. At present, an archaeologist is not allowed to start new research without first submitting the results of previous work. This precaution ensures that the excavation and conservation measures taken on a site follow institutional policy and that archaeologists adhere to the original plan (even when the project is the work of a single individual), since it is sponsored by INAH and endorsed by the Council on Archaeology. All the measures that we introduced in Cacaxtla received the approval of the Council.

The areas where the excavation work was concentrated were on two levels, one in the north and the other in the southwest. They were covered with shrubbery, a large quantity of long-rooted *nopal*,[1] and a few trees. An esplanade on level ground showed signs of having been cultivated for years, implying that there had been plowing and hence the growth of roots from crops over time, which could still be present among the archaeological remains. The villagers told us that on several occasions the local white pitted rock (*tepetate*)[2] had been used in the construction of the modern town. Both these circumstances alerted us to considerations we would have to bear in mind in the course of our conservation work.

1. Plants of the genus *Nopalea*, resembling somewhat the prickly pear.

2. Tephorite, a rock derived from volcanic ash.

Disposing of hundreds of cubic meters of excavated earth presented a different problem. We had to build immense ramps to bring the earth down to areas accessible to trucks. Our vehicles also made use of a path leading to the higher platform to carry away accumulated soil.

The excavation exposed an architectonic grouping characterized mainly by colonnaded structures built on small platforms, which surrounded patios of various sizes. In these structures, the "slanted panel configuration" peculiar to Cacaxtla turned up repeatedly. These panels of varying relief produced a harmonious play of chiaroscuro. The structures that were excavated to great depth revealed that this architectonic device had been used in this zone for a very long period without any abrupt alteration in design, proportion, or technique. The most significant change was the introduction of adobe, which replaced the wall construction of the oldest period.

Some of the debris that obscured the perimeters of the foundations was removed. Drill samples indicated that the foundations were 80 percent manmade. The excavation of architectural elements, a number of pictorial sets, and some clay reliefs provided researchers with enough iconography to hint at their ethnic origin and, generally, to suggest a wide range of cultural studies that could follow. The discovery of this material also created the need for its conservation; this enlarged the scope of research and discussion to include the participation of conservators who, in fact, had been there since our first visit to the site.

The theoretical basis of conservation methods

From the outset of the project, there were two schools of thought on the question of whether or not the Cacaxtla murals should remain in situ. One group of conservators maintained that the murals ought to be removed and replaced by reproductions, while the other group (which included us, the archaeologists in charge of the project) thought that the correct action was to leave the murals in their original context.

This lack of agreement meant we archaeologists had to establish our theoretical position. Each step we took, each solution we adopted, had to be justified. We had to expand our conservation theories to include the realm of social science.

We regard archaeology as a historical science that seeks to determine the characteristics of past societies; to understand the changes they underwent; and, based on the above, to discover some consistencies in the process. While archaeological study in general is concerned with the development of human societies over time, specific research may concentrate upon the material remains of societies and cultures, e.g., ceramics, implements, religious objects, and architectural remains. These remains may give us insight into social relationships and allow us to infer essential social content. They are documents—embodiments of the culture—and should be appreciated for more than their aesthetic or architectural value.

The concepts that we use to define and justify conservation and restoration in general depend upon our definition of archaeology and our understanding that archaeological remains are historical documents. Thus, material evidence is conserved (1) for the study and interpretation of laws that govern social change; (2) to affirm what is actually present (which is why we must not rebuild anything unless the reconstruction is based on accurate and authentic data, and why we must make evident our interventions); (3) to serve as a didactic tool for teaching our history; (4) for future generations to use in the dialectic process of our social evolution; and finally (5) to benefit the general and local populace, since the conservation of monuments affects directly the economic, political, and social development of the community. Indeed, evidence conserved in its own context is always a richer source of information, because we see it as part of a whole rather than as an element divorced from the social factors that generated it.

In areas where the language of conservators and archaeologists more or less coincides, our position was as follows: (1) we would adhere to the general outlines of the national and international documents that govern acts of excavation and restoration on archaeological sites; (2) regarding acts of site conservation that have a direct effect on the excavated remains, we would make use of restoration (in its modalities of consolidation and reintegration), protection, and the posting of signs; and (3) we would do maintenance work during our stay on the site and would avoid any unnecessary removal.

Article 9 of the *Charter of Venice* permits the relocation of entire structures or their parts in extreme cases only and, although this is a subjective judgment, we consider extreme only those cases in which the destruction of the cultural asset is imminent. Removal must be carried out after all other available means of on-site protection have been exhausted.

Removal in itself does not guarantee the conservation of the object. In fact, it subjects objects to major risks of deterioration during disassembly, transport, and reassembly. We now have the technology and the technicians to enable us to successfully remove cultural property from almost any site. However, it is not enough to possess adequate techniques: One must also have the proper exhibition or storage facilities. When an object is consigned to a warehouse, the storage conditions must guarantee its integrity (after all, the fact that its integrity had been threatened was the reason for removing the object in the first place). When an object is exhibited, it should be displayed in a manner that recognizes, if not replicates, its original context.

The above arguments have been set forth from our perspective as historians and conservators of the patrimony. As social scientists, however, we must address issues that transcend scientific curiosity: The impact upon the indigenous community must be considered.

From a sociological point of view (for inhabitants of San Miguel del Milagro in particular, and for the Tlaxcaltecans in general), the discovery of the murals provided the possibility of strengthening the precarious tourism-based infrastructure of the state. The township and the government believed that ruins = tourism = money = well-being. Within a few days after the start of conservation, the residents began to receive economic benefits from newly created jobs, and the town was on its way to renewal. The townspeople were aware of the social implications of removing the murals to another location. There were no museums in Tlaxcala that could house the paintings at that time. The relocation of the paintings to some distant site would have aggravated the Tlaxcaltecans. This strengthened our commitment to keep the Cacaxtla paintings and reliefs in situ.

The design of operations

The decision to keep the murals in their original context meant that we had to provide an acceptable alternative to removal. Actions had to be taken to prevent deterioration by the elements and by direct contact with people and animals. In order to accomplish this, alien elements—easily distinguishable modern materials—would have to be used. Such elements generally include protective covers—either roofs or dustproof enclosures— specifically designed to prevent exposure to the direct action of rain, wind, and sun. They can be fences or handrails that protect the objects from being touched by visitors. They can also be passageways and ramps with simple rubberized runners or elaborate structures that prevent visitors from walking on, and thereby contributing to the deterioration of the site.

As I have pointed out, these protective elements are conceptually and practically incompatible with the cultural objects and, therefore, they also have an adverse visual impact. Until better conservation measures are available, this is unavoidable. In the meantime, roofs, curtains, handrails, passageways, and dust-absorbing covers are preferable to the deterioration of the object or its removal. The challenge is to design elements that are as neutral and unobtrusive as possible.

Conservation operations

The majority of the structures at Cacaxtla were built with *tepetate* stones laid as double walls. On top of these walls were beams that held up the roofs. The walls were mortared together with mud, sometimes with lime mixed in. Their state of preservation was generally good. There was no need for major intervention, except in the case of some walls at the far ends of the groundworks. These were severely affected by earth movement. For example, the back wall of Edifice A was leaning out at a considerable angle; it was necessary to prevent its collapse because a mural was painted on its inside surface. To retain the wall, we built buttresses of *tepetate* stones that would blend in with the original material. In general, the restoration of the *tepetate* walls consisted of filling the butt-ends and wall facings with mortar and waterproofing the upper parts. We used the same treatment on the pillars that held up the porticoes of the buildings.

A noteworthy architectural feature of Cacaxtla is the slanted panel configuration on the walls of the rooms. This contrasts with other Mesoamerican archaeological sites, where the same configuration appears only on the platforms on which the rooms are built.

Throughout Cacaxtla, one observes the slanted panel configuration on the side walls and at the angle they form with the facade of the building; between these corners are a row of pillars placed at regular intervals, which form a part of the portico present in all the buildings of the area.

The panels were built in two ways and in both they acted as structural components of the exterior decoration. Sometimes the *tepetate* was carved to form different levels of relief; in other cases, this effect was achieved by allowing the *tepetate* stones constituting the panel to project beyond those in the lower part of the wall. The two techniques were often combined.

The mounds were built over a core of clay mixed with volcanic rock (*tezontle*), then covered with an additional coat of clay and smoothed out. Small holes were pricked into the surface or small pitted stones were embedded in it, and finally it was stuccoed over.

The walls of the apartments belonging to the ensemble we call the Palace were built of adobe. This represents an important transition in the use of materials; however, neither the architectural style nor the size of the modules changed very much. The builders of the adobe walls put ceramic fragments on the surface of the rows of mudbricks mortared with mud, probably to increase the adhesive power of the layer of stucco.

The Palace is located in an area that has been cultivated for many years, and the walls have been badly damaged by plowing and root growth. The walls we found had an average height of 60 centimeters and, as in similar situations all over the world, the conservation of these adobe elements presented one of our major problems.

As a first attempt at preservation, we cleaned the walls and used fungicides to arrest plant growth. We then applied a consolidant several times in various degrees of solution, and finally we sprayed on a waterproofing agent. The effects of this treatment lasted only a year. Erosion during the rainy season caused heavy damage, particularly to the tops of the adobe walls.

We then tested a number of products and decided on a coating that prolonged the life of the adobe but had the disadvantage of concealing the original material. The surface no longer looked like rows of adobe mudbricks, and gave instead the impression of a mud wall.

With the *tezontle*, we proceeded similarly, all the way to the consolidation stage. Once the original was consolidated, we covered the top and sides (stopping when we reached the stucco level) with a coat of mud impregnated with fungicide, a waterproofing agent, and a consolidant. When this coat dried, another thinner coat with a higher proportion of consolidant of the same mixture was applied. This time the coating was textured to approximate the original surface. This treatment was effective for more than three years, even though it never received the maintenance we recommended.

In Cacaxtla, there is a large area with stucco-covered walls and floors. These stuccoes are of very good quality and thus fairly well preserved. The major deterioration was from roots, from intentional destruction in the pre-Hispanic period when new buildings were erected using the old stones as a source of material, and from looting.

Restoration work consisted of trimming; filling gaps and cracks; and, where we found the stucco separating from the support of the wall, consolidating with a mixture of lime, sand, and consolidant. The plaza in front of the Mural of the Battle had lost more than 90 percent of its original stucco. To have left it alone, consolidating only the remains, would have caused serious problems for the paintings. The wind would have swept up dirt and deposited it on their surfaces, where it would have acted as an abrasive. We decided to fill the gaps with a mixture of lime, sand, and a waterproofing agent. We opened new drainage ditches and connected them with old ones to prevent flooding of the plaza, and some of them were covered with removable grates to facilitate clean-up operations.

When we were looking for the best way to drain the plaza, we discovered latticework, an architectural element that functions to separate spaces. Its appearance reminded us of the false lattices of the Maya area and of El Tajin. The design was based on a pattern of intertwined rhombuses on the outside, with a plain surface on the inside. Latticework made of adobe had a core of reeds within the body of the rhombuses, and there had been some kind of fiber in the core of the molding that crowned them. This element was not part of the original building but had been added to it later.

When we came across this kind of latticework, we immediately noticed how fragile it was, so we proceeded to combine restoration with excavation. The reed cores that should have been found in the rhombuses as well as in the cornice had disappeared; the structures were totally hollow. The earth that covered the lattice had kept it from collapsing, and removing the dirt would have caused damage if we had not taken proper precautions.

To prevent the lattice from being pulled down by its own weight, we cut rhombuses out of expandable polyethylene foam to fit each opening exactly. We cleared each opening this way, and inserted in place the material we had precut.

Once this was done we could proceed with consolidating the element. We made a kind of "chain" with a thin rod, which was found in the same openings that they used during the pre-Hispanic period. To this chain we joined the elements of the cornice that remained in situ; the replacement cores we inserted reinforced the rhombuses. The ends of the chain were soldered to give greater solidity to the element.

Then we introduced a wire rod into each of the openings; each of these reached the end of its respective rhombus—this without doubt gave a wide margin of security for the interventions that would follow and added to the life of the element already restored. The top ends of the wire rods that served as cores were soldered to the cornice chain. Subsequently, we injected into each of these channels a very fine mixture of lime, sand, and Primal® AC-33. We repaired the exterior cracks and we injected Primal wherever we found a trace of separation between the finished stucco and the adobe that gave support to the lattice.

To conclude the intervention, we constructed a plywood crate to fit the element perfectly, made on site. In it we mixed a cast of lime, sand, and Primal AC-33. We removed the crate after thirty days, proving the success of the operation.

The construction of exposed cores I consider to be a small contribution to local conservation. Actually, they are retention walls that keep the edifices and fillings (natural or artificial) from caving in. This type of wall can be absolutely vertical or slightly angled. Once raised, it is treated to approximate the color and textural features of the original core. Care is taken to leave appropriate blanks corresponding to what is known in painting restoration as lacunae.

We used the following techniques in the architectural restoration of Cacaxtla: consolidation, stucco binding, gap filling, the building of retention walls and exposed cores, and some reintegration. Our foremost criteria were respect for the pre-Hispanic work and the conviction that the least possible intervention sufficient to ensure the stability and preservation of the restored element was desirable. This attitude characterized the restoration work on each building; the specific details of the process are not relevant here.

The restoration of the paintings and the clay reliefs was done by the conservators, so I will confine myself to presenting general information on their activities and the materials they used. No mixtures, emulsions, or mortars were applied until they had been tested in various quantities and proportions on controlled areas.

The most ancient paintings discovered were a group found in the Staircase Room, applied to a smooth surface. They had been damaged by roots over time.

A second group we call the Mural of the Battle had been painted over smooth lime. This mural had been exposed to weather in pre-Hispanic times. Its entire surface was covered with a fine opaque film that resulted from the dissolving of the stucco while the paintings were exposed and the presence of carbonates. The central area was better preserved than the ends. The painting was covered with earth, but

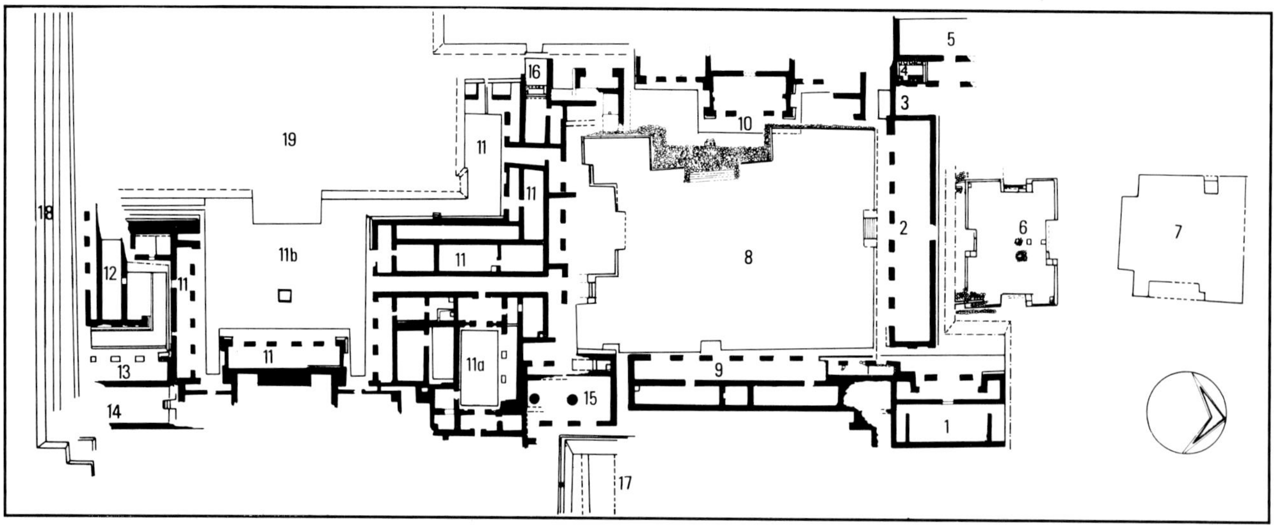

Figure 1. Plan of the Cacaxtla Site, with the structures excavated to the level of the great Platform.

1. *Edifice A*
2. *Edifice B*
3. *Passage of the vertical panels*
4. *The "rabbit warrens"*
5. *Edifice C*
6. *Sunken patio*
7. *Mound Y*
8. *North Plaza*
9. *Edifice D*
10. *Edifice E*
11. *The Palace*
11a. *Patio of the Rhombuses*
11b. *Patio of the Altars*
11c. *Doorway A*
12. *Edifice F*
13. *Doorway F*
14. *Staircase Room*
15. *Building with columns*
16. *The Latticework*
17. *Slanted panels, east*
18. *Slanted panels, south*

later a staircase was built and, in the extremities, the stones of the new mounds were placed directly on the painting.

A third ensemble was found in Edifice A. Four of the paintings were on the front wall and the door jambs. At the time of their obliteration the builder was careful to take measures to avoid destroying them, filling up the space in front of the paintings with finely ground *tepetate,* mostly free of impurities. The deterioration we detected was due to root penetration on the north mural and looting on the south one. The fifth mural was found on the back wall of the buildings. It had been painted on a smooth mud surface. At some point the painting was obliterated by a wash of mud. We must add that the wall that supported it was one of the most heavily damaged; it was full of gaps and cracks.

The painting techniques are true fresco and fresco *secco*. The colors are blue, red, yellow, green, pink, and various shades of ocher, white, and black—all of mineral origin except a certain black called "rubber" black. The restoration consisted generally of cleaning, consolidating pigments and supports, removing carbonates, eliminating fungi, and filling gaps and cracks by chemical or mechanical means as the situation required.

In Edifice A, we found a raw clay relief covered with red paint affixed to the north mural. Part of it had separated from the wall and there were numerous fissures requiring restoration. The excavation and the consolidation of the elements were carried out simultaneously by the restorer. He primed the relief with a consolidant, used adhesives where necessary, and made small repairs with paste, applying color to blend the patched areas with the original background. More or less the same work was done on the highly damaged relief, also made of raw clay, found on a pillar in Edifice D. It had been coated with stucco, which may have been polychromed.

During the project, there was always close interaction between the restorers and the archaeologists. None of us felt that our fields of action were being invaded; we continually exchanged information and had a common interest at heart—the conservation of our heritage.

To maintain the paintings and reliefs in their original context, it was necessary to build shelters to protect against direct sunlight, water, and as much as possible, wind and dust. The roofs or coverings had to be proportioned so as not to rob the architecture, or the space in general, of its stateliness. The supports for the shelters and covers could not interfere with the viewing of the motifs. The roofs had to blend in as much as possible with the structures to avoid creating an adverse impact on the environment. They could not be temporary, but they had to be reversible. We used high-quality materials, not short-lived solutions that could eventually endanger what they were intended to protect. Glass and acrylics could not be placed near the paintings because they could create a microclimate giving rise to such problems as fungi, dampness, or carbonate deposits.

These factors led us to erect individual roofs with four supports. In Edifice A, we modified the design and added a side lattice to protect the wall paintings from strong winds and driving rain. In Edifices A and B, curtains were appended to the fronts of the roofs to keep direct sunlight from striking the murals.

The excavated site at Cacaxtla was opened to the public after a series of walkways and handrails had been installed to prevent people from stepping on the stucco and the structures. Although this is not the first site where such measures have been used, here we emphasized the necessity of leading the visitor along fixed routes, protecting that which must remain out of reach but not out of sight. We found that visitors accepted the restrictions; they did not wish to destroy in a short time what had lasted for centuries.

Cacaxtla is one example, fortunately not the only one, of the effective application of in situ conservation measures and of the benefits of interdisciplinary work in preserving the cultural patrimony.

Joaquín García-Bárcena

CACAXTLA

Joaquín García-Bárcena is presently Director of Prehispanic Monuments, INAH, Mexico. He received the degree of Master of Archaeology from the National School of Anthropology and History in Mexico and has worked on the Fuentes de Obsidiana, Cuevas de Santa Marta, and Altos de Chiapas projects. He is a professor of archaeological methods and techniques and has published widely both nationally and internationally.

In this volume, Daniel Molina Feal describes the initial stages of the exploration of the Great Platform of Cacaxtla, emphasizing the restoration and conservation measures taken, as well as the criteria used in defining these measures. I will refer briefly to one operation mentioned in his paper that is now being carried out—that is, the roofing of the Great Platform—and then discuss the criteria used to reach the decision to install it, as well as the considerations taken into account when designing it.

The Great Platform is totally manmade. On the west side, it stands directly on top of the underlying rock; on the east, on 6 meters of archaeological deposits. Unlike other pre-Hispanic platforms, which were built as bases for temples, the Cacaxtla platforms are the result of many successive stages of construction of the so-called palaces. Each layer was partially destroyed, and earth, adobe, or rubble were used to fill up the spaces. A new platform was thus formed, and new edifices of adobe were erected (the floors and the walls were covered with stucco). As the height of the complex increased, sloped retention walls were built, partly of *tepetate* (a very brittle volcanic tuff) and partly of stone—the only stone used in the construction.

Mural paintings and bas-reliefs of raw clay were found on these walls. They are only partly visible today, but we have information indicating that on still-unexposed lower levels there are more mural paintings and various works in raw clay.

A few years ago, the south and east retention walls of the Great Platform were explored and consolidated. However, after the next rainy season, evidence of faulting appeared in the form of cracks parallel to the slopes on the top platform. This discovery, coupled with the obvious need to protect the edifice, made it clear that a partial or complete roof was necessary. It was decided that all the archaeological components should remain in their original location whenever possible. The monument as a whole—the buildings, the murals, and the bas-reliefs—was considered worthy of conservation, equal to the more notable or better-known elements.

The main causes of deterioration have been rain, intense light, wide variations in temperature, surface condensation and evaporation, and wind erosion. A roof protects a building from rainfall and direct sunlight and reduces (but does not eliminate) thermal variations, but it cannot prevent wind erosion. This is an important consideration in Cacaxtla because prevailing northwestern winds can (for short, rather frequent periods) reach speeds of up to 80 kilometers per hour. To completely control wind erosion, temperature, and humidity, one would need an enclosed and air-conditioned building. This was not considered

Figure 1 (opposite). Detail of wall painting. Cacaxtla, Tlaxcala, Mexico.

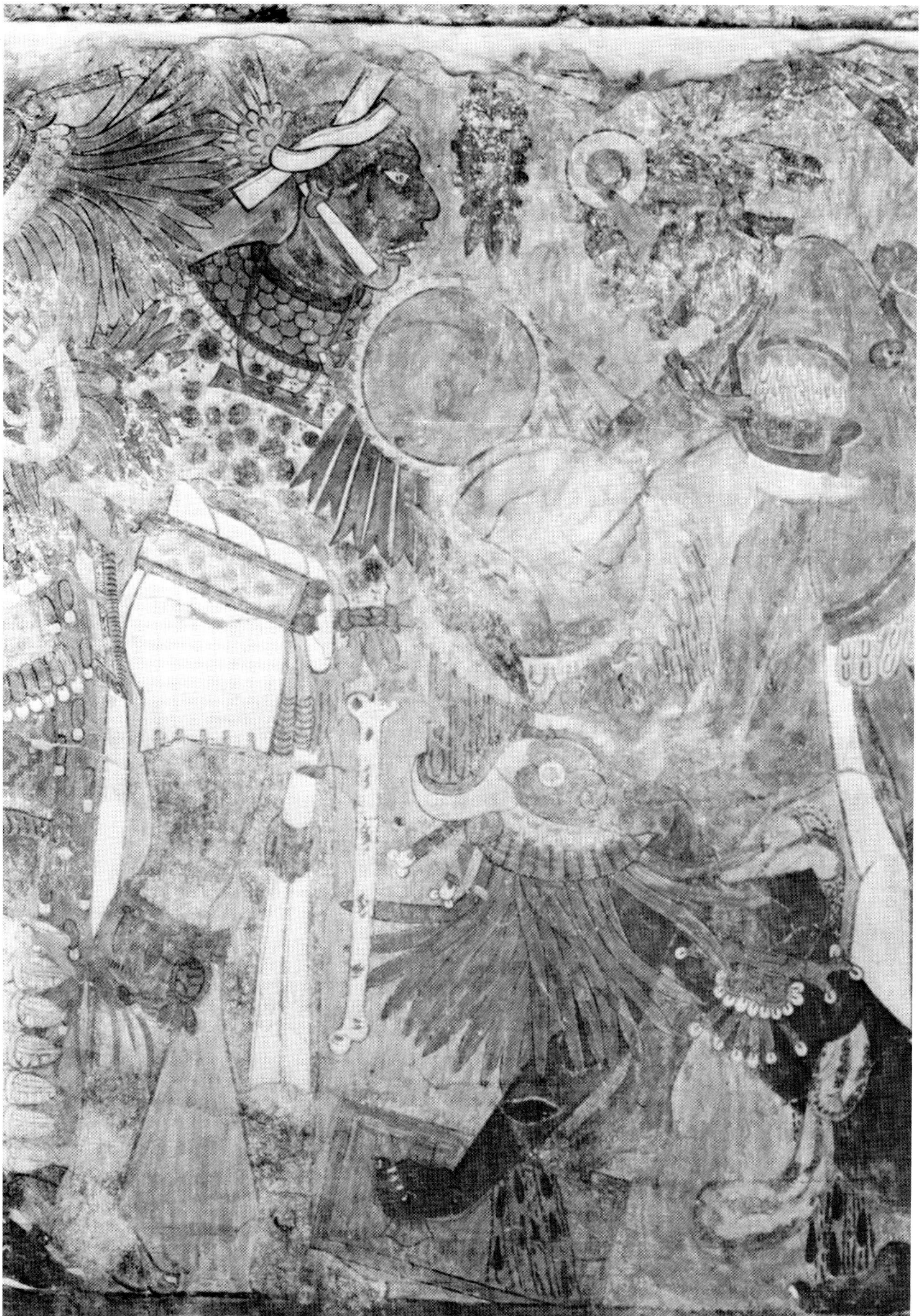

a viable option, however, given the maintenance requirements and the costs involved.

Designing an appropriate roof for an archaeological site involves the consideration of a number of factors: Conservation issues must be weighed against archaeological and architectural requirements, as well as against civil engineering needs. A detailed archaeological exploration and analysis of the whole monument would provide the most complete information. However, this would imply its disappearance as an integrated unit. From the conservation point of view, any intervention that protects the buildings for an extended period of time is desirable. From the architectural point of view, the introduction of new elements should be avoided whenever possible, as should modifications that would alter the interior spaces or the relationship of the buildings to the environment. And the engineering point of view recognizes the constraints on design and the selection of materials and techniques in order to achieve the necessary resistance to wind, earthquakes, and other causes of damage or destruction. Moreover, one must consider the economics of the situation, designing features wherever possible, aimed at keeping future conservation expenditures to a minimum.

Although any roof intrudes upon and alters the architectural spaces of the site and the immediate environment, the highest priority at Cacaxtla was the protection of the sloping walls of the Great Platform, the murals, and the associated components. This goal provided the basic criterion used to evaluate the various roof designs.

A preliminary analysis of the options found that three proposals seemed to address the multiple requirements of conservation, archaeology, architecture, and engineering.

1. *Reconstruction of the original volumes*. This would be accomplished by means of metallic sections, superimposed on the pre-Hispanic remains, to which protective covers would be added as necessary. The structures would not have foundations. This proposal was ultimately deemed unsatisfactory because (1) there was not enough evidence left to reconstruct the buildings, except volumetrically; (2) the sheer quantity of new building elements, far from enlightening the public as to how the edifices looked originally, would make the existing evidence more confusing; (3) even after placing protective covers over all these structures, the interior courtyards and the sloping walls of the Great Platform would remain exposed to the weather; and, (4) having no foundations, the structures would be aerodynamically unstable, and could not be anchored without gravely altering the substructures not yet excavated.

2. *Modular partial roofing, without foundations*. This proposal, though more satisfactory than the first, presented the following drawbacks: (1) the interior courtyards, the sloping walls of the Great Platform, and the unexcavated areas would remain exposed to the weather; (2) the ninety supports of the roof modules would not permit an unobstructed view of the area as a whole; and (3) the modules, having no foundations, would be aerodynamically unstable, and it would be impossible to anchor them without seriously affecting the existing substructures.

Another option, that of stabilizing the modules by loading them down, would require the use of plates to distribute the load. Such plates, according to soil mechanics studies, would occupy approximately 50 percent of the area covered, considerably altering the site.

3. *A single roof covering the Great Platform*. This would have foundations and would have no internal supports. This seemed to be the most viable option, because it was the only one that protected the whole of the Great Platform, including the edifices that crown it, the murals, and the other elements. It would also facilitate later explorations

of the unexcavated parts of the area. Then, too, it was the only proposal that did not present problems of aerodynamic instability. And it had the least visual impact when the visitors were inside the site, but the greatest impact of the three options from outside the site.

The single roof was the solution adopted with the following stipulations: (1) the excavations required for soil mechanics studies, as well as for the foundations, should be preceded by corresponding archaeological rescue operations; (2) the construction program should preclude, in so far as possible, the presence of assembly crews and materials on the Great Platform, and should proceed under the supervision of archaeologists; and (3) restoration specialists should be consulted to determine the appropriate measures necessary to ensure the best conservation of the archaeological site and its elements.

This archaeological rescue operation is presently being carried out at Cacaxtla together with the construction of the roof foundations.

Appendix: Proprietary Materials Cited in Texts

Proprietary Name	*Function*	*Major Chemical Composition*
Acritex	Adhesive	Acrylic emulsion
Acryloid B-72*	Consolidant/ Adhesive	Acrylic resin
Araldite	Adhesive	Epoxy resin
Bromacil	Biocide	Bromobutyl methyluracil
Calgon	Water softener	Sodium hexametaphosphate
Cetavlon	Biocide	Cetyl trimethyl ammonium bromide
Colmadur BV	Adhesive	Epoxy resin
Coroplast	—	Polyethylene
Crystic release agent	Release agent	Silicone mixture
Crystic resin 405	Castings	Polyester resin
Curasol AR	Adhesive	Polyvinyl alcohol
Diuron	Biocide	Dichlorophenyl dimethylurea
Duco cement	Adhesive	Cellulose nitrate solution
Endurol	Adhesive	Polyvinyl alcohol
Mowital B60H	Consolidant	Polyvinyl acetate
Mowilith (various grades)	Adhesive	Polyvinyl acetate emulsions
Multex	Adhesive	Acrylic emulsion
Orvis	Detergent	
Paraloid B-72	Consolidant	Acrylic copolymer
PEG-4000	Consolidant	Polyethylene glycol
Primal AC-33	Consolidant/ Adhesive	Acrylic copolymer emulsion
Resistol 850	Adhesive	Polyvinyl alcohol
Rhoplex AC-33	Consolidant	Acrylic copolymer emulsion
Saran Wrap	—	Polyvinylidene chloride
Simoniz wax	Polish	Silicone - wax mixture
Teflon	—	Tetrafluoroethylene
Tyvek	—	Spun polyethylene paper
UHU	Adhesive	Polyvinyl acetate solution
Wacker's 190L	Water repellant	Silicone resin
Wacker's OH	Consolidant	Tetraethyl orthosilicate
Zapon	Lacquer	Acrylic solution

*Also known as Paraloid B-72